I0490145

1,000+
SUDOKU

HARD PUZZLES

Collin Deloach

Contents

Your mission

is to solve the puzzle by filling in the empty cells with numbers from 1 to 9 without repetition in each row, column, and sub-grid.

The goal is to use logic and deduction to find the missing numbers and complete the puzzle.

Without repetition in each sub-grid

						3		2
			8			4	9	1
	1			3				
2			7	4	1	9		6
6			8	9	2			7
9		1	6	3	5			4
			4			8		
8	4	6		7				
7		9						

Without repetition in each column

						3		2
			8			4	9	1
	1				3	5		
2						9		6
6			8		2	1		7
9		1				8		4
			4			7	8	
8	4	6		7		2		
7		9				6		

Without repetition in each row

						3		2
				8		4	9	1
	1				3			
2						9		6
6			8		2			7
9		1						4
1	2	3	4	5	6	7	8	9
8	4	6		7				
7		9						

Enjoy !

HARD
Puzzles

Hard # 1

						4	2	
2		6	1				8	
		1	4		2			
3				5	8			
		5				2		
			7	4				3
			5		9	3		
	6				3	7		4
	9	3						

Hard # 2

6								
3	4			7	2			
2	5		1					
5			7	8			6	
		7				8		
	2			3	9			7
				7			1	9
			3	9			2	4
								3

Hard # 3

		6		9	5			2
	5		2				7	
				4				
2		1					5	
6		3				8		1
	7					6		9
				3				
	1				7		9	
5			4	6		7		

Hard # 4

	4					3		
5		2					9	
		8			1			2
		1		7				9
			4		5			
9				3		6		
3			7			9		
	2					1		5
		4					8	

Hard # 5

	7			6				9
	9	6			4	5		
1					5	7		
				8				
	6	3				8	7	
			5					
		7	1					2
		2	8			4	9	
3				9			8	

Hard # 6

	6		2					9
		8	3	9			1	
		1				8		
				4				7
	3	6				9	5	
2				6				
		4				5		
	8			3	1	4		
1				5			7	

Hard # 7

```
. . . | . 2 8 | . . 3
4 . . | . 5 . | 7 . .
5 . . | 4 . 9 | . . .
------+-------+------
6 . . | 2 . . | . 8 .
. . 2 | . . . | 7 . .
. 3 . | . . 4 | . . 5
------+-------+------
. . . | 9 . 1 | . . 4
. 9 . | 8 . . | . . 7
8 . . | 5 3 . | . . .
```

Hard # 8

```
. . . | . 5 . | 2 . .
. . . | 3 . . | . . 5
. 1 . | . 9 6 | 8 . .
------+-------+------
. . . | 2 . . | 7 . 4
5 . . | . . . | . . 8
4 . 9 | . . 3 | . . .
------+-------+------
. . 3 | 8 2 . | . 6 .
. 6 . | . . 4 | . . .
. . 7 | . 3 . | . . .
```

Hard # 9

```
. . . | 4 . . | . . 2
7 9 . | . 2 . | . 5 .
4 . 3 | . . . | 6 . .
------+-------+------
. . 4 | . . 7 | . . .
8 . . | . . . | . . 3
. . . | 1 . . | 9 . .
------+-------+------
. 5 . | . . . | 3 . 4
. 3 . | . 5 . | . 2 7
6 . . | . . 9 | . . .
```

Hard # 10

```
2 . . | . 7 9 | . . .
5 . 6 | 2 . . | . . .
. 8 . | 3 . . | . . .
------+-------+------
. . . | . . . | . 6 7
3 4 . | . 8 . | . 9 1
6 1 . | . . . | . . .
------+-------+------
. . . | . 5 . | 7 . .
. . . | . 7 5 | . . 8
. . . | 1 6 . | . . 9
```

Hard # 11

```
9 . . | 6 8 . | . . 3
. 3 . | . . . | . . .
. . . | 7 . . | 5 9 .
------+-------+------
2 . . | . . . | 7 . 9
. 8 . | 2 . 9 | . 3 .
3 . 7 | . . . | . . 8
------+-------+------
. 2 3 | . . 4 | . . .
. . . | . . . | . 5 .
8 . . | . 5 1 | . . 4
```

Hard # 12

```
. 8 2 | 5 . . | . . .
4 . 5 | . 1 . | . . .
. . . | 4 . . | . . .
------+-------+------
. 9 . | . 2 8 | . . 4
6 . 8 | . . . | 5 . 1
3 . . | 1 7 . | . 9 .
------+-------+------
. . . | . . 1 | . . .
. . . | 8 . . | 7 . 6
. . . | . 9 1 | 3 . .
```

Hard # 13

8			2	7				
		5			3	7		
	1					4		
	4					6	7	
			4	3	8			
	8	9					2	
	6						5	
		1	8			3		
				9	2			7

Hard # 14

	5							
		1		8	9			7
9				4		2		
3	2	4	8					
		6				5		
					6	4	2	3
		5			7			2
2			5	6		9		
							1	

Hard # 15

1				7		6	8	
		2	5				3	
	4			9				
				9		1	3	
			3		6			
3	2		7					
				7		5		
	6			4	1			
4	1		8					7

Hard # 16

			1	4	7			3
			2	6		4		1
		8						
4						5	1	
		6				8		
	3	1						6
						1		
9		4		3	1			
2			4	8	5			

Hard # 17

		3		7		6	5	
			4			1		
	6	4	5					
	5			3				1
2								9
8				9			6	
						7	8	1
		6			1			
	1	2		8		7		

Hard # 18

4				8	2			3
6			8	1	9			
		3			1		4	
	7						2	
	8		5			1		
				5	4	8		7
9				2	3			4

Hard # 19

8		1		2	3			
3			8	1				
9								
	9	7	6			8		
		4				5		
		3			4	2	7	
								5
				4	7			1
			3	6		9		8

Hard # 20

	8					5		
	7		3					
2		3	9					
				1	8	9	3	
7	5						1	6
	1	9	4	7				
					2	8		5
					3		4	
	7						9	

Hard # 21

				8			5	
		5		1			7	6
					3	4		
2					8		3	5
	3						6	
7	5		1					8
		7	9					
9	2			4		7		
	1			7				

Hard # 22

			5	1	6			
5	1			7		4		
								8
	4		2			3	7	
		3				5		
	7	9			5		4	
3								
		6		3			9	5
			7	5	2			

Hard # 23

		4		5	1			
2				8	7		6	
	6							9
	5	1			4			3
4				2		8	1	
3							8	
	9			7	2			5
			3	4		9		

Hard # 24

						9	2	
		4			6			5
1		8	7	2				
	8							4
			3	7	1			
7							9	
			3	7	1		6	
4			8			5		
	3	6						

Hard # 25

```
5 . . | 2 . . | . 7 .
4 . . | 5 . 7 | 6 . .
. . . | . . . | 2 . .
------+-------+------
. 8 9 | . 6 . | . . 1
. . . | . . . | . . .
6 . . | . 4 . | 3 8 .
------+-------+------
. . 8 | . . . | . . .
. . 2 | 3 . 4 | . . 5
. 6 . | . . 2 | . . 3
```

Hard # 26

```
3 . . | . . 5 | 6 . 7
. 9 8 | . . . | . 2 .
. . . | 7 . . | . . .
------+-------+------
1 . 3 | 4 . . | 5 . .
. . . | 1 . 6 | . . .
. . 4 | . . 7 | 3 . 8
------+-------+------
. . . | . 4 . | . . .
. 2 . | . . . | 7 8 .
8 . 9 | 3 . . | . . 4
```

Hard # 27

```
4 . . | . . . | . 5 6
. 2 . | 4 . . | . . .
. . . | 2 8 9 | . 4 .
------+-------+------
3 . . | . . . | 7 6 5
. . . | . . . | . . .
7 1 5 | . . . | . . 3
------+-------+------
. 4 . | 9 6 1 | . . .
. . . | . . 5 | . 8 .
2 5 . | . . . | . . 1
```

Hard # 28

```
4 . . | 3 7 . | . 1 .
. 3 . | . 6 . | 5 7 .
. . . | . . . | . . .
------+-------+------
. . 3 | 6 . 2 | 8 . .
. . 6 | . . . | 3 . .
. . 8 | 7 . 5 | 2 . .
------+-------+------
. . . | . . . | . . .
. 1 9 | . 5 . | . 3 .
. 2 . | . 1 3 | . . 4
```

Hard # 29

```
. . 9 | 6 . . | 7 . .
. . 7 | . . . | . . 1
2 . . | . 8 . | . 6 9
------+-------+------
. . . | . 7 6 | . . 8
. 4 . | . . . | . 9 .
7 . . | 4 5 . | . . .
------+-------+------
4 8 . | . 6 . | . . 5
6 . . | . . . | 4 . .
. . 5 | . . . | 7 8 .
```

Hard # 30

```
1 . . | . . . | . . .
. 2 . | 7 5 . | 8 . .
. . . | 2 . 3 | . 7 .
------+-------+------
. . . | . . . | 6 . 5
6 . 4 | . 7 . | 3 . 9
3 . 9 | . . . | . . .
------+-------+------
. 6 . | 1 . 4 | . . .
. 8 . | 6 9 . | 1 . .
. . . | . . . | . . 4
```

Hard # 31

					2	8	1	
			9	3				6
	4			5		2		7
	5					6		9
1		6					4	
5		8		7			2	
4				6	3			
	2	7	8					

Hard # 32

	8		7					
	2		5	1				
7					4	8		
	6		4			5		
	1			7		2		
	5				1		4	
	9	1						5
			9	3			7	
					7		1	

Hard # 33

3		9			5			
					7	9	3	6
8	6			9				
	4					8		
	5						7	
		1					6	
				6			8	7
7	9	4	1					
			3			5		4

Hard # 34

				7			6	2
						1	8	
		9		3	1			
9	6				8	5		
4								1
		1	5				2	6
			7	8		2		
	2	8						
1	3			5				

Hard # 35

					4	7	8	
			5					
	8			2	7			6
		7				1		2
9				3				5
2		5				4		
4			9	8			3	
					6			
	9	1	4					

Hard # 36

			3					
			1	4			3	6
5						9	7	
			2			8	9	
	1		9		3		5	
	7	4			1			
	5	9						2
6	8		1	3				
				5				

Hard # 37

				4			9	
	9	5				1	6	
	6				9			8
1			3	7				
	3						2	
				8	6			3
7			5				1	
	8	6				3	7	
	1			2				

Hard # 38

8		9						
3				7				
			8			4		1
		6	9			7		2
			2		5			
7		3			1	5		
4		2			8			
			1					5
						6		9

Hard # 39

	4				6			
			7			1		2
7		6	1					9
						9	2	
			7	3	9			
	7	4						
2					3	6		8
6		1		2				
			8				7	

Hard # 40

	2						5	6
			1			2	8	
	7				6	4		
			8	9	1			
2								9
			5	7	2			
		1	3			7		
	4	9		5				
7	3						9	

Hard # 41

			9			6		
3			6				1	4
		6	2				8	
8	6							
		1	7			6	4	
							3	6
	9					5	8	
6	5					7		2
		7			4			

Hard # 42

8	2		1	7		5		
						6		
			9	4	3	2		
6			5					
	1						8	
				2				4
	9	7	2	6				
	1							
		5		8	7		9	2

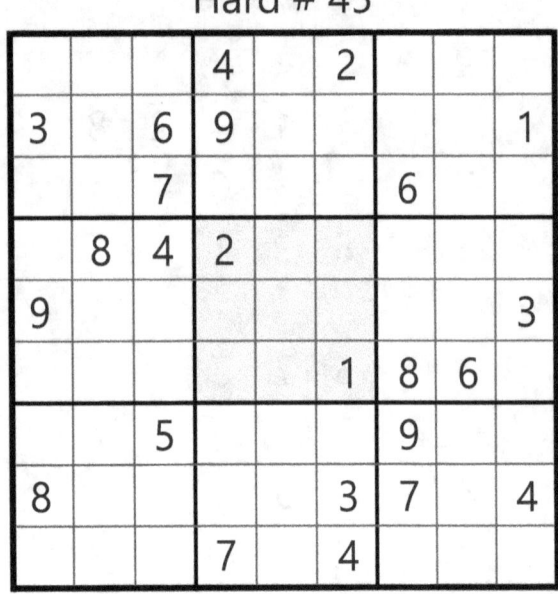

Hard # 43

```
. . 6 | 1 . . | . 9 .
. 9 7 | . 3 . | . . .
. . . | . . . | 7 1 4
------+-------+------
. . 4 | 3 5 . | . . .
9 . . | . . . | . . 8
. . . | 4 7 6 | . . .
------+-------+------
3 6 5 | . . . | . . .
. . . | . 8 . | 2 5 .
. 2 . | . . . | 6 4 .
```

Hard # 44

```
. 8 . | . . 3 | . 7 4
. . 1 | 7 . 2 | . . .
7 . . | . . . | . 5 .
------+-------+------
. 5 6 | . 9 . | 4 . .
. . . | . . . | . . .
. . 7 | . 8 . | 9 1 .
------+-------+------
. 6 . | . . . | . . 9
. . . | 8 . 5 | 2 . .
2 7 . | 3 . . | . 4 .
```

Hard # 45

```
. . . | 4 . 2 | . . .
3 . 6 | 9 . . | . . 1
. . 7 | . . 6 | . . .
------+-------+------
. 8 4 | 2 . . | . . .
9 . . | . . . | . . 3
. . . | . 1 8 | 6 . .
------+-------+------
. . 5 | . . 9 | . . .
8 . . | . 3 7 | . . 4
. . . | 7 . 4 | . . .
```

Hard # 46

```
8 . . | 1 9 . | 7 . .
. . 5 | . . . | 3 . .
. 6 . | . 7 . | . . .
------+-------+------
. 5 . | 2 . 8 | . 3 .
4 . . | . . . | . . 6
. 9 . | 6 . 5 | . 8 .
------+-------+------
. . . | 8 . . | . 1 .
. . 4 | . . . | 5 . .
. . 3 | . 5 6 | . . 4
```

Hard # 47

```
. 7 . | . 6 8 | . . 2
. . 9 | . . . | . 3 .
. . 2 | 4 . . | . . .
------+-------+------
. . 6 | . . 4 | 7 2 .
. . . | 6 . 9 | . . .
. 5 8 | 2 . . | 6 . .
------+-------+------
. . . | . . 7 | 8 . .
. 6 . | . . . | 3 . .
4 . . | 3 1 . | . 7 .
```

Hard # 48

```
5 . . | 9 . . | . . .
8 . 6 | . 1 . | . 3 .
. . . | . . . | 2 9 .
------+-------+------
. . 8 | 3 2 . | . . 6
. . . | . . . | . . .
4 . . | . 5 6 | 3 . .
------+-------+------
. 2 5 | . . . | . . .
. 1 . | . 9 . | 5 . 8
. . . | . . 3 | . . 9
```

Hard # 49

	5							
			7			4		
		5	4	1		9	3	
	8				6			1
	6	3				4	2	
4			3				7	
8	1		9	2	7			
	7			8				
						8		

Hard # 50

9	2		4				5	
				6		4		1
	7							8
				2	6			4
1								6
7			9	1				
2							4	
3		7			8			
	8				5		7	3

Hard # 51

					3			
	2			4		7	1	
					3			6
		9	7	5		2		
	7		2		4		8	
	1			9	8	4		
6			8					
	9	7		6		1		
		4						

Hard # 52

			2			8		
		2	3	9	7			
		6					2	5
	9	7						
2				4				1
						4	9	
6	3					5		
			9	5	6	7		
		4			3			

Hard # 53

3			9			6		
2						8	4	
					8			
					5	1		2
	2		7	6	1		9	
5			7	2				
			8					
	6	1						3
		3				2		6

Hard # 54

5					8			
		1	5	2	9			
7			3					
4		9			3		7	
		5				1		
	7		6			3		2
					2			8
		8	3	1	2			
			5					7

Hard # 55

4				7		5	3	
					6	1		
						7		
	8			2				6
	5	6	3		7	8	2	
7				6			5	
		1						
		9	4					
	7	4		8				3

Hard # 56

					7	8		1
3		9	6					
	8						5	
		3			9		7	
9								4
	7		4			6		
	4						1	
				4	9			8
7			6	3				

Hard # 57

7			5			8		
			4			6		9
			1	7		2		
6								2
	2	7				9	5	
9								4
		8		9	3			
2		1			4			
	9			2				6

Hard # 58

	4		5			9	1	
6								
	3		8			2		
2							3	
3	7			4			2	6
	5							4
		8		2		5		
								2
		5	4		7		8	

Hard # 59

		5		3	1			9
		4				2	8	
								6
	2			7				
1			6		5			3
				9			4	
6								
	8	1				7		
4			3	2		8		

Hard # 60

1				5				4
		7		6		2		
	2			7			9	
				6	7			
	9	3				7	4	
			8	9				
	5			3			1	
	6		1			9		
4				8				3

Hard # 61

```
. . . | 9 . 8 | 3 . 1
8 . . | . 6 7 | . . .
. . . | . . . | 2 . .
------+-------+------
. . 2 | 5 . . | . 7 3
3 . . | . . . | . . 9
4 6 . | . . 1 | 2 . .
------+-------+------
. 9 . | . . . | . . .
. . . | 8 7 . | . . 6
5 . 6 | 2 . 9 | . . .
```

Hard # 62

```
. 1 . | . . 7 | . 2 .
. 6 9 | 3 1 . | . . .
7 . 3 | . . . | . . .
------+-------+------
6 . 7 | . . 1 | . . .
. . . | 6 . 3 | . . .
. . . | 2 . . | 7 . 8
------+-------+------
. . . | . . . | 8 . 9
. . . | 2 9 3 | 4 . .
. 5 . | 8 . . | . 7 .
```

Hard # 63

```
2 1 . | 9 . . | . . .
8 . . | . 2 . | . . .
9 . . | . 3 . | 6 . .
------+-------+------
4 . 8 | . . 1 | . . .
. . 2 | 4 . 8 | 7 . .
. . . | 6 . . | 8 . 2
------+-------+------
. . 5 | . 7 . | . . 3
. . . | 2 . . | . . 6
. . . | . 9 . | . 7 4
```

Hard # 64

```
6 . . | 4 . . | . . 7
. . . | 3 5 6 | 4 . .
. . 4 | . . . | . 9 .
------+-------+------
. . 8 | 1 . . | 5 . .
4 . . | . . . | . . 5
. . . | 7 . . | 8 2 .
------+-------+------
. 7 . | . . . | . 3 .
. . 4 | 6 7 8 | . . .
2 . . | . . 3 | . . 9
```

Hard # 65

```
5 . . | . . . | . 6 .
. . . | 5 2 . | . 4 .
1 . . | . . 3 | . . .
------+-------+------
. . 2 | 8 . . | 5 . 6
. 8 . | 7 . 6 | . 2 .
6 . 5 | . . 9 | 7 . .
------+-------+------
. . 9 | . . . | . . 2
. 5 . | . . 1 | 8 . .
. 4 . | . . . | . . 8
```

Hard # 66

```
. 4 . | 1 . . | 3 7 .
1 . . | 6 . . | 8 . .
. . 7 | 4 . . | . . .
------+-------+------
. . . | 2 . . | . . .
4 . 1 | . . . | 6 . 7
. . . | . 3 . | . . .
------+-------+------
. . . | . 8 . | 7 . .
. . 8 | . 3 . | . . 6
. 7 9 | . . 5 | . 3 .
```

Hard # 67

7				2	6		8	
		9		5				
	5				8		4	
	2				9			3
9								7
8			4				5	
	3		9				7	
				8		5		
	7		6	4				1

Hard # 68

3		8		9				
7						4		
4			2	8	7			
9	2	3						
				5				
						1	7	2
			8	1	4			9
		9						3
				2		8		1

Hard # 69

4			2			9		
5		6			9			
				3				4
	3				7		2	
1		2				7		6
	9		5				3	
9				6				
			8			2		5
		4			1			3

Hard # 70

		9	5	6	4	2		
		1						9
2			3					
								1
3	8		4		9		5	6
5								
					8			7
6						1		
		5	1	2	6	8		

Hard # 71

2					1	5		4
	1				8			2
			6			1		
		2						1
		6		4		9		
8						2		
		7			6			
5			4				7	
4		8	3					5

Hard # 72

				8	7			4
		1	3				8	
4		9	5					
1	9							
	6		9		8		5	
							3	7
				2	3			6
	1			5	2			
9			4	1				

Hard # 73

	3					9		4
				4				
1		2	8			5		
		5		1			6	
2			3		7			5
	6			9		2		
		3			6	1		2
				3				
7		9					4	

Hard # 74

	8	2			4			6
9			7					
				2		9		
		4			2		7	
	3			5			1	
	6		4			3		
		5		4				
					1			7
1			9			8	3	

Hard # 75

	4		5				6	
		3				2		
	6		4	1		5		
			2			1	7	8
2	9	1			4			
		7		3	6		1	
		8				7		
	1				8		9	

Hard # 76

		8	5			7		
					9			3
3						1	4	
4			3	2				
		7	9		1	2		
				4	5			7
	8	3						1
1			6					
		2				4	6	

Hard # 77

		8		5			2	1
	4						3	
9		7						5
	8		6					
			3		4			
				2		9		
1						5		2
	5						1	
4	3			6		7		

Hard # 78

2		7		8				
					1	4	9	
		4						
5					9			2
8			7		6			1
9			8					4
						1		
	8	5	9					
			7			8		6

Hard # 79

9		7				5		
			8		2		1	
3					7			
4			2					5
	3	5				2	6	
2					6			4
			3					9
	9		6		8			
		8				3		7

Hard # 80

		7	1	3			2	
		9	4					
	2		6					8
	3		2					
2		7		5				6
		6		1				
3			9			7		
			6	2				
	9		7	1	5			

Hard # 81

4					2			
	7	6				9	1	
	2		9	1				
9			2	7				
		4				1		
			5	4				6
			8	6		2		
	1	5				8	7	
			1					4

Hard # 82

4		2	9					
	9	8	5					
3						5		
			4			1	6	
	4		7				8	
	7	1			9			
		4						9
					5	6	3	
					3	2		1

Hard # 83

	3		7		9	4		
5								1
			3		1			
					5	1		
7		3		2		6		8
	6		9					
			5		2			
2								5
	4		6		3		7	

Hard # 84

	3		9					
			1		6	8		
1	4							
	7	1	6				9	
		2		4		8		
	5				7	3	1	
						5	2	
9	7		2					
					1		4	

Hard # 85

							4	8
	8	7		5				6
6					9		1	
9				7				1
			6		4			
5				3				4
	3		9					7
2				4		3	8	
4	5							

Hard # 86

9			4	8				6
		4			6	3	9	
	6		7			9		5
1								2
2		7			8		6	
	2	3	1			7		
7				2	4			8

Hard # 87

7	3						9	8
		2	6					
	8				5	1		
			8					
4			1	3	6			7
			5					
		7	8			6		
				4	8			
8	9						5	4

Hard # 88

			4	9			7	
7	9							
1		2						3
2				5			1	
		3	1		2	4		
	7			3				2
6						7		8
							4	6
	3			6	5			

Hard # 89

	8	5			1		2	
								9
			2	4	8			7
			4	5		9		
4								3
		1		9	7			
1		3	7	6				
2								
	7		1			6	4	

Hard # 90

2					6			
			3		1		4	
	8			7				
3		8				4	1	
		4		5		8		
	6	1				2		3
				4			9	
	7		2		8			
			9					4

Hard # 91

		9		2				
3					6			1
	6		5	8				
9		3			5			
1								7
			7			4		6
				9	1		2	
6			3					4
				4		8		

Hard # 92

5		8	4					
6	4		1				8	
	2	3	8					
8							1	
4								9
	6							3
					5	6	7	
	7				2		5	4
					7	9		2

Hard # 93

	7			3				8
2		5						
	1		7		9			
6	4			7	8			
9								7
			4	6			9	2
			6		1		3	
						7		4
3				5			6	

Hard # 94

	7	5			8			
						9		4
			7	2			5	
6			1		5		8	2
5	2		3		4			7
	4			5	6			
7		6						
			9			4	2	

Hard # 95

	2				9			
3		1		8				
			1		6	7		3
4		9				8		
5								1
		3				9		5
2		4	7		5			
				3		6		2
			6				5	

Hard # 96

		6			3			
		8	7					
	3	5		8			6	1
					1	9		
6			8		4			7
		1	2					
	1	2		7			8	5
					5	7		
		6			4			

Hard # 97

		5	9				6	
4				6				
1			3		7			
	4				3	9		8
	7						1	
2		3	1				7	
		6			8			2
				3				4
	3				9	1		

Hard # 98

2			3				1	9
		4		1				8
9				7		3		
	4							7
		2				8		
3							6	
		3		6				1
5				8		6		
4	2				9			3

Hard # 99

			9	3				
9			6	2			7	
1								9
4		1				8	6	
	7						2	
	5	9				1		7
8								2
	3			9	6			1
				1	4			

Hard # 100

	5	7		1			6	
1			7			4	3	
			6					
8			5					
6	9						1	3
				4				7
			8					
	2	3		5				8
	8			4		7	5	

Hard # 101

			4		1	2		6
	6							9
5				9				
		2	7		4			
	5						8	
			5		8	1		
				6				4
1							2	
7		8	9		5			

Hard # 102

	9		3		2			
			7			2	3	
			6				8	4
1			5		7			
	5						2	
		7			1			8
9	4			5				
	2	5		3				
			6		7		9	

Hard # 103

9	5						1	
3		7			9			
				2	8			
			3			2	5	
			7	2	6			
	6	4			8			
		3	9					
			4			7		5
	7						9	3

Hard # 104

9	4		3					5
			4				8	
	2							1
		7	2	6		9		
		8		9	1	4		
7							9	
	1			2				
2				4			6	7

Hard # 105

			2				1	8
4				7				
	2	5	4					9
		7	5			8		1
1	9				8	5		
5					2	7	9	
			6					5
3	6				7			

Hard # 106

9	2		1	8				3
	4			3		5		
			5					
2		4	3					
8								6
					7	9		2
				3				
		7		1			3	
6				2	9		7	1

Hard # 107

1					3			
		7				8		
	5	9	7			4		1
				2		6	9	
			5		4			
	8	5		1				
2		4			1	3	6	
		1				7		
			4					2

Hard # 108

4		1			9	2		
3							6	7
			4			3		
			4	3	2			8
7			8	1	5			
		4		2				
6	9							2
		7	5			1		3

Hard # 109

1		5		3	9			
		9		1				
2	3				8			
6		7	8				5	
	8				6	7		9
			6				9	4
				8		6		
			4	5		8		7

Hard # 110

3	5		8			4		
				7	4			
	6					1	7	
		2		5	7			
	4						2	
		4	1		8			
9	3						1	
		1	4					
	5			6			4	3

Hard # 111

		6	5					
		2	8					
8	4				1		5	
		9		5			4	
	2		3		4		9	
	7			1		6		
	3		4				1	8
					9	2		
					3	4		

Hard # 112

		3		2			6	
		8			9			
	9		4				3	1
				5	4	3		2
2		5	7	6				
3	1				5		2	
			2			7		
	5			8		1		

Hard # 113

			4	8		3		
	1							
	6		1			5	8	
2	3		6					
9				5				3
					3		9	7
	8	4			9		5	
							7	
		9		6	4			

Hard # 114

				3		1		
	6	7			8		4	
9					2		8	
	2					8	3	
		4				5		
	7	1					6	
	9		4					3
	8		9			6	5	
		6		1				

Hard # 115

							4	
		7		1				
	3	1	2					5
	2		6		3		5	
	7						9	
	4		7		8		3	
9					1	3	8	
				2		9		
	6							

Hard # 116

5		8	4			7		
				6	7			
1			9					3
9			1			8		
	6						3	
		2			9			5
2					5			9
		7	1					
		3			6	1		4

Hard # 117

			4		5			
	8	4	9			6		
7				1				9
			7		1	5		
	3						2	
		9	2		4			
6				5				8
	4				6	1	9	
			1		7			

Hard # 118

		7	8	2				
					5	7		
1		5	3			8		2
3								
	9		2		8		1	
								9
5		1			2	3		8
		9	6					
				5	7	2		

Hard # 119

3	4					5		
		7						
			9	8				4
1	7				5			
		2	6		8	3		
			2				7	6
8				2	1			
						9		
	1						6	5

Hard # 120

8	4		5			9		
			4				8	
7					9			1
				4	7			
5		7				3		9
			9	3				
9			1					6
	3				6			
		8			3		1	2

Hard # 121

	8			3				
		7				5	9	
		9	1		4			
			4			7		2
3			2		6			9
8		2			3			
			8		7	1		
	1	4				3		
				1			2	

Hard # 122

9						5		
			4	5				
	1						6	3
		3	5	2	9	8		
7								2
		4	7	6	8	9		
3	5						2	
			3	1				
			1					8

Hard # 123

	7							
9				2	4			
	5		6	9		8		
		8		6	5			
		2		8		7		
			7	4		2		
		6		1	2		4	
			9	7				6
						3		

Hard # 124

		6						
8				6		9		1
		1	8		2		6	
9	3				4			
			3		5			
			6				2	4
	1		5		9	2		
3		9		4				6
						5		

Hard # 125

	1	4						
				7		6		3
			9			4	2	
8			1			3	4	
		5				2		
	7	2			3			9
	2	9			6			
3		7		4				
						7	9	

Hard # 126

			3			2	7	
4		7	9	8				
9				1				
	9				8		3	
		5				8		
	7		4				1	
				2				5
				4	3	7		1
	8	3			6			

Hard # 127

	8	1	9				2	
					5	1		
				8			9	
3	4		5	2				
		2				5		
			9	4			1	7
	2			1				
		4	8					
	1				7	9	8	

Hard # 128

	5				8	3		4
8		1				2		
			7			8		
2								9
	6		8		5		2	
5								6
		5			4			
		6					9	5
9		7	2				1	

Hard # 129

						6	1	
2						4	8	
			2	9	6			
	2			4				
9		3	8		1	7		6
				7			3	
			1	3	5			
	5	7						1
	8	2						

Hard # 130

			6				4	5
5	3				4			7
	9			1				4
		8	2	7	3	5		
1				5			8	
8			5				6	1
9	2				7			

Hard # 131

5		4	9			6		
				6				4
6		2		4				
	7		6	1		8		
		1		7	5		2	
			5			9		2
7			1					
		3			9	5		6

Hard # 132

					4	8		
	7	1			5			
5	4			7				9
			4	8			5	3
7	9				2	3		
4				5			2	1
		7				5	4	
		3	1					

Hard # 133

	5	2	3					
	6		8				7	
	8	9		7		5		
			9		6			
5								4
			4		7			
		4		3		8	2	
	2				8		5	
					2	7	1	

Hard # 134

			9					
		3		8	2			
8				5		4		3
						9	3	
2		9		7		5		1
	4	7						
6			2		9			5
			3	1		7		
					6			

Hard # 135

5	4		2	6		8		
	9		3					
		2			4			
	1			9				3
			8		3			
7				2			4	
			1			6		
					2		9	
		1		8	7		5	2

Hard # 136

	4		5			2		
7				2			6	1
				1				
	5		7				1	6
4								3
1	3				8		7	
				5				
3	1			9				2
		7			1		9	

Hard # 137

			3		8			6
1		4						5
					2			7
	2	8		1	5			
			6	2		8	1	
5			4					
9						5		3
6			5		7			

Hard # 138

7			5			4	1	
				7				3
		6		1		5		7
5						4		
			4		6			
	8							6
2		8		5		6		
9				4				
	4	7			8			2

Hard # 139

			9			6	5	
3	9		5					1
					2			7
8	5						7	
		2				4		
	4						9	8
7			8					
9					6		2	4
	8	5			1			

Hard # 140

1	5				6			2
						3		
		7		8	2			
	4	9	5			1		
7								6
		1			8	7	4	
			2	4		9		
		5						
2			1				7	3

Hard # 141

		9		5	1		3	
8						9		1
			8					
6				3			1	4
	7						2	
4	8			1				6
					4			
7		6						2
	2			1	8		3	

Hard # 142

				2				
					3		6	
3	7				8		9	1
1			4			8	5	
		9				1		
	8	7			5			6
7	6		5				3	9
	9		8					
				9				

Hard # 143

	5	2						8
	8			4			2	
				9				3
	2			7	5			
		5	4		9	6		
			6	1			4	
4				3				
	9			5			8	
5						3	7	

Hard # 144

	6					1		3
5	8							
			7		3			9
	3	8	6	9				
			8		7			
			1	4	2	9		
1			5		9			
							5	7
8		7				1		

Hard # 145

9	4	1						8
2				7				1
		7				2	5	
					6			
4			8		7			2
			9					
	9	5				1		
7				2				9
1						3	8	4

Hard # 146

			7					
9						4		
			9	5	6	3	8	
1	6		5					
		7	8			2	5	
					1		7	3
	8	3	4	6	9			
		6						2
					3			

Hard # 147

3					1			
	4	9					7	
	2		5			9		
2	6			4				
9								6
				3			5	7
		1			5		4	
	9					6	3	
			2					5

Hard # 148

		9	8					
6						7	2	
		8	5	4				3
	6			3	5	8		
		3	7	8			4	
	9			7	6	4		
	4	7						5
				4	2			

Hard # 149

5			2					
			6			8		5
	8			3				
		4		8		1	2	
6								9
	7	2		5		3		
				9			5	
1		3			4			
					1			4

Hard # 150

		1		9		8		
	2						3	1
4		9						
	5		4	1				6
			2		5			
8			3	7		2		
						7		5
3	9					1		
		7		4		2		

Hard # 151

		5	7		4			
	9			8	3		7	
		4						3
				3			5	8
4								1
5	7			1				
8						9		
	6		2	9			4	
			1		5	2		

Hard # 152

	4		3					2
		7						
	5		4	7				3
5			2			8		6
	1						2	
8		6			5			7
2				8	3		1	
						3		
9					6		5	

Hard # 153

8			4	7		2		
	2					4		
	5		1					
					8			
	1	8	3	9	5	6	4	
		3						
					3		9	
		7					1	
		6		5	8			4

Hard # 154

	3					1	7	
			6					3
		2			1		8	
					7	8	6	
	6		3		5		1	
	8	5	9					
	1		8			5		
8					6			
	9	4					2	

Hard # 155

	5					2		
			5		1		6	
9	4			7	2			
2	8							6
			6		3			
5							9	1
			1	8			3	4
	9		2		6			
		4				8		

Hard # 156

		3				2	1	4
7	1							
2				6				
3		7	6		9			
			2		9			
	4		5	1		6		
		8				4		
					1	2		
7	1	9			3			

Hard # 157

9	8		4	7	1			
			3			9		
				6				3
				3	4			9
	5				6			
3		4	7					
6			1					
	7			4				
			6	3	2		5	1

Hard # 158

			7				2	
	9				2		5	
		7		4	9	6		
8		6		3				
	4						6	
			9			1		5
		8	2	7		9		
	5		8				1	
	3			1				

Hard # 159

4								
			4		2	8		
			5		3	7		
	5		8			4		
	4		1	5	6		2	
	7			9		3		
	8	6		4				
	9	2		6				
								8

Hard # 160

2	4			3				9
			1					6
7				8				
				5			6	2
	6			1			3	
2	5		6					
			5				2	
9				4				
5			3			7	4	

Hard # 161

		4		5	3			1
			8					
6						5		7
	3		9			1		
4		8				6		2
		9			4		7	
3		6						8
					8			
1			2	4		7		

Hard # 162

	9		5			6		
2				1			3	
		5		9				
			6	3			4	
7	8						9	3
	4			5	9			
				6		2		
	2		9					4
		7			4		8	

Hard # 163

9					5			
8					1			2
1	6	3	7			8		
			8				7	
			9		4			
	1				2			
		9			8	7	2	3
3			5					9
		1						8

Hard # 164

			6				3	9
		3				1		
		2	1		9	6		4
	7	4						
5								7
						9	1	
3		1	7		2	5		
	7				4			
8	6			3				

Hard # 165

			1	7				5
	9							
8				2		4		
	3		2	9		5		
	2	8				9	6	
	5		1	8		2		
	9		7					3
					1			
7			2	5				

Hard # 166

5	1							
	7	8						
	9		6	2	4			
	8	5		4				9
6								3
9			7			2	8	
		6	4	2		3		
						6	1	
							6	7

Hard # 167

	8	4		6				7
		7					3	4
	3					5		
			8	3				
9				4				6
				7	6			
		8					5	
3	9					7		
5				2		1	9	

Hard # 168

	1							4
			8		9			
5		8				9	6	
9				3	6	7		
		4				5		
		2	5	7				1
	9	6				1		8
			7		2			
8							2	

Hard # 169

	7	1				2		
8					9			7
			5					8
				6		4		1
		8	9		1	3		
6		4		3				
9					2			
4			8					6
		2				5	8	

Hard # 170

			3		9		1	7
					4			6
	1			7			3	
2		8				6		
			8		2			
		6				5		2
	4			8			5	
8			9					
1	7		4		5			

Hard # 171

	2		9					
4			8	1				6
1		8		2				
							9	1
2	1						6	4
7	9							
			3			6		5
3			5	7				9
			4			3		

Hard # 172

			8	6		5		
		1	2				9	
				3				2
9	8					7		
4			5		6			9
		2					4	1
3			6					
	2				9	6		
		8		3	4			

Hard # 173

			1					3
	5	2				6		
6			8					
						7		1
	1	9	7		6	4	2	
5			3					
					8			6
		4				8	9	
9					4			

Hard # 174

	6			5		9		7
			1					
		3	2			8		
		2	4				6	8
1	8				6	3		
		4			5	1		
					7			
7		5		2			3	

Hard # 175

	7					1		
				2				7
1			5			2	4	6
3								
	2	6	1			4	9	7
								8
7	3	5		2				4
6			8					
		4					9	

Hard # 176

	5			1		9		
	7					6	5	
			2				4	1
	2		6		1			
	3						1	
			9		8		7	
4	6					3		
	7	8				1		
		2		5		8		

Hard # 177

3	2	1			7	9		
		4		5		2		
								1
	6		1	8				5
4				9	6		7	
1								
		8		6		7		
		2	7			8	4	9

Hard # 178

3								7
8	4		3		7			
7			8	1				
		2						1
		1	2		4	6		
5						3		
			3	8				2
		9		5			6	3
2								5

Hard # 179

8				1		9		
	4		5				1	
	7	1						
9			2		5		6	
6								4
	8		9		7			5
						8	7	
	9				4		2	
		2		8				3

Hard # 180

		6				4		3
	4	3						
			8			3		7
1		5			8		4	
2								1
	7		1			2		9
4				7		9		
							9	7
	3		2			1		

Hard # 181

	6		7				4	
4		3				2		
			5	3		6		
1	5							
	7						3	
							6	5
		4		2	8			
		6				5		8
	3				5		9	

Hard # 182

6					5			
			8					
		1		7	4		3	
	6			5	7	8		
9		2				5		3
		7	9	2			6	
	8		2	3		1		
				9				
		7						4

Hard # 183

	3		9					7
				1				
	7		2	8				6
		2					7	1
		7	3		6	8		
3	8					5		
9				5	1		3	
				9				
7					4		1	

Hard # 184

		4		5				3
5			4					
			1	2	8			
		9			7		1	4
	6					3		
4	1		5			9		
		5	7	6				
					3			2
7			1			4		

Hard # 185

	4		2					
7	1	8						
	5	3	9					
	2		6					7
3			8		7			1
6					3		2	
					8	7	6	
						5	8	4
					6		9	

Hard # 186

			1		6	7		
						1	3	2
	8	7						
9				8		5		
	4			5			1	
		5		2				3
						3	8	
5	6	9						
		2	7		4			

Hard # 187

	4	1			2	3		
	6							
				9		5	7	
4				3		9	5	
2	9			6				8
6	2		3					
						2		
		7	4			1	6	

Hard # 188

1	4							
		8	9		2			
			8	4			9	
7			2				8	
3			7		9			1
	2				6			9
	3			5	8			
			4		7	8		
							4	5

Hard # 189

					1	8		5
	5	8				4		
	7		4					3
			3		9	7		
7								6
		3	5		2			
8					3		1	
		7				3	4	
1			6	2				

Hard # 190

			6					
		7				2	4	1
				7	3			9
		5		2	9			
8	9						3	5
			1	8		6		
1			7	4				
4		8	2			7		
				6				

Hard # 191

	7			2	6			
	2		4			9		
		1		7				
8	1				5			
7	5						2	1
			8				5	3
				8		1		
		7			4		9	
			1	3			6	

Hard # 192

			4			1		
	9			8				
		5		3				7
		1	8			9		
3	5			6		9	7	1
		4			5	2		
1				5		3		
				7			6	
		6			4			

Hard # 193

						1		
	8	6			7		2	4
7		2				9		
			7	3				5
			5		6			
8				9	2			
	9					5		3
6	3		2			7	1	
		1						

Hard # 194

						9		
8				6	9	4		
1	3		4					
7			9		6			
	6		1		4		8	
			5		2			4
				8			4	1
		6	7	9				2
		5						

Hard # 195

2					1		6	
	6		7			5		
				3		8		
6			2				1	
		7	4		6	2		
	4				3			7
		1		6				
		4			2		9	
	9		5					3

Hard # 196

	1		6	2	3		8	
			8	5				4
		6						
	5		2	9				
9								7
			6	8		1		
						4		
2			1	5				
	8		7	3	6		5	

Hard # 197

	3		4					
8			1	5			6	
			9	7	3			1
						1		9
		9				3		
3		4						
4			5	6	7			
	5			2	9			6
				3			5	

Hard # 198

		9		4				
		6			1		9	
5	1					6		
7				2			3	
3			1		9			2
	5			8				4
		5					1	6
	9		3			2		
				6		7		

Hard # 199

```
+-------+-------+-------+
| 8 . . | . 6 . | . . 9 |
| . . 1 | . . . | . 7 . |
| 5 . 7 | . . 1 | 4 . . |
+-------+-------+-------+
| . . 6 | . 4 . | . . . |
| 3 . . | 2 . 8 | . . 5 |
| . . . | . 5 . | 3 . . |
+-------+-------+-------+
| . . 5 | 9 . . | 8 . 2 |
| . 4 . | . . . | 1 . . |
| 7 . . | . 3 . | . . 6 |
+-------+-------+-------+
```

Hard # 200

```
+-------+-------+-------+
| . . 8 | . 2 . | 1 . . |
| 1 3 . | . . . | 4 . . |
| . . 9 | . 3 6 | . . . |
+-------+-------+-------+
| . 8 . | 9 . 2 | . . . |
| 7 . . | . . . | . . 4 |
| . . . | 3 . 5 | . 7 . |
+-------+-------+-------+
| . . . | 2 1 . | 6 . . |
| . 1 . | . . . | . 5 3 |
| . 4 . | . 6 . | 2 . . |
+-------+-------+-------+
```

Hard # 201

```
+-------+-------+-------+
| . 2 . | 4 3 . | . 9 . |
| . . 3 | . . 1 | . . 2 |
| . . 9 | . . . | 5 . . |
+-------+-------+-------+
| . . . | 8 3 2 | . . . |
| 2 . . | . . . | . . 6 |
| . . 6 | 9 1 . | . . . |
+-------+-------+-------+
| . 3 . | . . 5 | . . . |
| 5 . . | 3 . . | 7 . . |
| . 8 . | . 6 5 | . 1 . |
+-------+-------+-------+
```

Hard # 202

```
+-------+-------+-------+
| 4 . 7 | 3 . . | 2 . . |
| . 2 . | . . . | . 1 . |
| . 5 . | 9 . . | . . . |
+-------+-------+-------+
| . 8 . | . . 1 | 5 . 4 |
| 1 . . | . . . | . . 8 |
| 9 . 6 | 5 . . | . 2 . |
+-------+-------+-------+
| . . . | 6 . . | . 3 . |
| . 6 . | . . . | . 8 . |
| . . 1 | . . 5 | 7 . 2 |
+-------+-------+-------+
```

Hard # 203

```
+-------+-------+-------+
| . . . | 5 . . | . . 6 |
| . 3 8 | . . . | . . 1 |
| 4 . . | 2 . . | 5 . . |
+-------+-------+-------+
| . 8 . | 4 5 1 | . . . |
| 9 . . | . . . | . . 7 |
| . . 5 | 6 2 . | . 4 . |
+-------+-------+-------+
| . . 6 | . . 2 | . . 5 |
| 8 . . | . . . | 9 2 . |
| 3 . . | . 8 . | . . . |
+-------+-------+-------+
```

Hard # 204

```
+-------+-------+-------+
| . 5 . | . 2 . | . 9 . |
| . . 6 | 9 . . | 7 . 4 |
| . . 7 | . 3 . | . . . |
+-------+-------+-------+
| . . 9 | . . . | . 5 2 |
| . 2 . | . . . | . 7 . |
| 1 7 . | . . . | 6 . . |
+-------+-------+-------+
| . . . | . 8 . | 3 . . |
| 8 . 4 | . . 6 | 9 . . |
| . 9 . | . 1 . | . 6 . |
+-------+-------+-------+
```

Hard # 205

					7			9
	1	8						
6		7	9					
5			4		3		8	
8								6
	4		1		5			7
					9	2		3
						7	1	
7			5					

Hard # 206

8			3		5			
		3	9			6		
			8			5	4	
		2					5	9
	9					7		
7	6					8		
	7	9		8				
		1		4	7			
			5		1			4

Hard # 207

9			8	5				2
	4			1				5
		8			7			
		6				1	5	
			5		3			
	9	3				2		
			6			9		
4				5			7	
7			9	3				6

Hard # 208

		9	7					
	1			9				5
			4			1	8	
4		1		6			5	
		7				3		
	5			8		6		4
	2	8			5			
9				7			1	
					6	5		

Hard # 209

	4		7			8		
	6		8		1	2		
	3							
7			2					1
2	9					6	4	
4			6					7
						9		
		5	6		3		4	
	8				4	5		

Hard # 210

2		3						
6	4	8		9				
			3				6	
5		4						2
1			2		9			5
8						6		4
	9				8			
			1			5	8	9
						4		1

Hard # 211

	5						9	8
3	2				6			
		8	1				7	
				6			8	4
		2				5		
8	9			4				
	3				4	9		
			2				5	1
2	8						6	

Hard # 212

8	7			3				
		1	4	2				
	3						4	1
		8		5	1			6
4			9	8		1		
5	8						9	
			4	2	3			
			6				5	2

Hard # 213

8	9			2		1		6
4								
			5	8				4
		4			9	8		
	2						9	
		3	5			4		
6			3	1				
								2
2		9		6			8	3

Hard # 214

7		5	8					
			3	1		8		
			5			7	4	
	4					6		1
	1						5	
8		2					4	
		7	6			4		
	3		9	1				
					2	9		8

Hard # 215

3	9	2						
	5	7	1		2		3	
			5					
			3			4		8
	7						5	
1		6			8			
				9				
	1		6			4	3	7
						9	2	1

Hard # 216

				2				1
5	9		3	1				
		8					7	
				2	4	1		
	5	4				3	8	
	8	1	4					
	6					5		
			4	7			3	6
3				9				

Hard # 217

```
2 . 7 | . 8 4 | . . 5
. . . | . . . | . . .
. . . | 1 2 . | 7 . 3
------+-------+------
. . . | 8 3 . | . 6 .
. . 9 | . . . | 1 . .
. 8 . | . 5 9 | . . .
------+-------+------
4 . 5 | . 6 8 | . . .
. . . | . . . | . . .
9 . . | 3 4 . | 5 . 1
```

Hard # 218

```
2 5 . | 7 . . | . . .
. . . | . 8 . | . . 3
8 . . | . 9 7 | . . .
------+-------+------
. 2 . | 4 . . | 5 . .
. 1 . | 8 . 6 | . 2 .
. . 9 | . . 1 | . 6 .
------+-------+------
. . 6 | 9 . . | . . 8
7 . . | . 3 . | . . .
. . . | . . 4 | . 7 2
```

Hard # 219

```
. . 2 | . . 9 | . . .
. . . | . 6 5 | 7 . .
3 . 4 | . 5 . | 8 . 2
------+-------+------
. . . | 1 . . | 3 . .
8 . . | . . . | . . 7
. . 3 | . . 7 | . . .
------+-------+------
7 . 6 | . 9 . | 1 . 3
. 2 5 | 8 . . | . . .
. . . | 6 . . | 2 . .
```

Hard # 220

```
6 . 8 | 7 1 . | . . .
. . . | . 5 . | . . 4
. 4 . | . . 8 | . 2 .
------+-------+------
. . . | . . . | 4 . 1
9 . . | . . . | . . 5
7 . 5 | . . . | . . .
------+-------+------
. 9 . | 5 . . | . 1 .
8 . . | . 6 . | . . .
. . . | 3 7 9 | . . 6
```

Hard # 221

```
. . . | 2 8 . | 3 7 .
. . . | 9 . . | . . .
. . 2 | . . . | 8 9 4
------+-------+------
. . 5 | 3 . 4 | . . .
. 9 . | . . . | . 5 .
. . . | 8 . 5 | 1 . .
------+-------+------
6 1 8 | . . . | 3 . .
. . . | . . 9 | . . .
2 4 . | 1 6 . | . . .
```

Hard # 222

```
. . . | 8 . 3 | . . .
. 6 . | . 4 5 | . 9 .
3 . . | . . . | . 8 2
------+-------+------
. . 1 | . 7 . | 8 . .
. . . | 9 . 1 | . . .
. . 9 | . 2 . | 5 . .
------+-------+------
9 5 . | . . . | . . 8
. 1 . | 7 8 . | . 3 .
. . . | 3 . 2 | . . .
```

Hard # 223

	3		2			4		
				4			7	
5		2			6			
		8			5			2
9	5						6	7
6			3			5		
			6			1		4
	8			3				
		1			8		9	

Hard # 224

9		5	2					
		8	5		6	1		
			1	4			5	
				1				
1		3				5		6
		4						
	7			2	4			
		6	8		9	2		
					3	8		1

Hard # 225

8				5				
				3		8	4	
				6	1	9	3	
		8	3				7	
9								8
	3			2	4			
3	4	7	6					
1	9		2					
			1					6

Hard # 226

3		9			1		2	5
		5	8					
	4						8	
			3					1
	2		5		8		6	
4					9			
	6						3	
				2	5			
5	1		4				6	7

Hard # 227

		6	9					
4			7		2		6	
				1	5	7	3	
								6
	9		8		7		5	
5								
	5	8	2	7				
	2		1		3			5
				6	9			

Hard # 228

4							8	
		3			7			
5					4		6	7
	9		1					8
1			8		2			3
7				5			9	
8	1		5					2
			2			7		
	4							9

Hard # 229

							9	
5		6		2	8			
	7							2
		7	3		2		6	
		3	1		5	9		
	6		7		9	1		
7							3	
			5	1		2		4
	8							

Hard # 230

9				3		5	8	
			6	5				2
1				9				
4		7						
		8		5		7		
						6		8
				8				1
6			9	4				
	7	9		1				5

Hard # 231

							3	1
				9	3			
		6			8		9	4
	9		7			2		
			3	1	5			
		3			6		7	
4	5		2			3		
			8	5				
9	1							

Hard # 232

			3	6		1		
	5		9			3	7	
	4				7			
						5		2
4	6						3	1
9		5						
			6				1	
	3	6			4		9	
		9		8	2			

Hard # 233

			8		6			
	6		3			5		
			7			3	6	1
			1				3	2
		1				7		
2	4			9				
7	9	3		5				
		8			7		4	
			2		9			

Hard # 234

			8	7	4	6		
							9	2
			6		1		3	
7	3				2			
		5				9		
			1				8	4
	1		5		3			
2	7							
	8	3	7	1				

Hard # 235

2	3			6	4		7	
		5				9		1
			2			4		
						4	8	9
5	8	7						
	4			5				
3		6			2			
	5		4	8			6	2

Hard # 236

1			3			4		5
			8			2		
		3				4	1	
	5				2			
8	3						4	2
			7				6	
		9	2			6		
		8		7				
3		4		1				9

Hard # 237

	6				5			
	5	8			6	2		
3		9						4
				2		1		
7			5		4			8
		3		7				
4						9		3
		2	1			4	8	
			3				2	

Hard # 238

	3	1				5		
4				9				
		9					1	2
			5		7		8	
		5	2		9	3		
	6		1		3			
1	2					8		
			2					1
		6				2	7	

Hard # 239

		1	6					7
					7		5	8
				1		9		
6			3	2		7		
	3					2		
	7		1	5				9
	8		9					
7	2		3					
1					6	7		

Hard # 240

	3		8			4		
6	2	7						
			1					
3	1				2	6	9	
	4						8	
	8	9	5				1	4
					5			
						1	3	5
		3			8		6	

Hard # 241

			3			7		
	7	5						
2			1				4	
	1		7	8			9	
4	2						6	5
	6			4	5		7	
	9				2			6
					4	8		
		6			1			

Hard # 242

	4					5		9
2						4		
		3			5		7	
			6			7	5	
9			8		4			3
	6	4		3				
	2		1			6		
		8						7
5		7					2	

Hard # 243

				7				1
		1	2		3		4	
7	3		9					
				2	4	3		
	5						7	
		6	7	8				
					1		2	3
	9			6		2	7	
6				3				

Hard # 244

		9	4	7				2
							9	
	4			8				3
						7		8
1	7						4	6
3		2						
4				9			8	
	5							
8				6	3	2		

Hard # 245

				1				
		6			7			
7	9			5			4	
8	3				6			
5			2		1			9
		9					2	8
	1			3			9	6
			7			2		
				4				

Hard # 246

1		4						
2					9	3		
			4	1		6		
3			7	8			2	6
8	5			6	2			3
		8		4	3			
	5	1						7
						8		9

Hard # 247

```
. 5 . | . . . | 2 . .
. . . | 6 3 1 | . . .
. . . | 2 4 . | . 6 5
------+-------+------
7 1 . | . . . | . 4 .
. . . | 3 . 2 | . . .
. 6 . | . . . | 9 2 .
------+-------+------
8 3 . | 2 7 . | . . .
. . 7 | 8 1 . | . . .
. . 6 | . . . | 8 . .
```

Hard # 248

```
7 . . | . . . | 3 . .
. 3 . | 1 . . | . . .
. . . | 5 . . | 7 1 .
------+-------+------
6 8 . | 4 7 . | . . 9
. . . | . . . | . . .
1 . . | . 9 5 | . 6 8
------+-------+------
. 6 2 | . 4 . | . . .
. . . | . 9 . | . 7 .
. . 4 | . . . | . . 6
```

Hard # 249

```
9 . 1 | 6 . . | . . 3
. . . | 4 . . | . . 9
. . . | . 9 . | . 4 2
------+-------+------
. . . | . . . | 6 . 5
. 8 . | 2 . 6 | . 9 .
1 . 9 | . . . | . . .
------+-------+------
8 9 . | . 6 . | . . .
5 . . | . . 7 | . . .
7 . . | . . . | 5 8 4
```

Hard # 250

```
. 7 . | 3 1 . | . . .
9 . . | . . . | 8 . .
4 8 . | 5 . 7 | . . .
------+-------+------
1 . 3 | . 2 . | . . .
. . . | 8 . 5 | . . .
. . . | 6 . . | 4 . 7
------+-------+------
. . . | 1 . 9 | . 2 8
. . 2 | . . . | . . 3
. . . | 5 3 . | 4 . .
```

Hard # 251

```
. . 1 | . . . | . . .
4 . . | 9 . . | 2 8 .
. . 9 | 3 5 . | . 4 .
------+-------+------
. . . | . 7 . | 5 . .
7 . . | 2 . 4 | . . 3
. 4 . | 6 . . | . . .
------+-------+------
. 9 . | . 1 3 | 6 . .
. 6 2 | . . 9 | . . 4
. . . | . . . | 1 . .
```

Hard # 252

```
. 7 . | . . . | 4 . .
5 . . | 8 . . | . 1 .
. . . | . 3 1 | . . .
------+-------+------
. . . | 6 . . | 5 2 .
7 . 2 | 9 . 4 | 6 . 1
. 6 5 | . . 3 | . . .
------+-------+------
. . . | 5 4 . | . . .
. 3 . | . . 6 | . . 9
. . 9 | . . . | 7 . .
```

Hard # 253

2	9					5	3	
5				2		8		
			4					
8			1			2		
	5	3				1	7	
		2		6				9
				7				
		1		9				2
		7	8				5	3

Hard # 254

1				8			7	
				5		2		
	5		1			3		
		3	5		8			
		9		7		5		
			6		9	4		
			3		7		9	
		7		2				
	3			1				6

Hard # 255

						9	1	
	8					5		
1			4	2		8		3
		2				7	6	
				8				
		9	5			7		
4		8		6	1			2
			2				4	
		7	9					

Hard # 256

3		1				9		
			9	5		7		
					1	4		
4	7			8	2		5	
	3		6	7			1	2
	2	3						
	1		9	5				
	6				5			7

Hard # 257

7		8		9				
	5		3		7			
6			1				7	
		6						9
1		2				6		4
9						8		
	4				6			3
			5		4		1	
				3		4		6

Hard # 258

			4					
2	1			3	8			
			7				6	5
9	8		6			5		
7								1
	5			7		9	8	
1	6		2					
			3	5			1	9
				5				

Hard # 259

			8		4		1	
				9				5
1	2						3	
	6				7	3		
		3				6		
		8	1				2	
	1						4	9
8				5				
	3		6		9			

Hard # 260

4				7			6	
	2		9		5			
		7			2			
6				4	9			8
		1				6		
7			5	2				1
			2			5		
			3		8		2	
	3			9				4

Hard # 261

		3			1		5	
6	9							3
7				4		6		1
								5
	6		3		8		9	
2								
1		6		9				7
3							4	6
	5			6		1		

Hard # 262

		7			2	8	6	
	9							
	8			6			3	
6			2		7			
3			8		6			2
			4		5			1
	1			5			7	
						9		
	4	5	6			1		

Hard # 263

5		2		1				
	4		7			1		
3				8				5
	3	4			2		1	
	8		4			6	3	
4			9					1
		1			7		2	
				6		4		9

Hard # 264

3					1	6		
		2		9			7	
7	1							3
			7				9	
8		1				3		7
	5				4			
5							8	4
	7			3		9		
		4	8					2

Hard # 265

	5		9				6	
6		4	3	1				
		3			5			8
	2	6						
7								1
						4	7	
1			4			7		
				5	6	1		2
	6				1		3	

Hard # 266

		2	7			1		4
		8		3		6		
					1		2	
					7			5
	1	7				3	6	
6			3					
	7		8					
		4		1		5		
8		5			2	9		

Hard # 267

		6	3					2
							9	
2	1			4		5		
				2				4
6		5				3		9
7				8				
		9		1			4	6
	7							
4					8	7		

Hard # 268

		3	9					
			4			8	5	2
				8			3	7
		5				6		
	6			7			4	
	4					2		
7	5		1					
8	1	6		9				
						4	6	

Hard # 269

				4				3
	7		3	1		2		
3	9				8			
	5	9	1					
8								6
					3	1	2	
			4				7	2
		6		2	5		4	
9				8				

Hard # 270

5		6				9	7	
				6				
3						4	8	1
	3	4				1		
			8			9		
			2			5	8	
	5	2	9					1
				1				
	8	3				6		2

Hard # 271

			6	3	9			
			4					8
		3				4	1	
2				1		3	4	
		8				5		
	5	4		9				2
	3	5				7		
4					5			
			7	2	3			

Hard # 272

					4		3	
		6	5		9			4
				2		7		5
6		3				9	5	
	8	9				2		6
8		2		6				
1			9			2	8	
	3			1				

Hard # 273

		7		2				
			4			1		
		3			1	4		2
9				8			5	
2			5		6			1
	5			9				7
4		6	8			9		
		5			2			
				1		5		

Hard # 274

		4			5		2	
	7							
9			6	3				8
			9		3		6	
1								5
	9		8		2			
4				9	8			7
						4		
	8		2			6		

Hard # 275

1		6	9			2		
			4					
	2		6			3		
			4					7
2		5	7		9	1		8
3					1			
		9			5		8	
			7					
	3			2		9		4

Hard # 276

					6	7	4	
6			2		8			
				9		2		
			7	6				9
	9		5		2		8	
1				4	9			
		5		2				
			6		7			4
	2	1	3					

Hard # 277

	5	3		7			9	1
7		2			8			
			3					
		4					8	
9	6						5	7
	1				9			
				7				
			8			5		4
4	3			9		6	1	

Hard # 278

		9	5				6	
	8		2			1		
			3				8	7
	7	6						
			6		5			
						8	4	
4	2				6			
		3			8		9	
	6				2	3		

Hard # 279

3			7	9		5		
	5		6			4		
		1		8				
	4	2					3	
		9			1			
	3				9	6		
			6			7		
		6			7		2	
		7		2	8			5

Hard # 280

		9						7
	4		8			1		
8				2				9
			8		9			6
9		2	7					3
3		1		6				8
1			9					8
	3			1		9		
2				5				

Hard # 281

						7		5
			4	5			1	
	4		2			6		
1		7	8		4			
				3				
			7		9	1		3
		1			3		2	
	6			4	5			
5		8						

Hard # 282

				4		8		
	7		9					1
1		9		2		7		
					8		6	
8		7				2		3
	9		2					
		5		6		3		2
9					4		5	
		1		8				

Hard # 283

7				9			5	
			8	1		6		
		2			5			4
8	2	4						
						1	9	2
1			9			7		
		8		7	4			
	9			3				8

Hard # 284

				5			9	
6		7	9	8				
	5			2		7		
3		1		4			2	
	8			5		1		3
		6		1			3	
			3	6	2			1
	7			2				

Hard # 285

4					3		1	
1		5		6				
				2		3	8	
		1	7	9		8		
	2			3	8	7		
7	1		9					
				1		8		6
	5		3					4

Hard # 286

8	4					3		
				6				7
6			1			4		5
		6	3			1		
			4		9			
		8			2	6		
3		2			4			8
5			9					
		1					5	4

Hard # 287

		9	5	4				
		8	2	1			9	
5					3			7
			6				3	
1								2
	5				9			
9			1					5
	3			6	5	2		
			8	2	6			

Hard # 288

4		7						
			9	8				
	5	1		6	7			
		8		1	3		4	
		3				5		
	1		5	4		6		
			1	5		3	6	
			7	2				
						9		2

Hard # 289

	9			6		4		
			5		3		9	
5					2			
	3		2	7	8			
4								1
		5	8	1		6		
		3						7
	6		7		9			
		9	8			1		

Hard # 290

	1	7			6			
		8		9				2
					4			7
	4		9		2		7	5
1	2		4		7		8	
8			6					
3				4		2		
			1			9	3	

Hard # 291

				4		6		2
			3		6			
	4				1			
5	6	4			8			3
1								8
3			9			4	6	1
			1				3	
			2		9			
9		7		6				

Hard # 292

3						1		
4					7	5		
	1	7			3			
	9				1		3	
7			9		2			6
	8		7				4	
			8			6	2	
		6	2					1
		5						4

Hard # 293

		6	4		9			
		4	5					6
1	9							
	8		2					9
7		5			8			2
6				1			7	
						9	3	
3				4	6			
			7		8	1		

Hard # 294

			8				5	9
				7				3
		9	4					
	3					5		8
1	6			3			9	2
9		7					1	
					2	9		
5				1				
4	2				8			

Hard # 295

7					3	8		
			4		2			3
				8			4	6
8		6	5				2	
	9				7	6		4
6	7			3				
1			7		8			
		5	2					9

Hard # 296

				2			5	8
	8	7	4	9				
4				6				
3	2						1	
			1		6			
	1						6	4
				8				7
			1	2	3	8		
9	3			5				

Hard # 297

5					8		4	
			4	9		8	3	
	7			3			1	
						6		3
			1		9			
2		7						
	8			5			9	
	2	6		7	3			
	5		8					7

Hard # 298

7						9		1
		3						
		3		8	1			5
		5				7		6
9			5		2			4
1		6				2		
6			7	5		1		
				4				
8		9						2

Hard # 299

	3		4	9		7		
	6			2			1	5
		7						
3	1							9
		6				8		
4							3	6
						5		
7	2			8			6	
		3		1	9		2	

Hard # 300

			1					
4	3			8	7			6
	7			4		2		
		6					3	1
			4		6			
5	1					8		
		8		3			9	
2			8	7			1	3
				2				

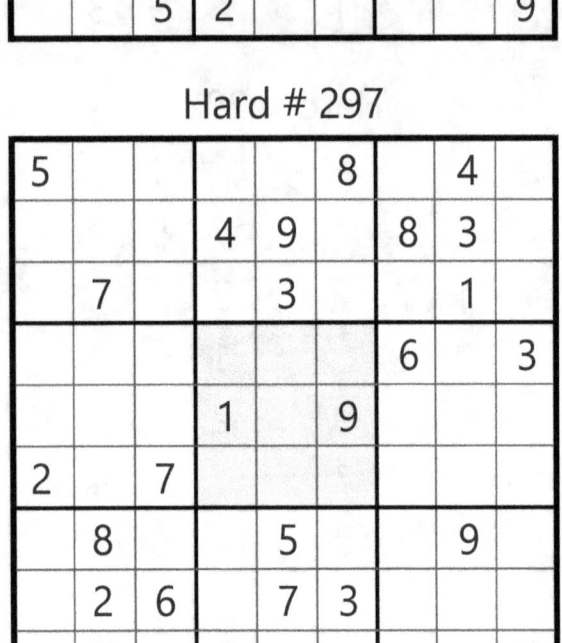

Hard # 301

	9		2					
			4			6		8
				6			5	1
9		5			4	7	3	
	1	2	3			4		9
3	4		9					
5		1				2		
					7		1	

Hard # 302

9						8	4	
	2			4		5	9	
				6			2	
			8		5		1	
5								7
	6		4		3			
	7		2					
	8	1		5			3	
		6	7					4

Hard # 303

4	5					8		9
				3	4			6
6				8				
5						2	6	
			6		2			
	1	6						4
				4				5
3			2	5				
8		5					9	1

Hard # 304

	9		6			1		
2	1		9					
		7		8				
5		6	3			7		
			4		5			
		3			1	5		2
			7			9		
				3			1	8
		4			9		6	

Hard # 305

	9						3	
4						9	6	
7		6		5				
		7		1	3		9	
		8				7		
	5		8	6		3		
			1			5		3
	8	1						2
	6						4	

Hard # 306

			2	9				
4							8	7
				5	3	6		
		4				2		
8			3		7			1
	1					9		
	3	6	7					
5	7							4
				1	9			

Hard # 307

				4				
2					1	4		7
			5	3		6		2
	6					3		
			1	7	3			
		9					4	
1		5		2	7			
8		2	3					4
				9				

Hard # 308

	6	2	3	7				
	4			1				
		8				3	5	
			1					6
3	2						4	1
6					9			
	7	4				2		
			9				8	
			8	2	5	7		

Hard # 309

					9			
			5	7		1		
9			3			2		4
8	2			1				
	4			8		7		
				5		3	2	
6		5			4			8
		2		9	5			
		1						

Hard # 310

	1				5		6	
			2			3		
		8		6			4	7
			3			7		1
	8						2	
6		9		7				
8	7			5		4		
		2			3			
		6		8			7	

Hard # 311

					9			8
6	7		5			4		1
			3	6				
	8			9			2	
1								6
	9			1			7	
			5	6				
4		5			2		6	7
2			7					

Hard # 312

		1						
8	7		6				2	
		9		5		1	7	
		2					5	9
				4				
3	5					2		
	3	5		1		9		
	4				7		8	5
			7					

Hard # 313

```
. . . | . . 1 | . 3 .
. 3 2 | . 5 . | . . .
. . 7 | . 2 . | 8 . .
------+-------+------
6 . . | . . 9 | . 4 .
3 . . | . . . | . . 5
. 9 . | 4 . . | . . 6
------+-------+------
. . 6 | . 4 . | 9 . .
. . . | . 7 . | . 1 8
. 7 . | 8 . . | . . .
```

Hard # 314

```
5 . . | . . . | 3 9 .
. . . | 3 . 1 | . . .
. . . | 6 9 2 | . . 8
------+-------+------
. . 4 | . . . | 8 . .
9 . . | 7 . 4 | . . 2
. 2 . | . . 5 | . . .
------+-------+------
8 . 7 | 1 5 . | . . .
. . . | 9 . 3 | . . .
. 1 2 | . . . | . . 9
```

Hard # 315

```
. 7 . | 6 . . | . . .
1 5 . | . 8 . | . . .
. 6 . | 9 . . | . 3 .
------+-------+------
. . 5 | 1 . . | . . 3
2 . . | 7 . 3 | . . 9
3 . . | . 9 . | 2 . .
------+-------+------
. 8 . | . . 6 | . 9 .
. . . | . 2 . | . 6 5
. . . | . . 7 | . 8 .
```

Hard # 316

```
4 . . | 3 . . | 1 . .
. . 7 | . . 3 | . . 8
. 2 . | . 5 . | . . .
------+-------+------
. 1 . | . 9 . | . 7 .
9 . . | . 8 . | . . 2
. 3 . | . 4 . | . 6 .
------+-------+------
. . . | . 2 . | . 4 .
3 . . | 5 . . | 7 . .
. . 4 | . . . | 2 . 9
```

Hard # 317

```
. 9 . | . 6 8 | . 1 .
2 . . | 1 3 . | . . .
4 . 6 | . . . | . . .
------+-------+------
6 . . | . . . | 7 . .
. 5 . | 6 . 9 | . 4 .
. . 3 | . . . | . . 5
------+-------+------
. . . | . . . | 4 . 1
. . . | 9 5 . | . . 7
. 3 . | 7 8 . | . 9 .
```

Hard # 318

```
. 1 . | 6 . 5 | . . .
. . 2 | 1 . 8 | . 6 .
4 3 . | . . . | . . .
------+-------+------
5 . . | . . . | 1 3 .
. . 9 | . . . | 6 . .
. 8 3 | . . . | . . 7
------+-------+------
. . . | . . . | . 1 9
. 4 . | 7 . 6 | 3 . .
. . . | 5 . 1 | . 4 .
```

Hard # 319

9				1	7	2		
		4						
	5			9	4			
2			1			4		8
1								6
4		5			2			9
			3	7			5	
						7		
		3	8	5				2

Hard # 320

2				6			3	
		3			9		8	
		9				1		
			5		1	3		8
				2				
7		6	3		4			
		8				6		
	3		4			7		
	5			3				2

Hard # 321

6			2					
	8		1		6	4	7	
	4						6	
3		4	5			7		
		1			7	8		9
	9					3		
	1	6	7		3		2	
				1				5

Hard # 322

			6			5	4	
3				4		7		
			8					
4		5			6		3	
2				7				9
	7		4			2		5
				2				
		1		3				6
	4	9			1			

Hard # 323

4				7				
1			6	3				8
			8			5	7	
	4					3		
		2			7			
	3					9		
7	1		4					
8			3	9				5
			2					4

Hard # 324

6		2					1	7
8			5				2	
	5							
	6		4					5
		5	9		2	7		
1				3		6		
							9	
	8			9				4
7	1					2		8

Hard # 325

				8		6		
1				3		9		
6	5			9				
4	1		7					9
		8				4		
3				4		1	2	
			4				7	5
	7		9					3
		6		2				

Hard # 326

7			4		5			
		5				6	4	
3								7
		4		6			7	
			1	3	8			
	2			9		5		
5								1
	9	8				2		
			8		3			6

Hard # 327

	5	2		6			3	
7		9				2		4
			5		3			
	3	1		4		8	9	
			9		6			
6		4				7		3
	1			9		4	8	

Hard # 328

			1					
1		7						3
	3			9	6			2
2	6	5	4					
			9		8			
				7		4	6	1
9			3	5			4	
8						5		7
				9				

Hard # 329

2				7	6	9		
	4			3	5			
9								3
4			1	5		9		
		7		3	9			2
5								9
		6	4			3		
	1	9	6					4

Hard # 330

	7	2	1					
	1		5					2
			6		3			
6		3					7	9
1								4
7	8					3		1
			9		4			
5				2			4	
				6	9	3		

Hard # 331

```
. . . | 8 . 1 | . 7 3
8 . 5 | 2 4 . | . . .
. . . | . . . | . . .
------+-------+------
. 9 . | . . 2 | 3 . .
7 4 . | . . . | . 8 1
. . 1 | 4 . . | 9 . .
------+-------+------
. . . | . . . | . . .
. . . | 2 8 7 | . 5 .
6 5 . | 1 . 9 | . . .
```

Hard # 332

```
. . . | 7 . 6 | . . .
. . . | 4 . . | 3 7 1
1 . . | . . 6 | 8 . .
------+-------+------
3 4 2 | 9 . . | . . .
. . . | . . . | . . .
. . . | . . 7 | 2 8 4
------+-------+------
. . 7 | 2 . . | . . 5
5 2 1 | . . . | 3 . .
. . 6 | . . 9 | . . .
```

Hard # 333

```
. . . | . 2 . | . . .
7 . . | 4 3 . | . . .
. 4 . | . 5 . | . 3 8
------+-------+------
1 7 . | . . 9 | 6 5 .
. . . | . . . | . . .
. 5 3 | 7 . . | . 8 9
------+-------+------
4 1 . | . 2 . | . 6 .
. . . | . 6 7 | . . 5
. . . | 1 . . | . . .
```

Hard # 334

```
. 7 . | . 4 . | . . .
2 . . | 9 . 5 | . 7 .
. 3 4 | . . . | . . .
------+-------+------
. . 8 | . 2 6 | . 5 .
9 . . | . . . | . . 6
. 5 . | 4 7 . | 3 . .
------+-------+------
. . . | . . . | 1 6 .
. 2 . | 7 . 4 | . . 8
. . . | 8 . . | 2 . .
```

Hard # 335

```
. . 4 | 2 . . | . 8 9
. 5 . | . . 7 | . . .
. . . | 1 8 . | . . .
------+-------+------
. . 1 | . . 9 | . 2 7
. . 6 | . . . | 3 . .
2 7 . | 3 . . | 8 . .
------+-------+------
. . . | 8 2 . | . . .
. . . | 6 . . | . 7 .
8 3 . | . . 1 | 5 . .
```

Hard # 336

```
. . . | . 5 . | 4 . .
4 1 2 | . . . | . 7 .
. . . | 7 . . | 9 . .
------+-------+------
7 . . | 9 . . | 8 . .
. . . | 3 6 1 | . . .
. . 1 | . 7 . | . . 6
------+-------+------
. . 4 | . 9 . | . . .
. 9 . | . . . | 3 1 7
. 8 . | 2 . . | . . .
```

Hard # 337

						2		
	8				6			
	7	2	1		3	8		6
9					1			7
3								4
8			6					1
7		1	4		2	6	3	
			7				5	
		9						

Hard # 338

5		3	6	2				
	1						8	6
6				8				
		2			7			
9	3						5	1
		7			4			
			5					8
2	9					4		
			3	9	1			2

Hard # 339

	9							2
8					4	7		
	3	4						
	8	5		4	6			
	1		9		8		7	
			7	3		4	5	
						9	1	
		8	1					3
1						6		

Hard # 340

	8		3					
		1	6	2			3	
			8					9
	7	2				3	1	
	6						2	
	3	9				5	4	
5				7				
	2			4	3	1		
				8			9	

Hard # 341

8		3		7				
7	4		8				2	
			4		5			
		1		5				3
				8				
9				2		4		
			5		2			
	5				6		9	2
				9		7		1

Hard # 342

3	2						9	
	6		4	1			5	
			3					
			2	6				4
1				7				6
8				3	5			
						3		
	3			4	6		2	
	9						4	8

Hard # 343

9			5				7	
4					5		9	
	5		2	6				
		8				6		
7				1				5
	9				4			
			4	2		5		
5		9						8
	2				1			6

Hard # 344

3		2	1			5		
			5			6		9
				2				
8	9			1				
		1		7		4		
				9			5	1
				5				
2		8			4			
		7			8	2		4

Hard # 345

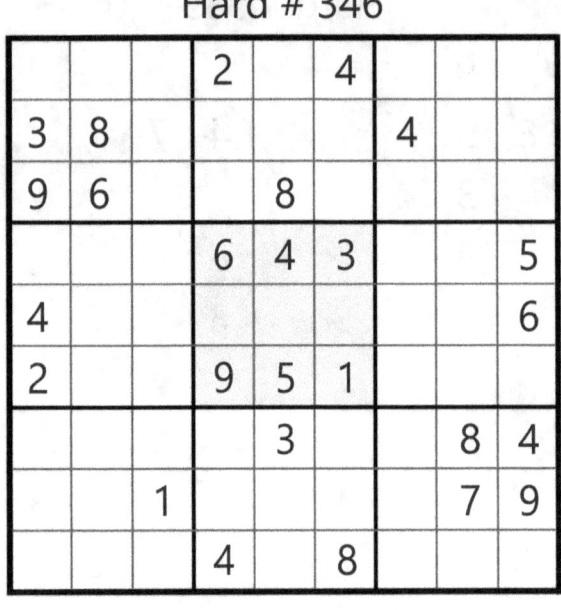

	4				8			
7				3				6
6							1	
			2	5		6		1
	2		8		9		5	
5		8		4	1			
	7							4
9				7				5
			5				3	

Hard # 346

			2		4			
3	8					4		
9	6			8				
			6	4	3			5
4								6
2			9	5	1			
				3			8	4
		1					7	9
			4		8			

Hard # 347

3				4			5	
		8	3		1	6		
	9				8	7		
				6			1	8
4	6			7				
		3	5				4	
		2	4		6	5		
	4			3				2

Hard # 348

		6		8	2			
5		7	3					
		8	9			3		
6	8			2				
3								1
				1			2	6
		2			8	4		
					6	8		9
		5	4		6			

Hard # 349

							9	6
	9		1			3		
2					8		1	
			7	1		9	6	
	2	3		8	6			
	3		2					1
		1			4		3	
8	7							

Hard # 350

		6		4			3	
1	2							9
7								
	9	8			2			7
			6		7			
6			1			3	2	
								6
3							4	8
	8			7		5		

Hard # 351

1			9					4
	3	4				5		
7				2			8	
4						5		6
	7						2	
8		1						3
	2			9				5
	6				7	4		
9					6			7

Hard # 352

			8	5				
	5			2	9	8		3
						4	1	
						1	2	
9	6						8	1
			4	7				
			5	6				
8			1	4	7		9	
				1	5			

Hard # 353

			5					1
8	1			7	2			
5					3	9		
	4	2	6	7				
			9	3	4	1		
	7	6						5
		4	1				7	2
1				5				

Hard # 354

				6				1
4	1	9		5		3		
5								
9	4			2				
	8						4	
			1				3	5
								6
	7		6			8	5	4
1			9					

Hard # 355

		5			9	2		8
	2							3
			3	2	7			
			8					5
3		8				1		4
6					3			
			7	4	2			
5							4	
9		3	1			8		

Hard # 356

	2							
	5			7	2			9
	3		1	4				7
	6				1			4
				9				
3			8				1	
4				5	8		9	
1			2	3			6	
							7	

Hard # 357

4	5	1			2			
					3			1
			1		4			
			3					8
9		2				7		4
8					9			
		6		3				
1			8					
			4			3	7	9

Hard # 358

8	9			4				
7		3						
	2		8		9			6
						1	2	
	7		1		3		5	
	5	2						
9			4		5		1	
						8		5
				7			4	9

Hard # 359

							1	6
	6	9						
2			4		9			
			8			1	5	
7	4			3			8	9
	2	8		1				
			2		1			8
						7	3	
4	8							

Hard # 360

5						6		4
		3		8	4			
	1				6			
8	5				9	2	7	
	7	6	8				9	3
			3				4	
			7	5		8		
3		9						5

Hard # 361

	1			8				4
			1	3			2	
	2				4	6		
4		2			7			
6								9
			8			2		3
		5	7				3	
	3			5	2			
7				4			1	

Hard # 362

	9			6	8			
		5	4					
	4						1	8
3	8			5	1	6		
		1	2	8			7	5
7	2						6	
				9	3			
			6	7			9	

Hard # 363

5		2	6		1			
	6					7		
	8		2		4			
4	9	6	8					
					6	4	5	9
			9		8		2	
		3					6	
			3		7	9		1

Hard # 364

	4				5	9		
	6			9			2	5
1				3				
			8				5	2
		4				3		
7	1				6			
				6				8
5	2			4			3	
			1	9			4	

Hard # 365

			1					2
		4			7			6
	3	5			6			
	4	3		1		9		5
7		1		2		4	8	
			9			5	2	
4			7			3		
3				1				

Hard # 366

4								
2	1		6	3				
3			9			7		
1				9		6		
	5						2	
		6		7				8
		4			3			5
				2	7		4	3
								2

Hard # 367

						7		3
		9		3				6
	7		4					5
				6	9		8	
		7				1		
	1		2	5				
7					1		9	
3				7		6		
8		1						

Hard # 368

				3				
7	8		5			1		
3					2			8
	6		2		5		7	
		9				4		
	3		7		9		2	
8			9					5
		5			3		6	9
				5				

Hard # 369

			7	8		9	1	2
				1				3
				2				
	2		4	9				
3								6
			7	6		5		
			9					
8			5					
9	6	1		3	8			

Hard # 370

			4	3				5
		5	7				8	
				9				3
3		4				9		
	2		9		6		3	
		1				5		7
8			1					
	6			9		8		
2				8	7			

Hard # 371

				2		8	1	
9	4		1					
	3	8		7				5
							8	1
4								7
6	9							
2				3		7	4	
				5			2	3
	6	4		8				

Hard # 372

	6		3					7
	1		9	2				
		5						
		7		6			9	
	2	8				7	1	
	4			9		8		
						4		
			3	4			2	
1				8			6	

Hard # 373

	3						8	6
		2	3					
			9			5		1
		5	6	3			9	
	7						5	
	8			7	5	1		
5		3			4			
					3	7		
8	6						1	

Hard # 374

4					3		5	2
		7		2	1			
8				6				
		8						3
7	4						2	8
3					9			
			3					1
			9	7		6		
9	3		6					4

Hard # 375

3		9			8			
8			5			6	4	
					3			
			6	9	8		7	2
9	2		7	3	4			
			2					
6	1			8				9
		3				4		6

Hard # 376

	2					6	7	4
			6		7		8	
	6			8			5	9
		4			1			
5	7			3			2	
	8		7		1			
3	4	9					1	

Hard # 377

			8		9	4	2	
						8	1	3
	4							
				4	9			6
	8			9			3	
2		7	3					
								8
1	6	4						
	5	3	4		6			

Hard # 378

	6		2			4		
			8	4				5
			3				2	
3	7					8	9	
			6		1			
	5	1					4	7
	3				9			
4			8	1				
		8			2		5	

Hard # 379

2		4				1		
				8				
1			2			6	5	
9				5	6			
	3						2	
			8	1				6
	1	9			5			3
				2				
		6				9		1

Hard # 380

9	5	6		3	8			
1						6	2	
5	4			1		9		
			5		9			
		9		4			6	5
	8	4						3
			9	7		2	8	6

Hard # 381

				8		4		3
				4	1			
		3		9		7		2
6			2					
	5	4				2	9	
				3				1
4		8		2		5		
		6	8					
2		7		3				

Hard # 382

	1			9				
	3		4			5		9
		5	1					
	6	1			2			
7								3
			3			9	2	
				5	8			
6		2		3		5		
			9			4		

Hard # 383

		7	6					
		1				4	6	
3			4		1	8		
		6		3			2	
			1		5			
	5			4		7		
		2	7		4			3
	3	5				1		
					9	6		

Hard # 384

		8			1			
7	2			3				6
						9		8
5	8			7			4	
			5		3			
	3			8			5	7
2		6						
8				4			7	2
			6			3		

Hard # 385

1								7
	6			2		9		
	3		9			1		2
			2	4	8		6	
	4		6	5	9			
4		3			5		2	
		6		7			3	
8								5

Hard # 386

							9	5
5				4	7			
	2						7	3
9	7			2				
		1	3		5	7		
				8			3	1
3	9					2		
			6	5				7
6	4							

Hard # 387

7			2			6		
		3		9	6			1
9							4	
1	7				3	9		
		9	6				5	3
	6							9
2			3	4		1		
		1			9			5

Hard # 388

		3		8				
			1		9	7		
	7				3			5
9	4						1	
		2				5		
	5						8	4
1			6				9	
		5	3		8			
			4			2		

Hard # 389

						2		
			7	1	3		5	
3		9			2	7		
5			6					
2			8		9			3
					1			6
		7	2			1		4
	6		1	9	5			
		3						

Hard # 390

6						3		
7					6	8		1
			4		3			
	7	8		5			1	
	5						2	
	9			7		5	3	
			9		8			
5		1	7					2
		9						7

Hard # 391

					6	3	1	
	3		4	7	9		6	
		9	5					
		4				8		6
5		6				2		
					7	1		
	6		8	3	2		5	
	2	8	9					

Hard # 392

	4	5		8	3			
6	3				4			
			2				1	
						2	6	3
	8						4	
9	6	2						
	2			6				
		4					3	5
		8	1			7	2	

Hard # 393

				3		2		
4	6		1					
3				5				8
5					8	1		
		8		6		4		
	7	4						2
8			7					3
				2		8	4	
	1		3					

Hard # 394

	8		1		4			
					3			
2			3		5			8
9						7		
		5	7		8	1		
	1							4
5		2		8				9
	3							
			5		2		6	

Hard # 395

			5	8		2		
	7					1		
4		5	3			8		
6				3				8
	8					1		
1				4				6
	5				6	3		2
		2				7		
	9		5	2				

Hard # 396

7			6	3			1	
							5	4
8						7		
		4	8		1		3	
		7				4		
	2		3		4	5		
		9						5
3	8							
	1			9	8			7

Hard # 397

					9			1
	4	7	8					
				3			8	2
	1				8			9
3		8				5		4
2			5				3	
1	5			9				
					3	2	4	
4			7					

Hard # 398

	3			8	1		9	
	7					5		2
1			4					3
		7	2					9
9				4	1			
2				6				4
7		3				2		
	9		1	2			6	

Hard # 399

			9					
	7		6		8			
4		6			3			1
9						5	4	
1	4						6	3
	2	8						9
3			8			2		5
			7		4		9	
				5				

Hard # 400

8		1					9	
		7	2					
				6	8			2
4				1	5		6	
				7				
	7		8	2				4
3			6	4				
				9	1			
	8					4		5

Hard # 401

				1		7	9	
4					8			
		2		7	9		3	
	6				3			
			1	8				
		3				5		
	9		7	4		1		
		5						4
6	1		2					

Hard # 402

	1		6			4		9
3		4	1					
			9					
		7	5	1		3		
5								4
		1		8	3	9		
				2				
						1	7	2
7		6			9		5	

Hard # 403

```
. . 3 | . 1 . | . . 9
. . . | 3 . . | 4 5 .
. . . | 2 . 5 | . . 3
------+-------+------
. . . | . . 2 | . 8 .
5 . 4 | . . . | 3 . 2
. 6 . | 7 . . | . . .
------+-------+------
6 . . | 1 . 7 | . . .
. 7 5 | . . 4 | . . .
8 . . | . 3 . | 1 . .
```

Hard # 404

```
. . . | . . . | . . 4
4 . . | . 7 6 | 1 . .
. . 8 | 9 . . | 2 . .
------+-------+------
1 . . | 4 . . | . 8 .
6 . . | 3 . 7 | . . 9
. 3 . | . . 8 | . . 6
------+-------+------
. . 6 | . . 4 | 3 . .
. . 5 | 7 9 . | . . 8
8 . . | . . . | . . .
```

Hard # 405

```
. . 3 | . . . | 8 5 .
1 . . | 3 . . | 6 . .
. 7 . | . . 8 | . . .
------+-------+------
8 . . | 2 . 1 | . 6 .
. . . | 4 . 5 | . . .
. 4 . | 8 . 7 | . . 9
------+-------+------
. . . | 1 . . | . 4 .
. . 9 | . . 2 | . . 8
. 1 2 | . . . | 3 . .
```

Hard # 406

```
. . 9 | . . 7 | . . .
2 . . | . . . | . . .
8 . 3 | . 6 . | 7 . .
------+-------+------
. 3 . | . 4 8 | . . 9
. . 2 | 5 . 9 | 6 . .
4 . . | 1 2 . | . 8 .
------+-------+------
. . 4 | . 1 . | 8 . 7
. . . | . . . | . . 5
. . 9 | . . . | 4 . .
```

Hard # 407

```
. 8 1 | 9 . . | . . .
. 6 . | . . . | . . 9
4 . 7 | . 2 1 | . . .
------+-------+------
. . . | . 1 . | . . 2
. . 5 | 3 . 7 | 8 . .
9 . . | . 4 . | . . .
------+-------+------
. . . | 5 6 . | 3 . 8
3 . . | . . . | 5 . .
. . . | . . 3 | 2 1 .
```

Hard # 408

```
5 . . | . . 3 | . . 6
. . . | 6 2 . | . 4 .
4 . 8 | . . . | . . 1
------+-------+------
. 7 . | 1 . . | . 3 .
. . . | . 9 . | . . .
. 9 . | . . 6 | . 7 .
------+-------+------
1 . . | . . . | 4 . 3
. 4 . | . 6 5 | . . .
7 . . | 2 . . | . . 8
```

Hard # 409

6						9		
	2		1	4				
3	8				6		2	4
				2		7	4	
	7	4		5				
8	3		9			6	2	
			3	1		7		
		2						9

Hard # 410

	2			6				5
		1	9				7	
	6			1		9		
2				5		3		
			8		4			
		7		2				9
		2			3		6	
	8				6	2		
1				9			8	

Hard # 411

	9					2	8	
4			2				3	
				1		4		
	6	8	4	5				
				3				
			2	9	3	6		
		2		7				
	8				6			1
	3	1				5		

Hard # 412

4							3	
		6		2				
		8	5	1			6	
		7			4		5	
2			8		1			3
	4		2			1		
	1		6	5	3			
			3		9			
5								8

Hard # 413

			7	2				1
5		1				9		
	3					8		
3			8	2		1		
	1					7		
	4		5	1				6
		9				6		
	2					5		8
4				6	8			

Hard # 414

				1		4		3
3			4	7		9		
		7	2			5		
	9					5		
			4		9			
		6				3		
	8			9		7		
	6		8	7				5
9		2		5				

Hard # 415

		9	2	3			4	8
				4	9	6		
	8							
			7					5
2		5				7		4
3				5				
							9	
	6	4	1					
9	7			6	8	5		

Hard # 416

				1	3			6
	7		5			9		
		3		9		2		
1		8			5			
		5				3		
			1			4		2
		9		6		1		
		2			7		3	
3				8	2			

Hard # 417

		7				2	4	
6				3				
3				4	5			7
	9		3	6				2
4				2	9		1	
2			6	5				8
				9				5
	3	8				4		

Hard # 418

2		8						
	9		7				2	
					6			3
			5	2		7	4	
3			1		4			8
4	2		9	3				
6			8					
	7				1		5	
						6		1

Hard # 419

	3	7			5			
				5				7
6				9			2	3
9			1					
3	1					4	2	
				4				9
1	5		2					6
7			8					
		4				2	3	

Hard # 420

			8					
	7		2				3	
8	2	6		5		4		
		3				9		4
	6						2	
2		7				5		
		4		2		3	8	5
	3				9		1	
					1			

Hard # 421

8			9		6			
				7			4	3
		3			8			
				1		9		
7			3		2			1
		5		9				
		8				2		
2	7			4				
			1		3			5

Hard # 422

3			4					
7							9	6
			9	1	4			
			5	8		3	1	
	5					8		
8	2		3	1				
		5	2	3				
2	6							7
				6				2

Hard # 423

					9		8	
	1		3	5				
9		2						
4				8			3	
1		8	2		5	6		4
	5			7				2
						7		1
			3	8		4		
	9		7					

Hard # 424

	7	3						
6			3		8		2	
		4		5		3		
2			5		9			
	9						8	
			1		6			2
		5		1		2		
	1		8		7			9
						8	6	

Hard # 425

	8						4	6
7					2			
		9	5	6				
1						4	3	7
6								9
4	9	5						8
				7	3	9		
			6					5
2	3						6	

Hard # 426

			1	6			7	
1							5	
		4			7			9
			2			5		
		3	7		4	9		
		5			6			
2			6			3		
	1							4
	6			4	8			

Hard # 427

		5					3	
		1		7	4			
			4		9		6	
1		7		2				6
3								2
2				9		8		7
	4		3		2			
		3	7			9		
	1					5		

Hard # 428

			4	2	9			
		6	5			4	7	
							2	
	3			8	5		1	
9								3
	2		3	1			4	
	8							
	9	3				8	7	
		2	4	7				

Hard # 429

2	8		5	4				
5	1							9
					7			
		6	7				5	
			3		2			
	4				9	2		
			8					
1						4	7	
				7	3		9	2

Hard # 430

	4		2	7				1
		2			6		4	
		8	5					
						8	9	
2		5				6		3
	7	4						
					7	5		
	1		3			4		
4				8	2		3	

Hard # 431

					7	4		
		7				3		6
3	6				1			
	3			9				
5		9	3		4	8		1
				1			2	
			1				8	2
8		4				5		
		3	9					

Hard # 432

2			8					
		1		3	9			
5	6						8	
7		9	5					
			6	1	7			
				2	7			5
	4						1	8
			7	6		5		
				3				9

Hard # 433

6		3			1			
	1	7	4	5				
	8	9						
		8			7	1		
	7						6	
		5	6			8		
						2	7	
			9	3	6	1		
			8			9		3

Hard # 434

1			6				3	
		6						
5			2	3	7			
	6				2	4	9	
7								2
	2	4	8				5	
			3	2	4			1
						8		
	7				9			4

Hard # 435

			5					2
	4			7			8	
				3	9			1
		3	1			4		7
2								3
8		1			5	2		
4			9	2				
	2			5			7	
3					4			

Hard # 436

		2			9			
	1		6		8	7		
3			2	7				
	3				4		9	5
1	4		9				6	
			4	2				8
		5	8		6		3	
			7			2		

Hard # 437

2				5			8	
	3	7		6	9			
	4			1	6			
					8			6
	2						1	
3			6					
		5	1				9	
		9	4		1	7		
	9			8				2

Hard # 438

6			8			4	9	
		5						
	7	2	4		3			
1				8		9		
			2		6			
		6		7				5
			5		1	3	2	
						7		
	4	3			8			1

Hard # 439

		7		6				
6	5		2					
3	1	8						
			8			1	5	7
			5		3			
7	9	5			2			
						7	1	9
				9			3	4
				1		8		

Hard # 440

					4	3		
				8				9
	3		9		7			
	2	5		4		1		
	9	4				8	6	
		3		5		9	2	
		1		2		7		
	5		3					
		1	5					

Hard # 441

		3		2	4	9		
		8						
5				1			2	
		7	2		6	3		
	4						8	
		1	8		9	2		
	9			8				3
					5			
		5	9	7		8		

Hard # 442

	5	3	9					
	9		8			5		
1						4		
3	4					2	1	
				5				
	8	5					7	3
		4						1
		6		4			9	
					6	3	4	

Hard # 443

			9	4				
	5				2			
1		4	7		6			
	8		9					5
	4	1				7	2	
2					3		1	
			8		9	5		1
		7					3	
			1	3				

Hard # 444

	2						9	7
			7	4			3	
3		8						
9			2		8	4		
			4		7			
		4	1		3			8
						6		3
	8			3	4			
2	9					4		

Hard # 445

							8	
3					8		9	7
	9	4					5	3
				1				2
	3		6		7		1	
7				2				
2	1					5	3	
5	8		7					6
	6							

Hard # 446

		5			8	7	2	
					1			9
6								
2			6		9		1	
5			1		7			8
	9		8		3			4
								7
9			3					
	6	3	5			9		

Hard # 447

		5		2				
4	1				9			
			5	4		2	3	
2							1	4
	6						8	
1	8							2
	3	9		1	4			
			2				5	9
				7		8		

Hard # 448

9			8				1	
			6	4			5	
3					1			8
		8			2	4		
		3				8		
		1	5			3		
8			9					2
	3			6	4			
	5				8			6

Hard # 449

			9	7			5	
		7		4				8
	4	5				9		
4	8							
6			8		3			7
						2	5	
		8				5	4	
1				9		6		
	9			2	8			

Hard # 450

			8				6	
8	4	5				9		
		7		2			4	
				9		2		
	3		6		1		5	
		4		5				
	7			6		4		
		8				1	3	9
	1			2				

Hard # 451

	6					3	8	
		4	9			7		
	7				5			1
	5			7		8		
			1		6			
		7		9			4	
6			8				1	
		5			2	6		
	1	3					2	

Hard # 452

								8
		1		5		3		
			8	9			7	4
	2				5	1		
7			3		2			6
		8	9				3	
8	5		6	2				
	6			3		7		
4								

Hard # 453

3			1		7			
	7		3			8		2
2				4				
		7	9			6		
	9							3
		6			3	9		
				1				9
8		3			4		5	
			5		2			6

Hard # 454

4	8		3	2	7			
								3
			6	4		5		
							6	5
9		7				3		1
3	6							
	3		1	9				
6								
			7	3	6		4	2

Hard # 455

	2		3	7				
		9						
6		4		1			9	
5					3	7		
3	6						5	9
		7	6					2
	4			9		2		3
						9		
				6	1		4	

Hard # 456

		5			9	4		
	7			6				1
		6				7		
			1	2	5		8	
		7				5		
	5		4	7	6			
		9				8		
6				4			7	
		8	3			6		

Hard # 457

.	4	.	5	.	.	.	.	.
5	8	.	.	3	.	.	.	.
2	.	.	.	.	1	.	.	3
.	.	.	4	.	.	7	.	1
.	.	.	2	.	3	.	.	.
6	.	9	.	.	7	.	.	.
8	.	.	1	.	.	.	.	2
.	.	.	.	6	.	.	3	9
.	.	.	.	5	.	.	8	.

Hard # 458

7	.	9	.	.	4	.	.	.
.	2	.	.	6	.	.	.	.
3	.	.	5	.	.	.	.	.
.	.	.	4	.	.	.	7	3
1	.	.	3	.	2	.	.	8
9	8	.	.	5	.	.	.	.
.	.	.	.	.	6	.	.	4
.	.	.	.	8	.	5	.	.
.	.	.	1	.	.	7	.	2

Hard # 459

.	.	3	.	.	.	.	.	7
7	2	.	3	.	.	8	.	.
.	.	.	.	.	.	.	3	1
6	.	4	9	8	.	.	.	.
.	.	.	.	6	.	.	.	.
.	.	.	7	4	1	.	.	5
5	6	.	.	.	.	.	.	.
.	.	9	.	.	1	.	7	6
8	.	.	.	.	9	.	.	.

Hard # 460

8	.	2	.	.	.	9	.	.
.	.	4	3	1	.	.	.	.
.	.	.	.	5	.	3	.	.
.	5	3	.	.	.	.	.	4
7	4	.	.	.	.	.	8	9
6	.	.	.	.	.	3	5	.
.	8	.	7	.	.	.	.	.
.	.	.	9	1	8	.	.	.
.	6	.	.	.	.	5	.	2

Hard # 461

.	4	.	5	.	.	.	.	.
.	.	2	.	9	6	.	.	8
6	7	.	.	.	1	.	.	.
.	9	7	.	.	.	.	6	.
2	.	.	.	.	.	.	.	9
.	6	.	.	.	.	3	2	.
.	.	.	7	.	.	.	3	2
1	.	.	8	2	.	5	.	.
.	.	.	.	.	9	.	4	.

Hard # 462

8	1	.	.	.	6	.	.	.
.	.	.	7	6	1	.	.	.
.	4	.	.	1	.	.	8	9
.	3	.	.	.	.	.	.	4
.	.	.	6	.	5	.	.	.
2	.	.	.	.	.	.	9	.
7	2	.	.	6	.	.	1	.
.	.	3	7	9	.	.	.	.
.	.	8	.	.	.	.	3	6

Hard # 463

```
5 2 . | 9 . . | . . .
. 9 . | . . . | . . 2
. . 7 | 2 4 . | . 9 .
------+-------+------
. . . | . 7 . | 1 . 9
6 . . | . . . | . . 4
8 . 1 | . 3 . | . . .
------+-------+------
. 4 . | 8 6 9 | . . .
7 . . | . . . | 5 . .
. . . | . . 4 | . 8 1
```

Hard # 464

```
. . . | . . . | 2 . 8
. . 7 | 2 9 . | . . .
2 . . | . . . | 9 3 .
------+-------+------
. . 4 | 9 . 3 | . . 6
8 . . | . . . | . . 5
1 . . | 7 . 6 | 4 . .
------+-------+------
. 6 5 | . . . | . . 3
. . . | 1 4 8 | . . .
7 . 8 | . . . | . . .
```

Hard # 465

```
. 1 . | . . 7 | 5 . 2
7 . . | . . . | 4 . .
. 8 . | 6 . . | 7 . .
------+-------+------
. . . | 5 . . | . . .
5 . 4 | . 9 . | 6 . 3
. . . | . 1 . | . . .
------+-------+------
. . 1 | . 3 . | 8 . .
. . 8 | . . . | . . 9
9 . 2 | 1 . . | . 4 .
```

Hard # 466

```
7 . . | . 5 . | . 4 3
. . 2 | 3 . . | . 9 .
. . . | 2 . 6 | . . .
------+-------+------
9 1 . | . . . | . . .
. . 5 | 8 . 1 | 9 . .
. . . | . . . | . 2 5
------+-------+------
. . 1 | . 8 . | . . .
. 8 . | . 9 7 | . . .
4 3 . | 5 . . | . . 9
```

Hard # 467

```
. . 2 | 6 . . | . . .
4 6 . | 8 . . | . . .
. . 5 | . 7 . | 3 4 .
------+-------+------
. . . | 7 . . | 9 . .
. 3 . | 4 6 . | 7 . .
. 5 . | 8 . . | . . .
------+-------+------
6 4 . | 3 . 5 | . . .
. . . | . 1 . | 4 6 .
. . . | 4 . 8 | . . .
```

Hard # 468

```
. . . | 3 . . | 5 . 9
. . . | 2 1 . | 3 . .
. 8 . | . . . | . . 7
------+-------+------
. 6 . | 5 3 . | 2 8 .
. 5 2 | . 4 8 | . 6 .
3 . . | . . . | . . 1
------+-------+------
. 2 . | 8 6 . | . . .
9 . 8 | . . 4 | . . .
```

Hard # 469

	9			4		3		
3				5	1			8
2			7					
		6				1	3	
			6		4			
	4	5				2		
					7			3
8			3	6				1
		2		9			5	

Hard # 470

5					9			
	3		8		1			
	9			3			2	
1	7							3
		8		4		1		
2							4	6
	8			6			7	
			3		5		9	
			2					4

Hard # 471

	6	3		5		8		
7		2						4
	5		1					
			9				1	
3		6				9		8
	7			2				
				6			8	
6						3		7
		8		4		2	5	

Hard # 472

6			7	2				1
			4			9	5	
	9							
			8	3		1		
1		3			7			4
	7		2	1				
						4		
9	8			5				
4			1	3				9

Hard # 473

				5		6	4	
			3			8		
		8		1	9		7	
		9					4	6
		4			9			
7	1				5			
	4		1	2		7		
		5			4			
3	8		6					

Hard # 474

		5		8	6			
8			4				2	
9	6		2					7
2		1				3		
		3				9		5
3				2			5	9
	8			7				4
			3	5		2		

Hard # 475

	2					1		
		8		3			2	9
	3	6				4		
			9	7				8
			3		4			
7				5	8			
		7				5	6	
8	4			2		9		
		3					7	

Hard # 476

		1						6
		4	1					3
8			2	3	6			
1				4			3	
		5				8		
	8			1				5
			7	2	1			4
6					4	2		
7						3		

Hard # 477

	5				6			
					4			
			4		5	9	3	
1			6	2	9			5
		3			2			
8		5	7	3				1
3	4	2		5				
		6						
			6				7	

Hard # 478

2		8		3				
		5	4					
		7			6	8	9	
			2	8			7	
	5						3	
	9			6	3			
	7	9	8			2		
					2	5		
				4		9		3

Hard # 479

			2					
	1					7	6	
5	4	8						3
			3	5	8			
	8		6		9		3	
		5	1	2				
6						9	5	8
	7	3					1	
				4				

Hard # 480

		9		6				
6			2	4	3			
3							1	
5	2					9		
8			3		9			7
		7					4	8
	5							9
			5	8	4			2
				2		3		

Hard # 481

		8	1	3				7
2						6	3	
					4			
9				6	8	3		
	2						7	
		4	3	1				6
			2					
	8	5						9
6				8	1	7		

Hard # 482

				7				4
			9		4			8
			8			2	9	
	6			9		7		
	9	5				1	6	
		1		6			5	
	5	4			9			
3			2		7			
8				5				

Hard # 483

		2	3					
	4	3		7	9			
		7		5				1
			6	9			2	3
6	1			3	8			
4				6		1		
			5	2		4	3	
					4	2		

Hard # 484

4			6				2	8
						9		
			1	9	2			
			4			2	5	
8			2		7			3
	1	2			3			
			5	1	6			
		7						
9	8				4			6

Hard # 485

2								8
			3	2			4	
			5	8				9
6	7			4				
4			5		9			2
				6			1	3
5			1	3				
	3			7	5			
7								6

Hard # 486

				8				
							3	1
5	9	3						
3	5		6		8			7
	8		4		9	5		
9			5		3		6	8
						2	9	6
4	7							
			6					

Hard # 487

```
. . 5 | . . 4 | . . .
. 9 2 | 7 1 . | 5 . .
. . . | 2 8 . | . . 6
------+-------+------
1 . . | . . . | . . 7
. 3 . | . . . | . 6 .
2 . . | . . . | . . 1
------+-------+------
5 . . | . 3 2 | . . .
. . 6 | . 5 9 | 4 2 .
. . . | 6 . . | 3 . .
```

Hard # 488

```
7 . . | 3 . . | . . .
9 . 1 | . 5 6 | . . .
. . . | . . . | . 1 .
------+-------+------
. 8 3 | 4 . . | 7 . .
4 6 . | . . . | . 2 1
. . 5 | . . 8 | 6 4 .
------+-------+------
. 5 . | . . . | . . .
. . . | 8 2 . | 1 . 5
. . . | . 7 . | . . 4
```

Hard # 489

```
. . 5 | 3 . . | . . 9
. . . | . 9 . | . 1 4
. . . | 8 . . | 5 . .
------+-------+------
. . . | 6 . . | 2 . .
9 3 . | . . . | . 6 1
. . 4 | . . 8 | . . .
------+-------+------
. 1 . | . . 2 | . . .
5 6 . | 9 . . | . . .
8 . . | . . 7 | 3 . .
```

Hard # 490

```
. . 4 | . 3 5 | . . .
. . 7 | . . . | 4 2 9
7 . 8 | . 5 . | . . .
------+-------+------
. 1 . | 3 . 9 | . 7 .
. . . | 7 . . | 1 . 4
4 5 1 | . . . | 2 . .
------+-------+------
. . . | 6 2 . | 7 . .
. . . | . . . | . . .
. . . | . . . | . . .
```

Hard # 491

```
. . 4 | . 9 6 | . . 5
. . 1 | . . 8 | . . .
3 . . | 2 7 . | . 1 .
------+-------+------
9 . . | . . . | . 8 .
. . 8 | . . . | 3 . .
. 6 . | . . . | . . 4
------+-------+------
. 1 . | . 8 5 | . . 9
. . . | 6 . . | 7 . .
6 . . | 4 1 . | 8 . .
```

Hard # 492

```
6 . . | 9 7 . | 4 . 1
4 . . | 1 . . | . . 2
. . . | . 8 . | . . .
------+-------+------
1 . 3 | . . . | . . .
. 5 4 | . . . | 1 7 .
. . . | . . . | 2 . 5
------+-------+------
. . . | 8 . . | . . .
3 . . | . 2 . | . . 9
8 . 9 | . 4 1 | . . 3
```

Hard # 493

9				8	6			
	7		3				2	
1						4		3
			4	5				2
5								6
7				6	3			
2		7						4
	6				4		5	
			1	2				8

Hard # 494

	3		2		8			7
	8	9	4		3		2	
	6	1			4			3
9			6			2	8	
	4		9			5	6	1
3			1			6		9

Hard # 495

		9	6					
	4	2		8		5		
	3				7			
3				7				8
	2	8				4	1	
4				1				2
			3				7	
		3		9		1	6	
					4	8		

Hard # 496

				1		8		
		9			8	5	7	
	7		2			9		
					5		1	
	6			4				5
	1		8					
		2			9		8	
9	7	6				2		
		6		7				

Hard # 497

	1					6		7
			4					1
7				1		4		
		4	8			3	5	
	6	7				2	1	
		8		6				1
	7			5				
6		3					8	

Hard # 498

				4	9	3		7
	5			2				
	9				1	5		
				1			7	3
		7				6		
3	2			8				
		2	1			5		
				3			6	
1			4	8	7			

Hard # 499

					1		2	
	3		8	5	9			
				6	5			4
	5	9						
3		1				2		7
						4	9	
9		4	6					
		7	1	2			6	
	1		5					

Hard # 500

	9	1	7					
6						2		
			6	9			4	5
			2				3	
5			4		1			7
	1				9			
9	3			1	2			
		4						8
					7	3	5	

Hard # 501

				9			5	7
5			7			1		
			3		2	4	9	
			4			9		
8								2
		7			3			
	8	4	2		7			
		2			8			9
3	6			4				

Hard # 502

2		9		6			5	
		3	8					7
	5							4
				3		1		
	1	9		2	6			
	4		6					
5						7		
6				8	5			
	3		4			1		6

Hard # 503

			5			9		
4			6			1		
	1		9			4		
		4	2			6		9
		2				8		
8		3			1	7		
	6			3			7	
		7			6			8
	8			2				

Hard # 504

			1		9		7	
			7					
5						4	1	6
	6				2	3		4
		7				8		
2		4	8				6	
6	1	8						2
			2					
	3		4		7			

Hard # 505

6		5						1
	4	9	7	5				
		8				5		
8	6				7			
		3				9		
			4				8	6
		7				8		
				7	2	3	6	
1						4		7

Hard # 506

			5	4				
			2				1	
3						5	2	8
		5					6	2
	7			1			3	
2	3					9		
8	2	4						6
	1				7			
			8	9				

Hard # 507

		2					6	8
7			5					
		1	6	7		9		
	9			5				
5			9		2			1
			8			5		
	8		6	4	1			
				5				6
1	7				5			

Hard # 508

		1	5					
			1	7		2	3	
		8		9		5		
		2		3		7		
	1					3		
	9			8		1		
	7			5		9		
4	2		8	6				
						2	5	

Hard # 509

3		5	1	9			2	8
			5			4		
	1							
	2			6			3	
		8				1		
	3			5			4	
							7	
		9			2			
7	8			4	3	6		2

Hard # 510

9	2				6		1	
	6		1	2				
				7				
2							7	9
6		3				5		1
4	9							6
			6					
			4	5			2	
	5		2				8	3

Hard # 511

		9		4	6		8	
	8					2		
3						4	5	
					8			9
4			6		2			8
5			3					
	4	1						5
		3					1	
	7		8	1		6		

Hard # 512

1		8		4	7			
3				6			4	
			9				2	
			2			5		
		4		5		2		
		9			6			
	4					3		
	3			9				1
			8	2		7		3

Hard # 513

5				9	3			
8			4				1	
7	9		1					
			6			7	3	
3								1
4	7		8					
				2		9	6	
	2		1					5
		3	5					8

Hard # 514

		9		4				5
		6		1		2		
	8			2	7			
				3			4	1
3								6
8	5		9					
		2	1				3	
	6		2		9			
9			5			1		

Hard # 515

	6				8			
8			1	4		3		
	4		6					8
2			7					
	5	7				2	3	
					4			1
7					3		8	
		3		5	1			6
			4				2	

Hard # 516

2				4	5			
			9				1	
6						5		2
						2	4	
	7	2	1			4	3	8
	4	8						
1		7						3
	5					6		
			3	5				9

Hard # 517

```
. . 7 | . . 3 | 9 . .
. 9 . | 1 . . | . . .
3 . . | . 9 . | . 7 .
------+-------+------
. . . | . . 9 | 5 . .
. . 3 | 4 . 7 | 1 . .
. . 6 | 3 . . | . . .
------+-------+------
. 1 . | . 6 . | . . 8
. . . | . 4 . | 2 . .
. . 8 | 2 . . | 4 . .
```

Hard # 518

```
. . 8 | 5 1 . | . . 9
. . . | . 6 . | . 7 .
6 . 9 | . . 7 | . . .
------+-------+------
7 3 4 | . . . | . . .
2 . . | . . . | . . 7
. . . | . . . | 8 2 1
------+-------+------
. . . | 6 . . | 5 . 3
. 4 . | . 3 . | . . .
3 . . | . 9 8 | 1 . .
```

Hard # 519

```
. . . | . 2 . | . . 9
. . . | 7 3 . | 8 5 .
5 . 1 | . . . | 7 . .
------+-------+------
6 . . | . . 4 | . 3 .
. 2 . | . . . | . 9 .
. 7 . | 6 . . | . . 4
------+-------+------
. . 6 | . . . | 3 . 1
. 3 5 | . 4 6 | . . .
8 . . | . 9 . | . . .
```

Hard # 520

```
. 8 . | . 9 . | . . .
. . . | 3 . 1 | . 9 5
2 . . | . . 6 | . . .
------+-------+------
. 2 6 | 9 . . | . . .
. . 9 | 4 . 5 | 3 . .
. . . | . 6 9 | 1 . .
------+-------+------
. . 2 | . . . | . . 8
5 3 . | 8 . 2 | . . .
. . . | 4 . . | 7 . .
```

Hard # 521

```
. . 1 | . . 8 | 2 . .
8 2 . | . 1 . | . . .
. . 9 | 5 . . | . . 4
------+-------+------
2 9 . | 7 . 4 | . . .
. . . | . . . | . . .
. . . | 2 . 3 | . 6 5
------+-------+------
4 . . | . . 9 | 3 . .
. . . | . 3 . | . 4 1
. . 6 | 8 . . | 7 . .
```

Hard # 522

```
. . . | 9 3 . | . . 2
. . . | . 8 3 | . . 6
. 3 . | . . . | . . 4
------+-------+------
. . . | 3 7 . | . 1 .
. . 9 | . . . | 5 . .
. . 8 | . 1 4 | . . .
------+-------+------
7 . . | . . . | . 2 .
6 . 5 | 1 . . | . . .
9 . . | . 4 3 | . . .
```

Hard # 523

			1					
	1	4		7		6	5	
	8	7		6				
1		5			4			
		4				3		
			8			5		4
				1		9	7	
3	9		7			6	4	
					3			

Hard # 524

	9			7			6	
		1	6					9
					9	2		8
			3	1		9		
	7						4	
		5		4	8			
5		4	1					
6					2	4		
	2			6			7	

Hard # 525

				4	7	6		
4							5	1
8		1						
	4			9	3		1	
		3				5		
	6		8	5			9	
					2			8
6	9							4
		4	6	3				

Hard # 526

1	8				2			
			9	5				8
6						3		
	4	7			3			
5		3				4		9
			1			2	7	
		2						7
9				2	1			
			8				3	2

Hard # 527

						9		
	7			4	2			3
				9				1
7	2		4				1	5
	6						7	
8	1				7		2	9
5			6					
4			3	7			8	
		6						

Hard # 528

6			8				3	1
2			3	9		4		
	1		5					
							8	6
			7		9			
	4	5						
					5		4	
		7		2	3			8
4	2				8			3

Hard # 529

			2	1		6	3	
		4						5
3		9		7				4
			9					
1		3				9		2
					5			
9				6		8		7
7						2		
	8	1		3	9			

Hard # 530

5		7			2			
	3			4	9			
		1						5
8	5				1			
	7	9				8	5	
			4				6	9
1						4		
			6	3			1	
			8			9		2

Hard # 531

			3			1		
		4		8	5			
6								3
7	2		5		3		9	
	9		8		7		1	5
1								7
			9	2		5		
		5			6			

Hard # 532

2								
	6		3	4			2	
5	1		9			8		
				8		1		
1			7		4			2
		7		3				
		6			2		9	3
	5			6	9		4	
								6

Hard # 533

		7	2			9		
	3					4		8
			4			2		
1			8	7				6
		6				7		
7			5	9				4
	5			1				
6		2					5	
		1			9	3		

Hard # 534

		5			7	6		8
8		2						
			8	3		1		
	2			4				
	8	1				3	5	
			6			4		
	4		2	7				
						8		1
6		9	8			4		

Hard # 535

2			3	4			1	5
			9		1			3
						9		
		1			8		2	
		2				7		
	7		5			1		
	9							
6			7		5			
7	5			1	3			8

Hard # 536

7		3				5		
	8	6		4				2
			7				9	
				1		6	4	7
8	1	7		6				
	7				2			
5				3		1	7	
		1				4		3

Hard # 537

2								1
		8	2			5		
	1		8	4				2
			3		6			9
		3			2			
8			9		2			
1			7	3		2		
		6			4	1		
9								5

Hard # 538

	3		4	6		8		
			1			2		4
	4					3		
2			7				4	
		9				8		
	8			9				1
	5						9	
3		8		6				
	2		9	3			5	

Hard # 539

			7		3		8	
	7			9				
9			6		5			
		1						2
	9	4	3		8	6	7	
6					8			
			5		1			9
				3			1	
1			7		4			

Hard # 540

	9				6			2
		7	2	1				
		4						5
				7		1	9	
5								3
	8	3		4				
9						2		
			8	1	9			
8			5				6	

Hard # 541

							1	
2			3		6		4	
4		5	2	1				
		3			4	5		
				7				
		9	5			7		
			3	7	1			8
	7		1		8			4
	5							

Hard # 542

			4	8		2	7	
						6	3	
7								9
1				3			4	
5	2						3	7
	8			6				5
6								3
		7	2					
	5	3		4	9			

Hard # 543

	6			8		5		7
2								
	5		4					6
				1		8	4	
	8		7			6		
4	3		8					
9				5		7		
								3
6		1		4			9	

Hard # 544

			6			7		2
3			4		8		9	
4								8
			5	8				
		8	7		1	2		
				3	2			
1								3
	9		6		3			4
8		4		7				

Hard # 545

	2			1		6		8
			8	2				7
			5					4
	4	2	3					9
9				8	4	3		
8				9				
2			1	3				
4		9		7			5	

Hard # 546

		3	6	9				
1				8				
4		5			1	6		
	7	4		3				8
6				1		3	5	
		9	1			2		4
			9					3
				4	2	7		

Hard # 547

8				9			2	
			8		3			
			5	2		8	6	
2	9							
	5	3				7	9	
							1	3
	6	4		8	9			
			1			5		
		7		4				6

Hard # 548

		1	3					
					8	2		7
			2				3	9
		5		8		6		
1		9				5		4
		7		2		3		
3	5				4			
9			6	5				
						7	9	

Hard # 549

		3	9					
					1		2	
	5	6		8	3		1	
					5	4		
6		4				2		5
		9	3					
	2		6	1		3	9	
	9		5					
					9	1		

Hard # 550

								3
				1	2	4		
	7		9	5				
	8	6		2	9	4		
		5				8		
		3	6	8		7	9	
				3	7		6	
	1	7	8					
9								

Hard # 551

								8
	6		1	7		5		
		9	5					3
		6		5			9	
		4				6		
	3			6		7		
5					2	4		
		2		8	4		1	
4								

Hard # 552

						1		
			4	7			5	
8	9		1			5		
3		4		8	7			
	1						6	
			6	1		7		2
			5			6	1	4
	8			4	2			
		5						

Hard # 553

```
. . . | . . 8 | . 7 6
. 7 . | 5 . 3 | 2 . .
. . . | . 7 . | . 1 .
------+-------+------
2 . 5 | . . . | . . .
9 . . | 6 . 7 | . . 3
. . . | . . . | 6 . 9
------+-------+------
. 5 . | . 8 . | . . .
. . 2 | 1 . 9 | . 5 .
3 9 . | 7 . . | . . .
```

Hard # 554

```
. . 2 | . . 5 | 7 3 .
. . . | . 6 . | . . .
. . 8 | . . 3 | . . 1
------+-------+------
5 . . | 7 . . | . 1 3
. . 4 | . . . | 5 . .
2 3 . | . . 8 | . . 9
------+-------+------
9 . . | 8 . . | 2 . .
. . . | . 7 . | . . .
. 6 5 | 2 . . | 1 . .
```

Hard # 555

```
. 9 . | 5 . 7 | . . .
. . . | . . . | 6 4 .
. . 6 | . . 4 | . . 3
------+-------+------
8 1 . | 6 2 . | . . .
. . . | . . . | . . .
. . . | 5 3 . | 1 8 .
------+-------+------
6 . . | 3 . . | 4 . .
4 5 . | . . . | . . .
. . . | 2 . 8 | . 9 .
```

Hard # 556

```
. . 8 | 2 . 1 | . . .
. . 3 | . . . | . . 1
. . 7 | . 4 . | . . 8
------+-------+------
. . . | . . . | . 4 9
7 1 . | . . . | . 6 3
8 2 . | . . . | . . .
------+-------+------
3 . . | 1 . . | 4 . .
2 . . | . 5 . | . . .
. . . | 6 . . | 2 9 .
```

Hard # 557

```
9 . 5 | . 3 . | 6 . .
3 4 . | 2 . . | 9 . .
. 6 . | . . . | . . .
------+-------+------
2 . . | . 3 . | . . .
. . 7 | 5 . 9 | 8 . .
. . . | 8 . . | . . 4
------+-------+------
. . . | . . . | 8 . .
. . 9 | . . 8 | . 3 5
. 3 . | . 9 . | 4 . 6
```

Hard # 558

```
. 5 . | . . . | . . .
. . 9 | . 4 . | 2 1 8
3 . . | . . 6 | 9 . .
------+-------+------
. . . | 4 . . | 7 . .
. . 1 | 9 . . | 3 5 .
. . 7 | . . 8 | . . .
------+-------+------
. . 3 | 6 . . | . . 9
6 7 2 | . 3 . | 1 . .
. . . | . . . | . 7 .
```

Hard # 559

```
 . 2 . | . . . | . 8 .
 4 . 1 | . . . | 7 . 3
 6 . . | 5 4 . | . . .
-------+-------+-------
 . . . | 4 . 7 | . 6 .
 8 . . | . . . | . . 1
 . 3 . | 8 . 5 | . . .
-------+-------+-------
 . . . | . 3 1 | . . 2
 1 . 4 | . . . | 6 . 5
 . 6 . | . . . | . 1 .
```

Hard # 560

```
 . . . | . . . | 5 . 9
 4 . 7 | . . . | . . .
 . . 6 | . 3 . | 2 7 8
-------+-------+-------
 . . 2 | 3 . . | . 8 .
 . . . | 2 . 5 | . . .
 . 7 . | . . 4 | 6 . .
-------+-------+-------
 2 1 8 | . 7 . | 3 . .
 . . . | . . . | 8 . 1
 3 . 9 | . . . | . . .
```

Hard # 561

```
 . . 5 | . . 4 | . . .
 . 9 2 | . . . | . 7 1
 . . . | 6 . . | 9 . .
-------+-------+-------
 . . . | 8 . 2 | . . .
 1 4 . | . . . | 5 9 .
 . . 5 | 3 . . | . . .
-------+-------+-------
 . 8 . | 1 . . | . . .
 2 1 . | . . . | 3 8 .
 . . 6 | . 4 . | . . .
```

Hard # 562

```
 . . . | 2 3 . | . . 1
 6 . . | . . . | 7 8 .
 . . 7 | 4 . . | . . .
-------+-------+-------
 9 7 . | 2 . . | . . .
 . 5 . | 4 . . | 3 . .
 . . . | . 9 . | 8 4 .
-------+-------+-------
 . . . | 3 8 . | . . .
 5 1 . | . . . | . . 3
 4 . 3 | 6 . . | . . .
```

Hard # 563

```
 . . 5 | . . . | 3 . .
 . . . | 6 5 7 | 8 . .
 . 9 . | 4 . . | . . .
-------+-------+-------
 3 . 9 | . . . | . . 6
 5 1 . | . . . | 9 4 .
 4 . . | . . 2 | . . 8
-------+-------+-------
 . . . | . . 8 | . 6 .
 . 7 . | 9 1 3 | . . .
 . 3 . | . . 8 | . . .
```

Hard # 564

```
 7 1 . | 3 . . | . . 8
 . 4 . | . 9 . | . . .
 . 5 . | . . . | . . 3
-------+-------+-------
 2 3 . | . 1 . | . . 5
 . . 7 | . . 2 | . . .
 9 . . | 8 . . | . 1 4
-------+-------+-------
 4 . . | . . . | . 3 .
 . . . | 6 . . | . 7 .
 5 . . | . . 2 | . 4 1
```

Hard # 565

					8		5	
	3	6	4			1		
9			2					
	8					6	2	
			5		1			
6	7					3		
				6				7
	4			3	6	9		
1		9						

Hard # 566

	6	7						
4					1	7	6	
9			8			3		
				5	9			8
				4				
5			7	8				
		2			4			6
	4	9	3					5
						9	7	

Hard # 567

		1		7		2		
								1
9			6		8		7	
	3	7		5	1			
2								7
	4	8		3	6			
	3		1		4			9
6								
	4		5		7			

Hard # 568

					4		3	8
			1	2				
2	8						7	
	1	5						
9	2		6		7		1	4
						9	2	
	7						8	3
			8	9				
1	6		4					

Hard # 569

		6		9		1		
4	7							
			8	2		5		
		3		5		1		
	1					2		
	6			7		9		
	4			6	5			
							8	2
		7		8		3		

Hard # 570

	1					5		
8			2		9			
			3		6			4
3			6				1	
4	2						7	6
	8					1		2
9			4		2			
			5		3			7
	7					3		

Hard # 571

			4			5	3	
	4		8				9	7
		3		7		2		
3							7	
	5						1	
	9							6
		7		5		6		
2	6				1		8	
	1	8			4			

Hard # 572

		5						
		9			7		8	
			2	8				5
			8				2	1
	2	3				9	6	
1	6				9			
7				6	2			
	8		5			1		
						7		

Hard # 573

2			1		6			
				9	2		8	4
			5			7		
	5	2						
		9	8			3	5	
						3	4	
		5			8			
1	2		9	3				
				1		4		7

Hard # 574

			5					4
4		6	7				2	
7	1				6	9		
				4	9		3	
	6		1	2				
	4	7					6	8
	2				3	1		9
8				4				

Hard # 575

	2	8		3				
			2	7	8	3		
7			6					
	8	1						4
			7		6			
4						9	2	
					4			8
		7	9	8	3			
				2		6	5	

Hard # 576

	7	1		4				9
						6	1	
			3				8	
1				4	5			
8			2		5			6
		5	8					1
	1				9			
	3	7						
2			5			7	9	

Hard # 577

	6	4	2				7	
8				3				
	7	3					8	2
			3	5				
	1						4	
				9	4			
7	5					6	1	
			8					5
	8				5	7	2	

Hard # 578

			6			7	2	
1						9		
		3		4	8			
	8	9	7					2
3								7
5					3	4	9	
			5	6		1		
		5						8
	3	6			1			

Hard # 579

	1		2				9	8
						5		
	4			6	1			
		4		2		8		
7			8		5			3
		3		4		9		
			7	8			5	
		6						
1	8				2		7	

Hard # 580

		4			8			
7				2			8	
3		2		7				
		7			5	9		4
		9				3		
4		1	6			2		
				6		7		5
	5			4				9
			9			6		

Hard # 581

		3				4		6
				3				
6		1	8				5	7
5			6		1		4	
	1		7		8			3
8	2				6	7		5
			5					
3		9				2		

Hard # 582

	6							
3		7		1		2	4	
		1	3					9
			2		7			3
	1						6	
6			1		9			
1					4	5		
	5	6		3		9		7
							2	

Hard # 583

	9	2				5	4	
8						9		
4						6		3
					7		4	
1		4				7		6
	2			1				
	4		6					8
			9					2
		6	8			5	7	

Hard # 584

			6					
	3							
	6	1				7	5	9
		3	9	7			8	1
			8		5			
2	7			6	3	9		
4	1	7				8	2	
							6	
					8			

Hard # 585

			5	7	4			
9	5						7	
						2		
			1	9			5	
3		4				1		8
	9			4	3			
		7						
	8						4	6
			2	1	7			

Hard # 586

2				7			8	
5	8						1	7
	6		5					
6				1		3		
	4						2	
		1		8				9
					9		6	
7	5						3	1
	9			5				2

Hard # 587

1			6			9		
		2					4	1
		4		9			8	
	5				2			
8		9				6		7
			4				9	
	9			2		5		
6	7					2		
		5			7			9

Hard # 588

	9	6	2					8
	3			6		7		
			8		5			
7				1		9		
		5		8				3
			7		2			
		2		9			1	
9					3	6	7	

Hard # 589

```
5 . . | 9 . . | . . .
. 1 . | 6 8 . | . . .
6 . . | . 3 . | . . .
------+-------+------
4 9 . | . . . | 6 . .
. 3 2 | . . . | 7 9 .
. . 1 | . . . | . 3 2
------+-------+------
. . . | 5 . . | . . 1
. . . | 1 9 . | 2 . .
. . . | . 7 . | . . 5
```

Hard # 590

```
. 4 . | 1 . . | . 7 .
. 5 . | 6 . . | 2 . .
. . . | . 2 3 | . . 6
------+-------+------
. . . | 7 . . | 8 . .
5 3 . | . . . | . 1 4
. 2 . | . 3 . | . . .
------+-------+------
6 . . | 4 8 . | . . .
. 3 . | . . 2 | . 4 .
. 7 . | . . 6 | . 2 .
```

Hard # 591

```
. 3 . | 8 . 5 | . . .
. . 1 | . 2 6 | . . .
. 5 8 | . 7 . | . . .
------+-------+------
2 1 5 | . . . | . . .
. 8 . | . . . | 6 . .
. . . | . . . | 5 4 9
------+-------+------
. . . | 3 . . | 9 1 .
. . . | 5 6 . | 4 . .
. . . | 7 . 9 | . 2 .
```

Hard # 592

```
. 8 . | 5 . . | . . .
. . . | . 9 4 | . . .
. 6 1 | . . . | . 7 9
------+-------+------
. 1 5 | . 7 3 | . . 6
3 . . | 8 2 . | 5 4 .
6 4 . | . . . | 7 3 .
------+-------+------
. . 2 | 9 . . | . . .
. . . | . 4 . | 6 . .
. . . | . . . | . . .
```

Hard # 593

```
. . . | . . . | 7 . .
. 5 . | . 3 7 | . . .
. 4 7 | . . . | . . 6
------+-------+------
. 9 . | 7 . . | 8 . 1
8 . . | 3 . 2 | . . 9
1 . 6 | . . 5 | . 4 .
------+-------+------
6 . . | . . . | 1 2 .
. . . | 1 5 . | 3 . .
. . 9 | . . . | . . .
```

Hard # 594

```
. . 4 | 2 . . | . 6 .
. . . | . . . | 5 . 7
8 . . | . 4 . | . . .
------+-------+------
. . 5 | . 9 8 | 2 . 4
6 . . | . . . | . . 3
4 . 2 | 7 3 . | 6 . .
------+-------+------
. . . | . 1 . | . . 9
3 . 1 | . . . | . . .
. 2 . | . . 5 | 7 . .
```

Hard # 595

				4				8
			8			6		
				5	9	7	1	
	9	2						3
5		6				4		2
4						6	5	
6	4	1	8					
	8		7					
2			3					

Hard # 596

6								8
4		2		9				
				3	4	2	9	
	1							
	4		7	8	9		1	
							5	
	6	9	3	4				
				7		5		6
1								7

Hard # 597

		9				2		1
5			4					
		2	9			5		
2			8		6		1	
	1						7	
	7		5		2			3
	4				3	6		
				2				8
8		6				1		

Hard # 598

	1				6		2	5
	5		2	4			1	
		3						
			6					1
	2	7				9	5	
5					3			
						2		
	4			6	7		9	
1	7		8				4	

Hard # 599

				6		1		
7			8					
6			2			4	7	
5				2		6		
8								3
		1		7				5
	8	9			5			2
					2			1
		4		9				3

Hard # 600

4			1		6	3		
5	8							
				9		7		
		5	8					
	1		5		7		8	
					3	4		
	1		2					
							6	2
		9	7		1			3

Hard # 601

			5	7				2
						5	6	
		9		8	1			3
		4					1	
8			7		2			6
	3					2		
2			1	9		6		
	5	6						
1				5	8			

Hard # 602

2						6		
	1	4			6		5	
			1	8				3
7		1						4
	9						6	
8						1		2
1			7	4				
	7		5			2	8	
		9						5

Hard # 603

			6		2			
		5						1
		5	2	1				
2	6			3			4	5
	9						3	
1	5			7			2	8
			2	9		6		
7					3			
		1		4				

Hard # 604

			2					5
			1		9	2		3
		7			8			4
		5				3	7	
	8						9	
	7	2				5		
7			6			8		
2		3	7		4			
8					1			

Hard # 605

	9							7
		6					9	4
		3		2			6	
			5			6	3	
2			6		3			8
	3	5			7			
	8			9		4		
9	6					1		
1						8		

Hard # 606

1					3		6	
	3		1		2			
	8	4			2			9
	5	8	2					
					9	8	2	
4			7			1	8	
		5		8		7		
	6		1					5

Hard # 607

			1				6	
6					5		3	
1			2		4			5
							4	9
		9	6		2	5		
7	2							
4			5		6			8
	9		4					1
	7			8				

Hard # 608

	4							9
5			3				1	8
				2	5			
	9	7			8	1		
1								6
		8	1			2	9	
			5	6				
4	1			9				3
6							8	

Hard # 609

7								
			5		6			4
		3	7			9		
4			1	3	2			
5								8
		2	6	4				1
	4				6	3		
8		9	3					
								9

Hard # 610

					2			8
					3		5	4
3		8			5		2	
		2				1		
			8		4			
		6				4		
	9		7			2		3
4	3		1					
2			5					

Hard # 611

5	2							8
		9					3	
	3		9	1		2	4	
			2	6	7			
			3	8	4			
	8	6		7	3		9	
	9				3			
7							5	1

Hard # 612

2		1	6			3		
	5				2			
4	9			3				
		6	4					2
9								8
3				5	4			
			9				4	6
		5					8	
	4			3	5			9

Hard # 613

	4					7		
		2	5	6			1	
		3	9					
7						3	2	
4				5				6
	1	8						4
					1	2		
	3			7	5	1		
		6					3	

Hard # 614

5			9		1	2		
	7						8	
1			3		5			
2			1	9				
	1						3	
				3	4			5
			6		9			8
	2						6	
		6	5		3			2

Hard # 615

			8			9	5	
9	8				1			
1				5	3			
		2						3
	5			3			8	
7					2			
		4	1					5
			8				4	7
	7	6		5				

Hard # 616

			3	2				7
	9		7		1			6
	7							
2				8			4	
1		8				5		3
	4			6				2
							1	
8			4		2		6	
5				3	8			

Hard # 617

	9	8		1		6		
1							5	
			9	7				
	1		4	8		9	5	
9	4		1	7		6		
			3	2				
	2							4
		4		8		1	7	

Hard # 618

						6	7	
			3	2				4
8			6		1	3		
	4	6		8				
	3					5		
			4			9	3	
		1	8		3			5
4			5	6				
3	2							

Hard # 619

				2	6		7	
		8			3			4
6						5		
	6		4	7		3		
9								5
		1		5	2		6	
	5							8
3			2			1		
	4		3	6				

Hard # 620

			4		1	3		
1			5				8	
			9		7			
7				4	5			2
	5						6	
3			8	7				4
		9		6				
	6			8				3
		5	3		9			

Hard # 621

	8		3		4	2		
			1		2	6		
		4						7
8	7	2						
9								8
						9	2	3
2						5		
		1	4		5			
		6	9		1		4	

Hard # 622

		7			8		9	
1				5			8	
6					2		4	1
					5			
	1	5				6	2	
			9					
5	9		1					3
	7			8				4
	8		6			1		

Hard # 623

				3	1			5
5	7						4	
		8					3	
		7	4			8	9	1
2	4	1			6	5		
	3					6		
	5						2	4
6				1	2			

Hard # 624

	7							
					2			6
1			4	5			7	
		3		2				8
	6		7		5		3	
4				9		2		
	9			8	1			5
3			9					
							9	

Hard # 625

2							6	
				8				5
	7		5	3				9
		6			7	8		
7		1				3		2
		8	9			5		
6				2	8		1	
9				7				
	1							7

Hard # 626

	8		7			6	1	
				6				
2		9	5					
1	7	8	9					
		2				3		
					5	7	8	1
					4	8		2
			7					
	2	4			6		5	

Hard # 627

9						2	3	
		7		3				5
		5			8		4	
2				9	4			
			2		1			
			3	5				8
	8		6			3		
4				1		9		
	3	2						6

Hard # 628

5				4				
	6	3		7				8
	8					6	2	
		9	6					
			3		8			
				1	3			
	3	2					7	
6			9			2	8	
			7					5

Hard # 629

4	7	9			2			
			6	1				
						9		5
	5		4				8	
2			7		8			4
	4				6		7	
5		4						
			6	2				
			8			3	6	1

Hard # 630

	6	7	8				4	
				4			7	
2		4	5	1				
3								8
			9		1			
4								5
				9	3	8		2
	8		4					
	1				5	4	9	

Hard # 631

1	4		2					
			4	1		6		9
8						3		
		7	6		9			2
3			8		5	9		
		1						5
6		4		5	2			
					6		1	7

Hard # 632

		9	6					
	2				7		1	6
		7		8			3	
	9			4	3			
		6				1		
			7	6			2	
	6			7		9		
4	1		9					5
					2	3		

Hard # 633

				4	7	1	8	
						4		3
4				5			7	
					8	6		
		1	2		5	3		
		2	6					
	2			6				1
6		8						
	9	5	4	7				

Hard # 634

		9	8			2	4	
	2		1				7	
8			6		7			
						1		8
			5		2			
1		4						
			7		9			5
	5				1		9	
	3	8			5	7		

Hard # 635

			1		2	5		
8					6	4		
1			4			8		
	9	4		7			6	
	7			2		1	4	
		3			9			1
		9	8					5
		8	6		7			

Hard # 636

				1				
	6		3					5
	3	4			5			7
3	4			1				
9		8				4		2
			2				3	8
2			4			7	9	
5				3			4	
			2					

Hard # 637

			2					8
	6	5			3		2	
		7		9				3
9			8	2			5	
	4			7	6			9
5				1		6		
	7		3				1	8
4					8			

Hard # 638

					3	6		
	4				7			9
			9	5			2	
						9	1	
6	9	8				5	4	3
	7	1						
	2			9	5			
8				1				6
		4	6					

Hard # 639

				9		4		
			7					9
	2	1		4				8
	9				3		4	5
		6				1		
3	8		1					2
8				5			7	6
7					9			
		3			8			

Hard # 640

5			7				2	
				9			6	
		1	4					
	9		2	5		6	3	
		4				8		
2	8			4	3		9	
						6	2	
1		9						
6						4		7

Hard # 641

	6							
	5	4				9	7	
3				7				6
				1				2
1			9		4			8
6				3				
4				2				1
		7	1			2	5	
							7	

Hard # 642

	2				4			
	7				5		9	
3				7				2
6	9							
7	1		5		9		2	6
							1	8
2				1				4
	4		7				8	
			3				5	

Hard # 643

6		5		3	7			
	7							3
		8	6	2				
						3	1	
	3		4		1		9	
4	5							
				7	3	2		
8							5	
			2	4		6		8

Hard # 644

								8
	6			8	2		5	
1				6	7			
5	4						2	
6		3				5		9
	2						4	3
			5	9				4
	9		3	1			6	
8								

Hard # 645

8	6		4					
	3	7		2	5			1
								5
					2			3
		4			7			
9			7					
6								
7			8	1		9	4	
					7		5	6

Hard # 646

				6				
		5					1	8
	4		7	5				3
			3			7		5
	3						8	
1		8			2			
4				1	7		9	
3	9					1		
				3				

Hard # 647

			3	5				
	1	8				3	7	
7								6
	6	1	2				9	
			8					
	7				9	4	5	
1								3
	9	4				1	6	
				9	2			

Hard # 648

7					2			
				4		9	6	
		1		6		2		
	3				8	5		
4			5		6			2
		8	2				1	
		2		9		4		
	5	4		2				
			3					8

Hard # 649

		5	8	9				
		1	7		3			
3	2							6
	7	3					1	8
5	9					2	4	
4							6	9
			9		5	8		
			8	6	5			

Hard # 650

7		1		2			6	
6			7			8	2	
								9
		7		8				
	8		9		3		4	
				7		9		
5								
	3	9			5			2
	2			9		6		5

Hard # 651

		6	8					
1								8
		2	4		1	9		
							2	6
	7	1				8	3	
6	3							
		9	3		6	1		
7								4
					2	7		

Hard # 652

	9	1	7			4		5
			8					
2	3							
		5		4		9		
3								7
		9		5		1		
							1	4
				5				
9		6			3	8	7	

Hard # 653

	9	8						7
			3				4	
		1	9		5			
9		5	7		4			1
	7		6		3	5		9
			2		1	3		
	6				7			
5						1	8	

Hard # 654

								6
		4	1	6				
					8	4		1
	8		5		1		3	4
	2						1	
3	5		7		9		2	
2		9	3					
			1	7	9			
6								

Hard # 655

3	9	2			7			
4				3				1
	1		6		9			
					1			5
	8						1	
2			3					
			1		8		7	
7				6				4
			4			8	3	9

Hard # 656

			7		5	2		
		9		2	1	4		
				4				1
7						6		9
				3				
3		5						8
5				6				
		7	8	1		9		
		8	9		2			

Hard # 657

	5	6					7	
2				7		3		
7			5		1			2
			8		2			1
9		3		2				
4			8		2			9
		8		6				3
	9					7	8	

Hard # 658

1			6			7	5	
5		9						2
							4	3
			1	7	8		2	
	8		9	6	5			
4	5							
8						5		1
	7	1			2			9

Hard # 659

6			1		8		7	
				9			5	
			4					3
	2			6		8		
8		5				1		4
		1		5			2	
3					6			
	4			1				
	8		7		2			6

Hard # 660

		1		7		3		
9				4				1
			8	1			5	
		9		3	2		7	
	1		4	8		5		
	5			6	9			
3			7					8
		2		4		7		

Hard # 661

					6		3	
				9		2		
4	3		7					5
5			4					1
		8				7		
7					3			4
1					4		7	9
		2		8				
	8		1					

Hard # 662

				6		8	9	7
		8						
9			1					
		9	5			2		
6		5				7		4
		2			4	1		
						9		8
						4		
2	4	1		7				

Hard # 663

1		9						
	2		8		5			3
			7				8	9
			2			7		
2			3		8			1
		3			4			
5	4				7			
7			9		2		6	
						1		5

Hard # 664

4		9				7		
			3					
7	1		6					5
	3	7			9	5		
	9						2	
		8	7			3	9	
9					4		5	8
				5				
		1				6		2

Hard # 665

		1		2		4		7
4		8						
		9	4				5	8
				5				
6			8		9			1
			1					
8	3				1	2		
						5		3
9		2		3		8		

Hard # 666

		6			9			2
8	4			1		7		
					7		5	
	9		4	3		6		
		1		8	2		9	
	6		7					
	3			2			4	6
2			3			1		

Hard # 667

					7			8
8			2				5	
	9	7						
	8				1	6	3	
			8		2			
	5	9	7				4	
						7	6	
	3			6				5
9			1					

Hard # 668

	1	7			9			
			7		8		4	
5								3
				4	2		7	
9		8				2		5
	7		5	8				
7								2
	5		6			3		
			2			1	5	

Hard # 669

		3		6	1		8	
				3			1	7
			2					
		6	3					
	5	4				8	3	
					9	5		
					6			
2	4			5				
	7		8	9		3		

Hard # 670

6			9					
	2	5		4	1			
4			2		3			
2						9	5	
		4				1		
	5	8						3
			6		5			9
			8	1		4	7	
				2				6

Hard # 671

	6				1	8		
4	7			8	3			
			2		3			
3					1		6	
			6		2			
6		8						5
	4		5					
			9	3		1	4	
		1	2			9		

Hard # 672

5				1			7	9
			5				6	8
2					8			
	3			6	4			
		6				4		
			2	3			1	
			8					5
6	4			7				
1	5			2				4

Hard # 673

			7					
2		6	8		3			
1		4			9			7
		1	3					
6	9						4	3
						1	5	
7			6			3		8
			7		5	6		2
				9				

Hard # 674

4		3				1		5
1				8	7		4	
					2			
	3				6			4
9								1
6			4			3		
		2						
	4		1	5				2
8		9				5		3

Hard # 675

		1	7			9		
6				1	8		2	3
	3					6		
		5						
	1		6		9		5	
						2		
		8					6	
3	5		7	6				2
	6			4		7		

Hard # 676

	6				5		1	
			8					
			2	6	9			5
				9				6
8	3						5	7
2			7					
7		5	1	6				
				9				
	9		7				3	

Hard # 677

8				3				
3		4	9		6	1		
	7							
9		6			3			4
				5				
1			4			2		3
						4		
		9	1		4	6		8
			8					2

Hard # 678

				9				1
	5	8		7				
			5			2		9
			8			7		
5		4		2		3		8
		1		6				
2		5		3				
			1			5	3	
4				6				

Hard # 679

```
7 . . | . 1 3 | . 8 .
3 . 5 | . . 6 | . . .
. 1 . | 2 . . | . . 5
------+-------+------
6 . . | . 4 . | . . .
. . . | 7 . . | . . .
. . . | 3 . . | . . 9
------+-------+------
2 . . | . 8 . | 7 . .
. . 6 | . . . | 8 . 4
. 8 . | 1 5 . | . . 6
```

Hard # 680

```
2 . . | . 7 8 | 3 . .
. 8 . | . 6 . | . . .
. 5 . | 1 2 . | . . .
------+-------+------
. . . | 8 . . | . . 3
8 2 . | . . . | . 7 9
3 . . | . 2 . | . . .
------+-------+------
. . . | 9 6 . | . 1 .
. . . | 7 . . | . 9 .
. 6 7 | 3 . . | . . 4
```

Hard # 681

```
7 . . | 5 3 . | . 2 9
. . . | 2 4 . | 8 . .
1 . . | . . 8 | . . .
------+-------+------
. . 6 | . . . | . 5 .
. . . | 9 . 5 | . . .
. 5 . | . . . | 1 . .
------+-------+------
. . . | 8 . . | . . 4
. . 9 | . 6 4 | . . .
3 4 . | . 5 7 | . . 8
```

Hard # 682

```
. . 7 | 2 3 . | 8 . .
1 8 . | . 9 . | . . .
. 9 . | . . 1 | . . .
------+-------+------
. . . | . . . | 7 . 6
2 . . | 9 . 6 | . . 3
6 . 1 | . . . | . . .
------+-------+------
. . . | 7 . . | . 6 .
. . . | 1 . . | . 3 2
. . 5 | . 2 3 | 4 . .
```

Hard # 683

```
. . 9 | . . . | . 2 .
. 6 . | . 3 . | 9 . .
3 . 4 | 8 . . | . . 6
------+-------+------
. 7 . | 9 1 . | . . .
. . 6 | . . 4 | . . .
. . . | 8 6 . | 3 . .
------+-------+------
9 . . | . 8 3 | . . 5
. . 7 | . 4 . | . 1 .
. 1 . | . 8 . | . . .
```

Hard # 684

```
. . 2 | . 3 . | . . .
. . 7 | . 6 . | . 8 .
5 8 . | . 2 . | . . 1
------+-------+------
9 2 . | 7 8 . | . . .
. . . | . . . | . . .
. . . | 1 9 . | . 3 5
------+-------+------
3 . . | 1 . . | . 2 7
. 7 . | 6 . . | 1 . .
. . . | 4 . 5 | . . .
```

Hard # 685

2			4	6		9	1	
			7					
	9	8			5			
		3				6		
8			1		4			2
		4				5		
			6			7	4	
				7				
	4	9		2	3			5

Hard # 686

	9		6					3
	2		7	9				
1				8				
	4	2						6
	7		5		9		3	
6						7	1	
			3					7
			2	6			9	
5				4			6	

Hard # 687

		6					4	
	5		2				8	
	9	7	8					1
				9	7	3	1	
	1	9	6	8				
5					6	9	3	
	2				5		7	
	7				5			

Hard # 688

		6		4				
	4			1	9	2	6	
		5	6					
2						4		
			9		1			
		3						5
				3	8			
	3	8	4	2			7	
				8		1		

Hard # 689

		5		7	9		2	8
4		7	2					5
			6					
	2				6	7		
		6	3				1	
				4				
8					5	6		2
9	3		1	6		8		

Hard # 690

				1			5	
9		4	2	7				
						2		
				2		3	7	
7	2		3		8		6	1
	6	1		9				
		2						
			5	1	9			8
	7			6				

Hard # 691

			7				2	1
	8							
5			3	2	1			
8			5		4	6		
		9				3		
		5	2		9			4
			4	1	3			5
							3	
4	7				5			

Hard # 692

		2					7	
9			6	1				8
		1		7	4			
						7	3	9
1	5	6						
			9	4		1		
4				5	8			6
	6					8		

Hard # 693

		2		8				
8			2		4			7
		5			6			
		7		5		6		
5			4		1			9
		1		3		4		
			5			1		
2			1		9			5
				2		3		

Hard # 694

	6	1	4					7
	8				3	4		
4							9	
	1			2	7			
		8				7		
			3	9			1	
	5							4
		2	9				5	
6					4	8	2	

Hard # 695

	2			1			8	
			9		8	2	7	5
3					5			7
2		5				1		4
7				4				3
4	8	7	5		6			
	5				9		4	

Hard # 696

			3	2			1	
3	5							
1				5	9		3	
		7			8			
2	3						6	4
			2			7		
	7			5	4			1
							2	9
	4			8	1			

Hard # 697

				6		1		4
3			2	4		8	7	
					5			
				8		2		
8				7				9
	3		1					
			7					
	5	9		8	6			2
7		4		1				

Hard # 698

	6			2	9			
			5			6	9	
1			7					
9								3
4	2		5		8		6	9
7								8
					6			2
	8	2		7				
			9	8			4	

Hard # 699

	9			7	4			
1		8					2	9
	5		9			1		
	1			3				
		4				9		
				5			7	
		1			5		9	
7	8					4		2
			2	1			3	

Hard # 700

1				4	7		3	
8			3					
	7							2
		5		1		7		
7			5		9			4
		1		6		2		
4							5	
					6			3
	3		2	7				8

Hard # 701

		4				2		
2		8		3	1			
9	1			5				
			3	6			1	
		3				4		
	6			9	5			
				1			9	5
			7	8		1		3
		9				7		

Hard # 702

4						5		
		8		3			7	
5			7	8		2		
				2			3	1
			9		1			
	7	1			4			
		3			5	8		1
	6			8		4		
		2						9

Hard # 703

```
3 . . | . . . | . . 9
. 2 . | . . . | 7 . .
. 8 7 | 5 . . | 2 6 .
------+-------+------
. . . | 4 2 . | . . 1
. . 8 | . . . | 3 . .
4 . . | . 7 5 | . . .
------+-------+------
. 9 6 | . . 1 | 4 3 .
. . 4 | . . . | . 5 .
1 . . | . . . | . . 8
```

Hard # 704

```
. . . | . . . | 3 2 .
. 1 7 | . . . | 9 . .
. 8 . | . . . | . 4 5
------+-------+------
6 4 . | 8 9 . | . . .
. . . | . . . | . . .
. . . | . 5 2 | . 4 6
------+-------+------
7 . 2 | . . . | . 1 .
. . . | 9 . . | 8 6 .
. . 1 | 3 . . | . . .
```

Hard # 705

```
. 6 . | 1 . . | 4 . .
8 . . | . . . | 2 . .
. . 5 | 4 7 . | 9 . 8
------+-------+------
. . . | 3 6 . | . 7 .
. . . | . . . | . . .
. 3 . | . 8 4 | . . .
------+-------+------
1 . 8 | . 2 3 | 7 . .
. . 9 | . . . | . . 4
. . 3 | . . 1 | . 8 .
```

Hard # 706

```
. . 7 | 2 . . | . 1 .
9 . 2 | . . . | . . .
6 . . | 7 4 9 | . . .
------+-------+------
. . 4 | . . 1 | . 5 .
. . . | . 2 . | . . .
. 3 . | 5 . . | 9 . .
------+-------+------
. . . | 9 8 6 | . . 5
. . . | . . . | 6 . 8
. 5 . | . . 7 | 1 . .
```

Hard # 707

```
. 3 . | . . 4 | . . 7
5 . . | . 3 . | . . .
. 1 9 | . . 8 | . . .
------+-------+------
2 9 . | 6 . . | . 1 .
. . 7 | . . 8 | . . .
. 4 . | . . 7 | . 5 6
------+-------+------
. . . | 1 . . | 6 7 .
. . . | . 2 . | . . 3
4 . . | 3 . . | . 9 .
```

Hard # 708

```
1 . . | 8 . . | . 5 .
. . . | . 5 . | . . .
. 7 . | . 9 . | . 3 8
------+-------+------
. . 5 | . . 6 | 8 . .
7 8 . | . . . | . 1 3
. . 3 | 4 . . | 5 . .
------+-------+------
9 4 . | 1 . . | . 7 .
. . 7 | . . . | . . .
. 6 . | . 9 . | . . 5
```

Hard # 709

				1				8
5			8		4			
	4							9
6	3					9		7
			3		7			
8		5					4	6
2							1	
			7		8			5
4				9				

Hard # 710

			3	8	5			
3						1		
5	6	9						
			9	3		4	7	
	3						8	
	2	4	1	5				
						9	1	5
		7						3
			9	1	6			

Hard # 711

	2	4	8			3		
	5							6
1	7		4					8
				2	5			
		9				1		
			6	9				
2					4		8	1
4							9	
		8			9	7	4	

Hard # 712

				1		2		
	2	7	9					1
3						7		
	8		5			1	4	
				2				
	7	9			3		8	
		4						8
2				5		4	7	
		3		8				

Hard # 713

		8	2			5		
				5	7	9		
							4	8
	3		4	7			8	1
8	2			1	6		5	
3	8							
		5	7	4				
		2			1	7		

Hard # 714

					3	4		
6				5				9
						1	7	5
			8		7		1	2
	6						9	
8	3		9		2			
1	4	8						
3				1				6
		6	2					

Hard # 715

		5					9	
7	4		3			1		
			2	1				
			9	4	7	2		
8								1
		3	8	6	1			
				8	5			
		1			2		6	9
	8					3		

Hard # 716

	9		5					8
6	3						9	
				9		3	6	
8			6	4				
		1				8		
			7	1				5
	2	4		5				
	7						4	9
3				4			5	

Hard # 717

2	9				8	7		
		5	3	9				
7	4							5
			7					3
		4			9			
1				9				
9						8	6	
			8	6	2			
		7	4			5	9	

Hard # 718

7	6	1			2			
5				6	7			
4	2							
	9	2	3					
			2		1			
				4	3	8		
							6	3
			5	7				8
			6			4	9	5

Hard # 719

9	3			4				
		6					1	7
					2			
5		8		3			7	
			5		6			
	6			9		3		5
		1						
2	8					1		
				5			3	6

Hard # 720

		2	3		1			
		9		5		1		
			4		5			9
6		1				9		
			5		4			
	8					6		2
3		7		2				
	6		4			9		
			7		6	3		

Hard # 721

3								7
			9	7		4		
5					6	3		
	5		1					2
	1						5	
2					8		1	
		3	6					1
		4		8	7			
8								3

Hard # 722

					2		8	
	9			3	6			
			6				4	5
5	4						3	
8				1				6
	6						5	2
3	8				4			
		6	7				9	
	2		8					

Hard # 723

		4						3
9		6	1	7				8
					9			
		1		9			8	
6		5				7		9
	7			1		3		
			6					
3				8	5	9		6
5					2			

Hard # 724

		6		5	8	7		
3		8			7			
	7				6			
	9					3		1
	5						2	
6		3					4	
			5				3	
			3			2		6
		9	4	6		5		

Hard # 725

5			8			6		9
4	7	6			1			
				3				
			4			2		7
	5						6	
1		2			7			
				5				
			2			7	3	8
6		4			9			5

Hard # 726

			6				8	
				1		5	7	
7		1	9					6
		9		8				7
3								8
4			2			3		
6						9	4	5
	7	4	1					
	5			6				

Hard # 727

2			1	5	6	9		
							8	
	9	4		8				
	1							4
4			3		7			2
5							6	
				3		7	2	
	5							
		6	2	1	9			8

Hard # 728

							9	5
3			7				1	
	4			1	6			3
	2		9			3	5	
	6	7			1		8	
8			5	6			4	
	7				2			1
2	3							

Hard # 729

	1		7	9		8		
			3	8				6
8					1	2		
		9						3
	4						1	
2					6			
		4	7					8
3				1	4			
	6		2	3			7	

Hard # 730

8				4		7		
		4		5			3	
5	9	3						
					4		6	
1				9				8
	2		1					
						6	7	5
	5			3		2		
		6		2				4

Hard # 731

		4	7		3	9		
						6		
5	9			2		3		
7				9				
		6	1			7	5	
				8				2
		2		7			8	5
		7						
		5	9		1	2		

Hard # 732

				1	6	2		
2			4			5		
	9						8	
6				9	1	3		
5								7
		1	5	6				4
	8					4		
		5		4				3
		4	6	7				

Hard # 733

		6	9		7		2	
7				6				
		4			3			
		1					8	
6	9			2			3	1
	8					6		
			5			9		
				4				5
	3		1		9	4		

Hard # 734

			4			6		
		1		9	2	8		
		9	6				2	
	5				8		3	7
1	4		7				6	
	9				5	2		
		8	2	7		1		
		5			4			

Hard # 735

1					5			
9		8	1					
	5							7
	1			8		7	9	
6				4				3
	2	5		9			4	
8						3		
				4	2		5	
			8					4

Hard # 736

9		8	2	7				
								9
	3			9		6		
	8			3		2		
6		2				9		1
	4		6			5		
	2		5		6			
1								
			6	2	7		3	

Hard # 737

				5	9			
7		3		4	6			
		9		1				8
	6					4		2
9								5
3		8					9	
6				2		8		
		7	8			2		3
		5	3					

Hard # 738

			3			5		
3	4		8			2		
9		6			1			
		7						9
			3	9	7			
6						4		
			4			1		2
	3				2		9	4
	7			6				

Hard # 739

	6					8		5
						9		
			7	5	8		1	
9	7				5			
3			6		9			8
			2				9	1
	8		1	2	3			
		1						
5		7					2	

Hard # 740

6			4				8	
			5			7		4
		8	7					1
			2	8		1	9	
	6	7		1	9			
9					4	6		
4		6			2			
	5				6			2

Hard # 741

		6	4				3	
							1	
		1	2			7		
	9			8	1		5	
	3						8	
	7		3	6			9	
		9			7	5		
		3						
	1				5	9		

Hard # 742

7				6	3			2
			4	2		9		7
	5							
	8		2			4		
		9				5		
		7			9		8	
							4	
6		5		1	2			
8			6	4				5

Hard # 743

5			2			7	9	
	2							
8			9		7			
		3		4			2	
4			5		2			7
	5			8		1		
			8		1			6
							4	
	7	2			4			8

Hard # 744

1			2	7				4
				4	8	3		
		5				1		
9		4			5			
		6				4		
		9				3		8
		2				6		
	8	3	1					
4				3	7			1

Hard # 745

5			9			7	4	
		8		2			3	9
			4	7				
						6	5	
	3						1	
	4	2						
			7	8				
3	5			1		8		
	8	9			3			6

Hard # 746

8			5					
	2	1	3	8	4			
	9						3	
	8			6				
1		2				4		5
			2				9	
	3						5	
			7	4	1	9	2	
					3			6

Hard # 747

3	4	6		1				
				8				
	1	7	5	3				
2	3							
5		1				7		9
							1	6
			9	2	8	4		
			4					
			8			1	9	5

Hard # 748

				2		3		9
		7			4	1		
	1						5	6
	2		1					
6			8					2
			5			1		
1	6					3		
	5	2			6			
9		8		7				

Hard # 749

		8	4					
	1		5	7	2		9	
				9	1			
	7					3	6	5
9	4	3					2	
		9	2					
	2		1	3	7		8	
				8	7			

Hard # 750

9								
		1				4	3	
7	5		4			8	1	
				2		6		
	9		8		3		4	
		5		1				
	3	4			5		9	7
	1	7				5		
								4

Hard # 751

	3	7			2			
		4			6			
5		2						1
		5			8		9	4
			7		5			
7	9		1			5		
4						6		7
			5			2		
			8			3	1	

Hard # 752

	4	3			8	6		1
7				2				
								5
			9	3		2		8
				5				
1		8		7	4			
3								
				1				2
6		2	3			9	4	

Hard # 753

			6					8
			9	7		6	5	
							4	7
1			2	5			9	
		6				2		
	2			3	9			5
2	8							
	3	4		6	1			
7					3			

Hard # 754

9					6		3	
	3	7				4	8	
1			4					
	6		3	2				
			8		9			
			4	7			9	
				4				3
	7	5				9	2	
	8		5					6

Hard # 755

				4	1			
			5				1	7
				8		4	3	
3		6			2		8	
		1				6		
	9		1			3		2
	3	8		1				
1	5				7			
			6	2				

Hard # 756

		1		8	4	7		
	3		1			5		
	7		3			1		
2							6	
5								8
	6							5
		5		8		7		
		4		3		5		
		3	6	2		9		

Hard # 757

			2			3	9	
4			1	9	5		2	
								4
9		1				8		
			3		1			
		3				6		7
3								
	6		9	8	7			2
	8	9			3			

Hard # 758

1	4		5	6				
					7			9
		8			1			
7		4						
2	6						9	5
						1		2
			8			2		
5			7					
				1	2		7	6

Hard # 759

	2							
8		5			3	4		6
				5	4			3
	3			7		2		1
1		8		9			4	
9			5	8				
5		2	6			7		9
							1	

Hard # 760

	4		1		9	5		
8								
	2			5			4	
6				7			5	
	5		4		2		8	
	7			1				6
	6			4			2	
								4
		2	5		8		1	

Hard # 761

					1			
	7				3	2	8	
8				6		1		
					8			5
	5		8	7	6		4	
4		9						
		4		9				1
	2	3	4				9	
			6					

Hard # 762

1					6			8
	2		1	5				7
	4	9		3	8	5		
		5				9		
		8	9	1		2	4	
3				9	4		6	
9			6					4

Hard # 763

9	3	2	7					
		1			5			
								9
		9		4	1	2		
		5	9		8	6		
		4	6	2		3		
2								
			8			7		
					9	1	4	3

Hard # 764

4								
		6		5				2
	7			9	8	6		1
		4		1			6	
	8						1	
	3			2		8		
1		9	4	5			7	
3			7		1			
								4

Hard # 765

		8	9	5				
				2			9	
9				8		3	4	
7		3						
5		4				8		1
						2		3
	3	5		2				6
	9		6					
				4	7	1		

Hard # 766

			7			4		
	5		8					
	4	7		9	1			
8				2				7
1	3						2	4
7			4					6
		5	7			2	3	
				3			5	
	3			9				

Hard # 767

	1		5					9
3				7				
5		7		6		3		
		2					6	
4			8		7			2
	3					4		
		4		3		9		1
				4				3
6					2		8	

Hard # 768

		7		4				
	1						5	9
3			5	1		4		
6		1				3	8	
	8	3				5		1
		2		9	5			7
4	7					3		
				2		6		

Hard # 769

				3				1
	9							
2				7	5		9	
8			7			2		
		7	3	5	8	4		
		6			9			7
	7			4	2			3
							8	
4				1				

Hard # 770

	4				3		8	
		9						
5			7		1	9		3
						3		2
			6		7			
2		6						
3		1	9		5			4
						2		
	8		1				5	

Hard # 771

5		1			2			9
		2			6			
	7		4	3				
			3				6	
2		5				7		4
	9				1			
			9	3		8		
			7			4		
6			2			1		7

Hard # 772

			2	5		8		
5								
4			8				2	7
			4	5		8		2
1								9
3		8		9	2			
2	9				7			1
								8
	3			6	1			

Hard # 773

						8		3
								6
	2	1		5	8		9	
		4	2					8
		7		4		5		
5					6	1		
	1		9	6		7	5	
	3							
9		2						

Hard # 774

		7			2	5	6	
		9		8				
8				1			2	
					1		7	5
			4					
6	8		2					
	5			3				9
			9			8		
	9	3	5			7		

Hard # 775

	5	6	9	3				
			7				2	6
		2			1			
				3	4			
2				9				7
		1	2					
			3			9		
1	6			2				
				5	4	2	3	

Hard # 776

	8	2		4			5	
5					7			
	9			6			2	
	5	7			4			
3								9
			3			5	6	
		9			5		4	
			9					6
	1				2		9	7

Hard # 777

		4						
	2				8		9	1
			7	3	9			
5	7			8				9
			9		5			
6				7			2	8
			4	5	1			
9	8		6				4	
					7			

Hard # 778

						4	9	2
			9		3	5		
5				2		1		
4				1		8	2	
	6	8		5				4
		4		6				9
		6	2		4			
9	8	5						

Hard # 779

							5	
				2		1		3
			3	9	4		6	
5						7		4
	9	8				6	1	
1		2						9
	2		8	4	1			
9		7		6				
	5							

Hard # 780

9		5			6			
	4			2				3
8			5					
3				1	8		7	
		7				3		
	9		6	7				8
				5				4
2				3			8	
		1				6		9

Hard # 781

	7		2				8	
		3		8	4			
2				3				1
5		9	3					
	4						5	
					6	2		7
8				2				5
		4	6			7		
	5				7		2	

Hard # 782

								5
5			8	2				
1			3		6		9	
	5			9		6		2
	2						5	
9		3		7			1	
	4		2		5			6
				8	7			4
6								

Hard # 783

		2	8	7	5			
1		4		2				7
9								
4		9						
		3	9		7	8		
						9		5
								6
5				8			1	4
			2	5	1	3		

Hard # 784

			8			4		3
				6				1
		5	9					7
5				7	4			
		1				8		
			3	6				2
8					5	6		
4			7					
9		2			1			

Hard # 785

		7			2			3
	6				8			
3			6	7				
				5		2		4
		8	4		3	5		
2		4		1				
				9	7			5
		9					7	
1			5			4		

Hard # 786

			9		4			5
		2		3				
			6	1		7		9
	6		4					
3	2						1	7
				1			4	
8		5		2	9			
			1			9		
1			8		6			

Hard # 787

8			2				6	
		9	5		6			
1	7				4			
		2			8			4
7								3
5			1			7		
			6				3	9
			3		1	8		
	6				9			2

Hard # 788

	6	7	8		3			9
	5	8		4				
						7		
				6			8	
		9	4		7	2		
	8			3				
		3						
				2		5	6	
8			6		1	9	3	

Hard # 789

	8			9		1		
		7			1			
			5		4			2
						8	2	
5			4		9			3
	3	9						
9		6		7				
			6			3		
		1		2			8	

Hard # 790

							9	
8				6	1			
		2			4		5	
6			9				7	
		3	4			6	5	
	2				7			3
	6		8			2		
			5	4				7
	1							

Hard # 791

4		2			7			5
	9			2			7	
		7	3					
		6			5	2		
1								3
		9	6			8		
				4	6			
	2			3			5	
3				9			7	4

Hard # 792

3						7	1	
			1				9	4
		6			7			
					1	9	2	
6								8
	4	1	5					
			7			4		
8	7				3			
	3	2						7

Hard # 793

			3					
	6	2					9	
4	7	1						8
	3		5	7				9
6								2
5				6	8		1	
2						9	3	5
	9					6	8	
					4			

Hard # 794

								4
4			1	5		8		
	3			2			6	1
			6	1				7
	2						9	
6				4	3			
1	7			9			3	
		2		6	8			5
5								

Hard # 795

	4			8			9	
					5	2		
2			4			6		
9	5			1		4		
4								8
		3		6			7	2
		5			9			7
		9	1					
	8			3			6	

Hard # 796

	5			8			9	
2		1			6			
	8				4			6
			7	4				
6		4				5		9
			2	9				
8			4			3		
			3			2		5
	7			1			6	

Hard # 797

2				1			5	8
		1			7	4		
	9				8			6
		7			1			
				8				
			2			5		
9			5				6	
		3	1			9		
4	5			3				2

Hard # 798

	3					4		
1		8						
	6		8		2			1
	9		2		4			
		3				6		
			5		3		4	
4				9		1		8
						5		2
		7					3	

Hard # 799

```
9 1 . | 8 . . | . 3 .
. . . | . . . | 6 . .
6 5 . | . . . | 1 . 7
------+-------+------
. 2 . | . . 3 | . . .
8 . . | 7 . 5 | . . 4
. . . | 2 . . | . 6 .
------+-------+------
4 . 8 | . . . | . 2 6
. . 5 | . . . | . . .
. 3 . | . . 9 | . 4 8
```

Hard # 800

```
. . 2 | 8 . . | . . .
. 7 . | 2 4 . | . 8 .
. . 9 | . . . | . 7 .
------+-------+------
1 . . | . 6 5 | . . .
2 6 . | . . . | . 9 7
. . . | 7 2 . | . . 4
------+-------+------
. 1 . | . . . | 9 . .
. 9 . | . 3 2 | . 1 .
. . . | . . 8 | 6 . .
```

Hard # 801

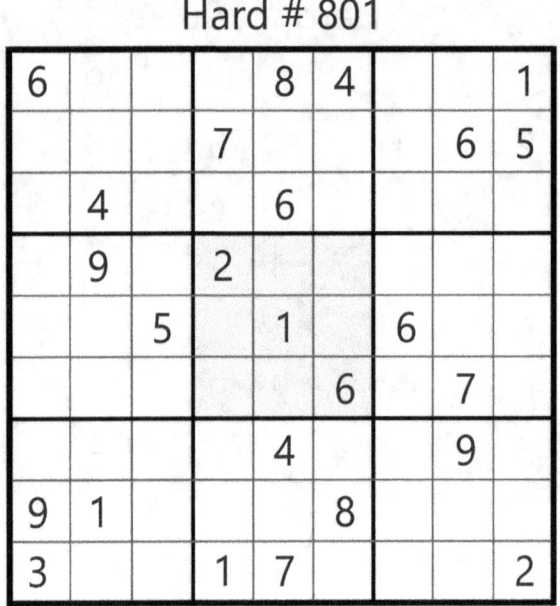

```
6 . . | . 8 4 | . . 1
. . . | 7 . . | . 6 5
. 4 . | . 6 . | . . .
------+-------+------
. 9 . | 2 . . | . . .
. . 5 | . 1 . | 6 . .
. . . | . . 6 | . 7 .
------+-------+------
. . . | . 4 . | . 9 .
9 1 . | . . 8 | . . .
3 . . | 1 7 . | . . 2
```

Hard # 802

```
2 . 9 | 4 . . | 5 . .
4 . . | . 8 . | . 9 .
. 3 . | 5 . . | 4 . .
------+-------+------
. . . | . . . | . 6 .
9 . . | 7 . 6 | . . 8
. 4 . | . . . | . . .
------+-------+------
. . 7 | . . 2 | . 5 .
. 6 . | . 7 . | . . 3
. . 4 | . . 1 | 9 . 7
```

Hard # 803

```
. . . | 8 . . | 4 . 2
. . 9 | . 4 . | 1 . 8
8 . . | . . 9 | . . .
------+-------+------
. . . | . 1 . | 7 6 .
. . . | 4 . 3 | . . .
. 9 7 | . 8 . | . . .
------+-------+------
. . . | 6 . . | . . 1
1 . . | 5 . 2 | . 9 .
9 . . | 8 . . | 7 . .
```

Hard # 804

```
. . . | . . . | . . .
. . 6 | . 3 . | . 8 2
. . 8 | 6 . 2 | . . 9
------+-------+------
. 3 . | 5 . . | . . 4
. 4 . | 8 . 3 | . 2 .
1 . . | . 9 . | . 3 .
------+-------+------
9 . . | 2 . . | 1 7 .
7 1 . | . 8 . | . 9 .
. . . | . . . | . . .
```

Hard # 805

		7	9				4	
	2				5		9	
			6					1
		1					6	
		8	5	1	6	9		
	3					4		
7					2			
	9		4				8	
	5			3		6		

Hard # 806

			3		9			8
								4
3		7					1	
			8	1	7			5
		9				1		
1		2	5	9				
	9					6		3
4								
5			4		6			

Hard # 807

							4	1
7			5					
2	6				8		9	
			4		6			2
1		9				6		3
8			9		1			
	8		1				3	9
					7			6
6	5							

Hard # 808

	8	2		6			4	
				7	3			
	1						8	
1			6		8			
	5		7			3		
			9		2			5
	6						1	
		8	4					
	4			8		9	7	

Hard # 809

5			9					
6		1		4		5		
	4		5	8			1	
9							7	
		2				4		
	6							1
	8			1	3		6	
		9		6			8	2
				8				4

Hard # 810

		2	3				7	8
	4		8					
				6	1			
6		8		9		1		
9								6
		3		8		5		2
			9	2				
			5			3		
2	5				4	8		

Hard # 811

```
. . 6 5 . 7 . . .
. . . . 9 6 7 . .
. . . 8 . . . . 3
5 . 9 . . . 2 3 .
. 3 . . . . . 4 .
. 6 4 . . . 5 . 8
2 . . . 6 . . . .
. 9 7 8 . . . . .
. . . 4 . 2 9 . .
```

Hard # 812

```
. 8 . . . . 1 . .
. . 4 . . . 2 6 .
. . . . 3 9 . 8 .
. . 7 . 5 4 . . .
4 . . 7 . 1 . . 2
. . . 8 2 . 6 . .
. 4 . . 1 6 . . .
. 7 5 . . . 9 . .
. . 2 . . . . 1 .
```

Hard # 813

```
. . 5 . . . 3 . 8
. 9 . . 7 . . . .
. 3 . . . . 5 4 .
. . . 2 6 . 4 9 .
5 . . . . . . . 2
. 4 7 . 5 1 . . .
. 7 3 . . . . 8 .
. . . . 4 . . 5 .
1 . 4 . . 6 . . .
```

Hard # 814

```
. . . 6 . 5 . . .
. 8 3 7 . . . . .
9 . . 2 3 . . 1 .
5 . 2 . . . . . 7
. 9 . . . . 8 . .
4 . . . . 6 . . 2
. 1 . . 5 3 . . 9
. . . . . 9 3 7 .
. . . 4 . 2 . . .
```

Hard # 815

```
. 1 5 . 8 3 . . 9
. . 2 . 7 3 . . .
7 . . . . . 8 . .
. . . 3 . 8 . 5 .
. . . . . . . . .
. 9 . 6 . 5 . . .
. 2 . . . . . . 4
. . 7 8 . 5 . . .
6 . . 7 5 . 1 9 .
```

Hard # 816

```
. . 9 . 2 . . 7 .
4 . . 9 . 8 . . 3
. 7 . . . . 8 . .
. 2 . . . 9 5 . .
6 . . . . . . . 8
. . 8 4 . . . 1 .
. 5 . . . . . 9 .
2 . . 7 . 5 . . 6
. 6 . . 4 . 3 . .
```

Hard # 817

6		9					1	
			5					
	7				6	2	3	
9	6			7				4
		5				9		
4				8			5	1
	5	8	7				4	
					8			
	1					3		9

Hard # 818

	5			8	9			1
							3	
4			3		1			
9			1		4	2		
		6					1	
		4	8		3			6
			4		8			9
	7							
2			6	9			8	

Hard # 819

2		1	7				3	
		6		8				
					1			
	8			5	6		9	
5			3		2			1
	3		8	1			6	
		5						
				9		4		
	7				4	5		2

Hard # 820

			7					
			2			8	3	9
4				5		7		
		1				4	9	
8								3
	5	7			6			
	4		8					2
6	9	2		3				
				6				

Hard # 821

	4	9		3				
	3	1	5	4				
	6				8			
		6			7	1		
5								3
		8	4			7		
			9				1	
			2	6	9	7		
			1		8	2		

Hard # 822

8						9		
2				8	7			
		6	9	4		8		
	3					6		7
			1		4			
1		5					4	
		9		5	3	1		
		6	1					9
		2						4

Hard # 823

	7				3			
6			2					
			9	6		5	8	
	8			7				
3			4		2			7
				1			4	
	2	4		8	6			
					5			6
		5					1	

Hard # 824

			7				1	
1			3			8	5	6
6					8	7		
	2		5					9
9			7			3		
		7	8					1
3	8	9				2		4
	4			7				

Hard # 825

1			8			4		
9				4			1	
				6		8		
				8	4			2
		2				3		
8			7	1				
		5		3				
	6			7				3
		1				5		8

Hard # 826

		9		8		2	1	
		6	3					8
1			9					
	4	7		1		3		
		3		6		5	7	
				4				6
2				7		9		
	3	4		9		1		

Hard # 827

		8			5	3		
			4					8
			3	1				2
6					4	8		3
	9						4	
1		3	8					5
9				2	3			
2					6			
		4	5			7		

Hard # 828

	7		2					
					6	2	5	
		6		7		1		
			8	3				1
5								9
6				1	4			
		3		4		6		
	5	9	3					
					2		7	

Hard # 829

		8			5			
		1	4	9				5
	7	4		3				
7					1			
	6	2				9	8	
			2					7
				2		6	3	
4				5	6	2		
			3			1		

Hard # 830

						4	8	
		1		2	6			
	2		5					7
7	6					4		
1								5
		4					6	8
4				8			1	
			4	3		2		
		5	1					

Hard # 831

	2							
8	7		2			9		
			9	6		8		2
				7			5	
4			1		2			7
	1			4				
1		8		5	4			
		7			9		1	5
							6	

Hard # 832

9			5			1		4
				1			5	
	6						7	
		6	3	7		4		8
8		7		9	6	3		
	4						6	
	9			8				
6		8			2			7

Hard # 833

	3			7			5	
		7	5					
	8	1			6			9
7						5	8	
			8		3			
	4	6						2
9			4			7	6	
						5	1	
	5			6			2	

Hard # 834

	4		5					
	6					2		
		8	9	2	7			
4			3				7	
		9	7			4	1	
	7					5		6
			2	4	3	7		
		7					3	
				9			2	

Hard # 835

	3				4	8		2
		7	3	2				
	9	5						
	5		9					
	1	3				4	5	
					7		3	
						6	8	
			3	6	2			
8		2	7				1	

Hard # 836

			2			5		
	6	2					1	9
5			6		9			
			5			7		1
				7				
8		9			6			
			7		3			6
3	1					8	4	
		7			4			

Hard # 837

	3					9	1	
9		6		8				5
		5	1			2		
			8	6				
5								3
				7	4			
		8			2	3		
2				9		6		4
6	9						5	

Hard # 838

	5	6				9		
		1	4					6
					7	5		4
	3		7					5
			3		1			
6				5			2	
9		3	5					
1				6		8		
		2				7	3	

Hard # 839

5		8			3			
			6			3	1	
		9						4
			4			3		
	2	3				7	6	
	7		1					
7						1		
	5	4		2				
			6			8		3

Hard # 840

			2					5
		4	5	6		9		8
				8			1	2
		8		1				
			4		6			
			5			3		
8	6			7				
5		9		4	2	6		
7				3				

Hard # 841

```
8 . . | . 9 5 | . . .
. . . | . . 3 | 5 4 .
. . . | 6 . . | 2 8 .
------+-------+------
3 . . | . . . | 6 . .
. 9 4 | . . 6 | 1 . .
. 7 . | . . . | . . 3
------+-------+------
. 9 6 | . 2 . | . . .
. 5 1 | 6 . . | . . .
. . . | 1 5 . | . . 2
```

Hard # 842

```
. . 4 | . . . | . . 1
2 . 5 | . . 9 | . 6 .
. 3 . | . 8 . | . . .
------+-------+------
. . 6 | 7 . . | . 2 .
. 9 1 | . . . | 7 3 .
. 2 . | . . 5 | 1 . .
------+-------+------
. . . | 6 . . | 8 . .
. 6 . | 8 . . | . 9 7
5 . . | . . . | 6 . .
```

Hard # 843

```
. . . | . 6 . | 9 . .
. 2 . | . 3 1 | . . .
1 . . | . . . | . . 4
------+-------+------
. 3 . | . . 5 | 8 . .
. 1 . | . . . | . 2 .
. . 6 | 4 . . | . 3 .
------+-------+------
3 . . | . . . | . . 9
. . . | 5 9 . | . 4 .
. 8 . | 7 . . | . . .
```

Hard # 844

```
. . . | . 8 . | . . 5
. . . | . 4 . | 3 2 .
5 . . | . . 3 | . . .
------+-------+------
6 9 . | . . 7 | . . 2
. . 3 | . . 5 | . . .
1 . . | 9 . . | . 4 3
------+-------+------
. . . | 5 . . | . . 9
. 6 1 | . 2 . | . . .
8 . . | 1 . . | . . .
```

Hard # 845

```
. 2 . | . . . | 1 . 8
1 . . | 4 . . | . 7 .
. . 7 | 2 . . | . 9 .
------+-------+------
. . . | . 1 . | 3 . 7
. . . | 3 . 8 | . . .
9 . 8 | . 5 . | . . .
------+-------+------
. 6 . | . . 9 | 7 . .
. 1 . | . . 2 | . . 6
5 . 9 | . . . | . 3 .
```

Hard # 846

```
. . 4 | . 6 . | . 5 .
. . . | . 1 . | . 7 .
7 . . | 3 . . | 4 8 .
------+-------+------
. 5 6 | . . . | . . 9
. . . | . 2 . | . . .
9 . . | . . . | 1 4 .
------+-------+------
. 1 7 | . . 9 | . . 6
. 8 . | 6 . . | . . .
. 3 . | . 5 . | 8 . .
```

Hard # 847

			4	9		2		6
					8			
		5					4	7
6	4				7			
		7	2		9	4		
			8				7	2
8	5					7		
			3					
3		1		2	5			

Hard # 848

1						8		
	9	4		2		5	3	
		3	5					
			6	1	2			
2								4
		1	3	9				
						7	9	
	7	6		3		4	5	
		8						3

Hard # 849

						4		7
		5		4	6			
		1				9		
3	2		7			6	8	
8	7				6		2	5
	8				2			
		9	3		8			
6		1						

Hard # 850

			5					8
9		2			4			
	4				2	7		
		5	1			9		3
7								6
1		6			3	4		
		8	6				9	
		8				3		7
2					5			

Hard # 851

			2					
							3	1
		7			4		6	8
			5	8		3		
	1	4		3		6	5	
		3		4	7			
1	2		4			9		
6	3							
					1			

Hard # 852

		1				6		
3			9		7			
	8				3			
	6			7		8	5	
8			6		5			2
	9	3		1		6		
			5				9	
			8		6			4
		4				5		

Hard # 853

```
. . 2 | 7 5 . | . . .
. 7 . | . . . | . 6 1
. 4 . | 6 . . | 8 . .
. 6 8 | 2 . . | 7 . .
. . . | . . . | . . .
. . 4 | . . 3 | 9 8 .
. . 5 | . . 2 | . 3 .
3 2 . | . . . | . 9 .
. . . | . 3 9 | 2 . .
```

Hard # 854

```
9 . . | 2 . 7 | . . .
. 6 . | 5 1 . | . . .
. . 8 | . . . | 6 . 1
. . 4 | . . 5 | 2 . 3
. . . | . . . | . . .
2 . 3 | 6 . . | 8 . .
1 . 6 | . . . | 7 . .
. . . | 4 3 . | 9 . .
. . 1 | . 6 . | . . 4
```

Hard # 855

```
. . . | 4 . . | 9 . 1
. . . | 7 . . | . . 8
8 1 . | . . . | 5 . .
. . . | 5 4 8 | . . 9
. 8 . | . . 5 | . . .
5 . 1 | 3 8 . | . . .
. 6 . | . . . | . 4 7
7 . . | . 2 . | . . .
2 . 9 | . . 3 | . . .
```

Hard # 856

```
. 1 . | . . . | . 5 .
7 9 3 | . 6 . | . . .
8 . . | 2 . . | . . .
. . 4 | 9 . 5 | . . .
3 . . | . 2 . | . . 1
. . . | 6 . 3 | 7 . .
. . . | . 8 . | . . 2
. . . | . 3 . | 6 8 4
. 2 . | . . . | 3 . .
```

Hard # 857

```
2 . 1 | . 5 . | 4 . .
. . 4 | . . . | . . 7
9 . 5 | . . 2 | . . .
. . . | . . 9 | . . 2
. 5 . | 8 . 1 | . 7 .
7 . . | 2 . . | . . .
. . . | 7 . . | 1 . 9
4 . . | . . . | 3 . .
. . 6 | . 3 . | 7 . 5
```

Hard # 858

```
9 2 . | . . . | . . 3
6 7 . | . 1 . | . . .
. 1 . | . 3 . | . . .
2 . . | 3 . 4 | . . .
. . 1 | 5 . 7 | 3 . .
. . . | 9 . 1 | . . 8
. . . | 4 . . | . 3 .
. . . | 9 . . | . 7 2
8 . . | . . . | . 9 4
```

Hard # 859

	5	9	2					6
3				9			5	
					5	2		
2							8	
	6	8				9	2	
	1							5
		3	8					
	7			4				1
4					9	7	3	

Hard # 860

7					5	1		4
				1	3		6	
		5	6					
		2	1					
1				9				8
					8	9		
					9	8		
	6		4	7				
9		4	5					2

Hard # 861

					4		3	
	3	1				2		
				2		1	6	9
	9		8		1			
	8						7	
			6		3		1	
5	2	3		1				
		8				6	5	
	4		9					

Hard # 862

2	7			8				
			2				5	7
3		5			9			
			5		3		2	
				9				
	9		8		1			
			1			3		5
7	4				6			
				7			8	4

Hard # 863

	4	8		3				
1	3		9		8			
						7		
	1		7	4	6			
4								5
			5	1	9		6	
	2							
			1		4		9	3
				9		6	5	

Hard # 864

					7			9
					2			1
9	7					5	3	2
2		4		7				
			3		5			
			1		2			7
3	1	2					9	6
6			2					
4			8					

Hard # 865

			4					7
				3	7	8	6	
	9				1			
	2					3		
		1	6	8	4	2		
		6					7	
			1				4	
	4	2	3	6				
9					2			

Hard # 866

								5
5			6			4	2	
8	3	4	7					
			8				9	
9		6				8		3
	1				9			
					4	6	3	7
	5	3			8			2
4								

Hard # 867

					7		6	
	8	9		3	1			
1				5		3		
4								2
	1	3				5	9	
6								8
	2		5					3
			3	7		4	2	
		6	9					

Hard # 868

6			1			4		3
			2				7	
7	8			9		1		
				6			5	
3								7
	7			2				
		4		3			9	6
	6				9			
9		5			4			8

Hard # 869

8	3		9		4			
5			1		6			2
6								
	2							1
	5	7				2	4	
3						5		
								6
7			6		5			4
			4		2		1	5

Hard # 870

3				6				
4	2		8			1		
1			4		9			
7	5							
	8	9				4	1	
							2	5
			3		5			2
		7			6		5	8
			9					1

Hard # 871

	7			6		1		5
			5	8		4		
					4			2
5	4				6			
		1				2		
			9				4	6
8			2					
		9		4	8			
4		7		1			3	

Hard # 872

					9			
5				1		4		
		6			4	3	8	9
	4					7	3	
7								1
	1	9				2		
3	2	1	9			6		
		7		5				2
			6					

Hard # 873

6	9			2				
	5	4	3					6
		8						
8				6		9	7	
			4		1			
	1	5		9				2
						2		
5					9	7	1	
				4			5	3

Hard # 874

		9		5	8	4		
6	1		3					
						1		
			4	3			2	
	7						5	
	9			8	6			
		5						
				7			6	2
		2	8	9		5		

Hard # 875

		1		7				
		2	4				3	
		3			1		6	9
	2		7	3			5	
	7			1	9		4	
1	8		3			5		
	9				8	2		
				4		6		

Hard # 876

	8			6		9		
		6						8
				2	1			
8	9							6
	7		5		6		8	
4							9	5
		1	7					
7						6		
		8		3			5	

Hard # 877

```
. . . | . . . | 5 . 2
. 1 . | . . 2 | . . .
. 2 . | 4 6 . | 8 . .
------+-------+------
. . . | 9 8 . | . 1 .
8 . . | . . . | . . 7
. 9 . | . 1 3 | . . .
------+-------+------
. . 5 | . 7 4 | . 9 .
. . . | 8 . . | . 6 .
7 . 3 | . . . | . . .
```

Hard # 878

```
. 5 . | . . 7 | . . .
. 8 5 | . . . | . . .
. . 1 | 6 4 8 | . . 5
------+-------+------
. . 7 | . 8 . | . . .
. 2 8 | . . . | 9 4 .
. . . | . 3 . | 1 . .
------+-------+------
3 . . | 8 7 9 | 2 . .
. . . | . 5 . | . 6 .
. . . | 4 . . | . 1 .
```

Hard # 879

```
4 . 7 | . . . | 6 . .
. 9 . | . 5 . | . . .
. 8 2 | 9 . . | 7 . .
------+-------+------
5 2 . | 3 . . | 4 . .
. . . | . . . | . . .
. . 8 | . . 2 | . 1 7
------+-------+------
. . 4 | . . 8 | 3 7 .
. . . | 4 . . | . 8 .
. . 1 | . . . | 2 . 6
```

Hard # 880

```
. . 3 | 6 2 . | . . .
. . . | 4 . . | . 3 5
. . . | 7 . . | 1 . .
------+-------+------
. . . | 2 . . | 8 5 .
2 . 5 | . . . | 6 . 9
. 4 8 | . . 7 | . . .
------+-------+------
. . 4 | . 6 . | . . .
5 2 . | . . 4 | . . .
. . . | 1 5 3 | . . .
```

Hard # 881

```
. 4 . | . 6 2 | . . 1
. 6 . | 4 . . | 7 . .
. . . | . . . | . . 8
------+-------+------
4 . . | . . 6 | . 1 .
. 1 . | 7 . 5 | . 3 .
. 8 . | 3 . . | . . 6
------+-------+------
1 . . | . . . | . . .
. . 7 | . . 4 | . 5 .
9 . . | 6 7 . | . . 8
```

Hard # 882

```
. 4 . | . . . | . 3 2
. . . | . 4 . | . . 8
1 . 7 | . 8 . | . . .
------+-------+------
5 . 2 | 7 . . | . . .
6 . 9 | . . . | 8 . 3
. . . | 9 . . | 2 . 6
------+-------+------
. . . | 3 . . | 6 . 4
7 . . | 6 . . | . . .
3 8 . | . . . | 2 . .
```

Hard # 883

7	9				3	1		
							9	
	4		5			3		7
	8	9			6	2		
		2	4			9	1	
8		5		7			3	
	6							
		3	6				7	2

Hard # 884

	4	9					3	
7				1				
	1					4		6
9	2		6					7
			7		8			
8				9		6	4	
2		3				4		
			4					8
	6					5	7	

Hard # 885

		8				1		
	5		6	4	8			
	3			7		2		
8			9		6			
			7					
		1	8					4
	8		9			6		
	5	2	1			3		
	9				2			

Hard # 886

8		6						
7	2		4					
	9	3		1	7			
5					9	1		
			7		8			
	8		1					2
			5	3		8	4	
				7			6	9
						2		5

Hard # 887

		2		5	4			
			2	4	6			1
9	2				1	5		4
8								3
1		4	8				2	6
6			3	1	9			
		8	5			9		

Hard # 888

	6					5		
		1		8				2
5		8	4					
	8				3	5		9
			6		4			
9		7	1			2		
					2	1		7
6			3			8		
	5						3	

Hard # 889

```
1 . 5 | 3 . . | 6 . .
4 . . | . 9 8 | . . .
. . . | . . . | 2 . 8
------+-------+------
3 . . | 5 . . | 9 . .
. 6 . | . . . | . 3 .
. . 2 | . . 7 | . . 4
------+-------+------
8 . 1 | . . . | . . .
. . . | 2 1 . | . . 7
. . 3 | . . 4 | 5 . 6
```

Hard # 890

```
. 7 . | . . . | . . 1
6 . . | . . 3 | . . .
. . . | 8 . . | 9 5 .
------+-------+------
1 . . | . 6 . | 5 3 .
. . 3 | . 7 . | . . .
. 9 6 | . 1 . | . . 8
------+-------+------
. 4 8 | . 5 . | . . .
. . 6 | . . . | . . 7
2 . . | . . . | . 4 .
```

Hard # 891

```
. . . | . 7 5 | . . 9
. . 1 | . 2 . | . . .
. 3 . | 9 . 8 | 7 . .
------+-------+------
. . . | . . 2 | 1 . 6
1 . . | . . . | . . 5
2 . 8 | 4 . . | . . .
------+-------+------
. . 4 | 1 . 3 | . 6 .
. . . | . 9 . | 3 . .
7 . . | 3 2 . | . . .
```

Hard # 892

```
7 9 . | 8 . . | . 5 .
. 1 8 | . . 4 | 6 . .
6 . 5 | . . 9 | . . 3
------+-------+------
. 7 . | . . . | . 9 .
4 . . | 7 . . | 1 . 8
. . 1 | 5 . . | 9 3 .
------+-------+------
. 4 . | . . 1 | . 2 5
. . . | . . . | . . .
. . . | . . . | . . .
```

Hard # 893

```
. . . | . 6 . | . . .
. 4 9 | 3 . 1 | . . .
1 2 3 | 8 . . | . . .
------+-------+------
. . . | 1 . . | . 7 .
9 . 6 | . . . | 2 . 1
. 7 . | . . 9 | . . .
------+-------+------
. . . | . . 3 | 1 2 5
. . . | 7 . 2 | 4 8 .
. . . | 4 . . | . . .
```

Hard # 894

```
. 9 . | . . . | . . .
. . 7 | . 5 8 | 4 . .
8 . . | 4 . 1 | . 5 .
------+-------+------
. 6 9 | . . . | . 8 3
. . . | . . . | . . .
3 7 . | . . . | 6 1 .
------+-------+------
. 1 . | 7 . 6 | . . 4
. . 3 | 1 8 . | 2 . .
. . . | . . . | . 9 .
```

Hard # 895

9				6			5	
			9			3		2
			2			7		9
		8	5				2	
3								8
	4				1	6		
7		5		8				
4		2		5				
	9			3				1

Hard # 896

6	3		9				1	
8								6
		7		4	6			
			4				2	
	9			5			8	
	2				3			
			7	8		6		
9								1
	8				2		4	5

Hard # 897

							1	
	5		6				8	3
2		3						
9	2			1			7	
			7	4	8			
	7			2			6	1
					6			5
8	1				2		9	
	4							

Hard # 898

			7					5
9			5				7	1
		5				8	9	
4	3				6			
1								6
			8				1	3
	9	6				2		
5	2				4			9
3				7				

Hard # 899

		2				8	3	
		6	7	8		4	9	
			6			2		
4			5	1				
				9	6			8
		9			7			
	5	4		2	8	9		
	8	3				1		

Hard # 900

		6			3	5	7	
		9		5	1			
3		2						
	9				7			
5		3				4		7
			8				9	
						2		3
			9	4		7		
	8	7	3			1		

Hard # 901

		3			1			
8	1			9	7			
			1	2			6	4
	9							8
	6					4		
3						7		
9	5			1	4			
			8	7			9	1
		2			3			

Hard # 902

							4	
2		5	4		1	3		
4				3				9
		3		1			8	
			8		2			
	5			7		4		
6					8			3
		1	6		5	8		4
	7							

Hard # 903

					6	7		
		9	8	7			5	1
	7			9				6
4		3						
			7		3			
						2		5
1				6			2	
7	3			4	8	5		
		5	3					

Hard # 904

		3	2	5				6
	7							
2	5							1
8			1				9	
	2		6		3			
	6				2			7
5							1	9
						4		
3				8	4	7		

Hard # 905

			4	9				
8	7				5	4		
9	5							
	9		2		6	5	4	
	6	5	7		4		1	
							5	1
		7	1				3	8
				6	7			

Hard # 906

				7	9			8
		9	3		5		4	
	1		2					
	2			6		3		
6								9
		7		3			1	
					8		3	
	5		1		3	7		
8			7	4				

Hard # 907

			1					
6				7	9		4	
					6		8	7
	5			6	8		3	
2								8
	1		3	4			6	
7	9		5					
	4		7	9				6
					1			

Hard # 908

	7		6	2			9	5
							8	
			7		1			6
		1				9		
	4	3				2	7	
		9				6		
4			2		3			
	5							
1	3			6	8		4	

Hard # 909

		6		5			9	
1			8					
				9	6			
		8		7			6	1
2		4				5		7
7	5			3		9		
		7	2					
					8			4
	2			6		3		

Hard # 910

		4						
	1		9	7			8	
6			5		4			7
			6				7	
	9			2			5	
	2				3			
8			7		9			5
	6			1	2		3	
						7		

Hard # 911

2		1						
				7	5			
4			2			7		9
1		4	3					
		8	7		9	4		
					1	6		2
5		9			6			3
			9	8				
						1		6

Hard # 912

3		6			5	8		
	2			3				
	9			7		5		4
							4	6
		8				7		
2	3							
7		4		2			1	
				1			6	
		2	4			9		8

Hard # 913

							6	
		3	4		1	2		5
5		1	7					
	4	6		5	7			
			1	9		4	3	
					5	6		3
2		9	8		6	1		
	7							

Hard # 914

			9	7				
	9					6		2
		1	6	2				
	6					5	2	9
9								3
2	1	8					6	
			1	4	7			
5		7					8	
			9	7				

Hard # 915

3			1					
		4			5			7
		9		8	6			
		1			7		3	5
9								1
6	2		5			8		
			4	7		2		
7			6			1		
					2			4

Hard # 916

	9		8					
7	5				2			
			4	7		1	6	
4			1				9	
		7				2		
	2				7			6
	1	5		3	4			
			7				3	9
				8			1	

Hard # 917

			8					
7				3			4	9
	9	6	7			5		
6			1			2		
		1				6		
		9			2			4
		2			1	3	6	
1	5		3					8
				9				

Hard # 918

		2		6			8	
						7	6	4
		8	4			5		
3			8					
9				2				3
					6			5
	6				9	2		
8	7	9						
	2			7		9		

Hard # 919

4					6	5		
			4	9			2	
2					8			
		6		4			9	
8			5		1			6
	1			8		4		
			8					3
	3			7	4			
		4	1					8

Hard # 920

	3						9	
	5		9			7		4
			1			2		
			2				8	
8		2	4		7	5		6
	6		3					
		9	4					
3		1		9			2	
	4						7	

Hard # 921

			6					9
	3					7		1
			1	2		5		
	4				6	2		
		8	4		7	3		
		6	8				9	
		9		8	4			
4		1					8	
7				5				

Hard # 922

	2				5	8		
							6	
7			4		8		5	
	4			2	6		9	
	1						4	
	3		5	7			2	
	7		6		1			4
	8							
		6	9				7	

Hard # 923

	1		5				2	
		8						9
			7			4	6	5
1				2			5	
			4		9			
	8			5				4
5	9	4				3		
8						1		
	6				4		9	

Hard # 924

	3				4	2		
			1	9				3
5						4		
		6		5			8	9
	7						2	
9	1			6		5		
		4						8
6				2	3			
		7	8				5	

Hard # 925

```
. . 4 | 1 . 3 | . . .
. . . | . 2 . | 9 . .
. 6 . | 8 . . | . . 7
------+-------+------
. . . | 9 . 3 | . . 2
. . 6 | . . . | 5 . .
8 . . | 4 . 7 | . . .
------+-------+------
3 . . | . . 8 | . 5 .
. 5 . | 7 . . | . . .
. . 7 | . 3 . | 2 . .
```

Hard # 926

```
3 . . | 7 . . | 5 . .
9 . . | . . 1 | 6 . .
5 . . | . 4 2 | . . .
------+-------+------
7 . . | . . . | 3 8 .
. . 6 | . . . | 9 . .
. 2 9 | . . . | . . 7
------+-------+------
. . . | 2 9 . | . . 4
. . . | 5 1 . | . . 9
. . . | 8 . . | 4 . 6
```

Hard # 927

```
1 . 9 | 7 . . | . . .
. . . | . 9 . | 5 4 .
. 2 . | 1 . . | . . .
------+-------+------
8 . . | 6 . . | . . .
7 4 . | . . . | 8 9 .
. . . | . 4 . | . . 3
------+-------+------
. . . | . 3 . | 4 . .
5 9 . | 8 . . | . . .
. . . | 4 . . | 6 . 1
```

Hard # 928

```
1 . . | 2 . . | 7 . .
4 . . | . . . | 8 3 .
. . . | 7 9 . | . 5 .
------+-------+------
9 . . | . . . | 3 . .
. . . | 1 . 8 | . . .
. . 2 | . . . | . . 7
------+-------+------
. 1 . | 9 5 . | . . .
. 8 3 | . . . | . . 6
. . 9 | . . 7 | . . 5
```

Hard # 929

```
. 2 7 | . . . | . . .
8 5 . | . . 7 | . 9 .
. . . | . 5 . | . . 2
------+-------+------
6 . 4 | . . . | 9 . .
. . . | 4 8 9 | . . .
. . 1 | . . . | 4 . 3
------+-------+------
7 . . | 9 . . | . . .
. 9 . | 5 . . | . 7 8
. . . | . . . | 6 4 .
```

Hard # 930

```
. 4 9 | . 7 . | 6 . .
. 2 . | 6 1 . | . 9 .
. . . | . 2 . | . . .
------+-------+------
2 3 . | 5 . . | . . .
9 . . | . . . | . . 5
. . . | . 8 . | . 4 6
------+-------+------
. . . | 9 . . | . . .
. 5 . | . 3 7 | . 2 .
. . 7 | . 8 . | . 1 5
```

Hard # 931

	9	6		2				
							3	
5	3				9	7		
1	4				3			
		9	6		8	2		
			7				4	8
		5	1				2	6
	1							
				7		8	5	

Hard # 932

	3		5	1		2		
					6	3		8
			2				6	
3				6				
4	8						9	7
				9				1
	5			8				
8		2	6					
		4		3	9		2	

Hard # 933

								3
	2			9	3			
		7	8	6			2	
1	6	5				9	3	
	9	8				7	6	1
	3			2	7	5		
			5	4			1	
4								

Hard # 934

				9			5	3
		1	5			9		
	2							8
2			8					5
6		4				1		2
1				3				6
7						4		
		5			2	8		
8	1		6					

Hard # 935

		7				8		
		5						3
	2			9		7		1
7		4			8			
			3		1			
			6			2		9
4		9		5			3	
6					2			
	5					6		

Hard # 936

		1		2	5			
5				3		6		1
							4	
	4		3			7		
	8	5				1	9	
		6			8		5	
	5							
2		3		1				6
			6	9		2		

Hard # 937

				8	4			6
		4	2				7	
				1			3	8
	5	6				1		2
3		1				7	4	
9	4			3				
	7				6	5		
1			5	7				

Hard # 938

4				9			1	
5		2						
			1		6			
	8	6			1	7		
1				4				3
		5	3			9	6	
	7		5					
				6				2
	5			3				4

Hard # 939

				1		8		
	5	6	3					7
2	4			7				
	2	5						
	3		4		5		6	
						4	2	
			8				7	6
5					2	9	1	
	6		1					

Hard # 940

						6	4	
	9			1				8
		1					2	7
9					4		1	
1			8		7			5
	2		1					3
2	6					1		
5				4			7	
	3	9						

Hard # 941

				1				
4							1	
3		1	2					5
		2	3	5				7
	1		7		2		9	
7				4	8	2		
5					4	9		3
	7							8
				3				

Hard # 942

		7	4	5		6		3
							7	
			1		3	2		
5		1				7		
			3		9			
		3				1		4
		5	6		2			
	6							
1		8			4	7	3	

Hard # 943

		2	8	5			9	
		3						
		7	3				4	1
				1		8		5
2								3
6		8		2				
4	9				6	7		
						9		
	2			8	9	1		

Hard # 944

3	2							
			6	3			7	1
	7					5		
	9		5					7
		1	9			4	6	
7				6			4	
		3					1	
2	6			4	9			
							6	8

Hard # 945

			9			2		
	1				3			5
8	2				1			6
		8						
3	6	4				9	5	2
					3			
5			8				2	9
6			2				7	
		7			6			

Hard # 946

		5	9				8	4
9	8							
4	1		8	6				
	9					1		
5								2
		1					6	
				2	5		4	3
							7	1
7	5					3	2	

Hard # 947

3					2			
			2	6				3
			3	1		5		
1	3							5
2		5				7		6
7						2		1
	4			7	5			
5			4	6				
		3						9

Hard # 948

			1	5				3
	9			8		4		7
4								1
	7				1			
8			4		7			6
			8			4		
7								2
3		5		9			1	
6				1	2			

Hard # 949

	3							
1			5					7
		6	8	7	3			
			5			2	4	
		7	1		4	6		
	6	4		2				
		9	4	7		8		
3					2			6
							5	

Hard # 950

							8	5
4				1	5			
		6	2					
				7		6	1	
7		3				9		8
	8	2	9					
				4	2			
			7	5				3
3	1							

Hard # 951

	1	3						5
9	2		4					3
	6		9					
				5				1
	3	5				7	6	
1				4				
				9		1		
2				7			4	8
6						3	5	

Hard # 952

	4	8		5			3	
			4					7
9	5		8					
			3			2		9
	6						1	
2		3		8				
					7		4	5
6				5				
	7			9		8	2	

Hard # 953

						6	3	
	7			6				8
			3		4	1		
			1			9	2	5
1								7
6	2	7		5				
		9	7		8			
7				3			4	
	8	2						

Hard # 954

4			8	5	6		1	
				2	6	9		
8	5					4		
	1	7				3	6	
		6					2	1
	7	1	3					
	3		4	1	7			2

Hard # 955

6			4	1				
5				2				
		8	9				1	
		5			7			
	1	4		5		8	7	
			2			6		
	3				8	2		
				9				5
				3	4			1

Hard # 956

	1		2		5			8
5	3	4			9		7	
							9	
2				4				
		7				8		
				2				5
	4							
	2		3			7	8	9
7			5		8		6	

Hard # 957

								2
		5						7
	7			1	8		4	
2			4			3		
		4				2		
		1			7			9
	9		3	5			6	
6						9		
8								

Hard # 958

6			4					
				7			3	4
		5					1	
			8		1		2	
4		8		2		7		1
	9		7		5			
	7					3		
8	5			6				
					4			6

Hard # 959

5		9						7
4								5
	2	8	4				3	
		5		2				
	9		5		1		8	
			8		6			
	1				3	7	9	
9								8
2						1		3

Hard # 960

8							3	
	4	6		8				
	5			1				6
					2	7		9
4			6		8			5
7		2	4					
9				4			7	
			2			5	6	
	7							8

Hard # 961

8			1			6		
5								8
	3	9					4	
7			8			3		
6			2		5			4
		4			7			1
	6					2	1	
9								5
		8			6			3

Hard # 962

			1			5		
7			9	6			1	
	2	5	8					
	8		9					
	2					7		
			4			6		
			7			8	9	
	3		6	1				4
		6		9				

Hard # 963

		9	6			5		
	6			3	4		8	
	2							
		7						6
4			8		1			7
5					2			
							2	
	8		4	6			3	
		1			2	4		

Hard # 964

	6		3					5
			4		6	2	3	
	4			2			6	
		5	8					
7								9
					1	5		
	3			1			8	
	8	6	7		3			
9					8		7	

Hard # 965

	7		5		6		8	
	9	8						
		1	3	4				
9	4			3		2		
		2		8			5	9
			2	7	3			
						8	4	
	5		8		3		1	

Hard # 966

	3	7		2		8		
		5		6		2		
6								7
7			6		1		3	
	8		4		3			1
9								5
	3		5			6		
	6		3			9	7	

Hard # 967

					6		3	1
	9		5			8		
		8	2					
	4		6			3	8	
1								7
	3	7			8		5	
					9	7		
		5			3		2	
4	6		8					

Hard # 968

3		9						8
		7	2	1				
4						6		
2	1	5			3			
		5				7	9	2
	4							5
			4	7	9			
9						2		3

Hard # 969

	7		4	6				9
5				8				
	4	3					6	
						3	8	
			7	1	3			
	6	2						
	2					8	7	
				9				2
6				3	4		9	

Hard # 970

	5		3	9				6
8	1				6	4		3
				7				
	3	8						5
7						9	2	
				3				
3		4	6				9	8
9				8	2		3	

Hard # 971

		8	5	6				2
	5		9			7	8	
								6
	4	6						9
			3		8			
2						5	1	
7								
	1	4			7		9	
9				1	4	2		

Hard # 972

	1		6		5		3	
4			2					
			7	1				
		7		3		8		
	8		7		6		1	
		5		8		6		
			8	6				
				2				3
	5		9		3		2	

Hard # 973

		1			2			
	7		9			6		
				4		3		9
	9		7					
	4			2			5	
					3		9	
8		2		1				
		6			8		4	
			6			2		

Hard # 974

4		8	9				1	
					8			
6	3		5					
	6	3		9				
1		7				4		8
		5				6	3	
				7			4	5
	6							
	8				5	2		9

Hard # 975

	2	4					8	
		1			8			
			7		9	2		
				5		7		9
	6						3	
7		3		1				
		5	6		2			
		5				3		
	1					8	6	

Hard # 976

	9	3	4	1				
4			9		7			8
	5						9	2
	4	8				3	5	
3	2						6	
7			2		9			6
				5	8	2	3	

Hard # 977

			5		9	3		
5				7	6		8	
2			3					
	9					4		7
				8				
6		5				9		
				8				6
	8		5	2				3
	2	7		3				

Hard # 978

				2		6	8	
						2	5	
	5			4				
3		7		2				
			1		9			
				3		6		4
				7			1	
	6	8						
2	9		8					

Hard # 979

	6	9				7	3	
	4		6	1		9		
2								
9		8		5				
			2		8			
			6		2			9
								4
		6		7	9		2	
		7	8			5	6	

Hard # 980

4		2	1	9				
3	9		5					
						3	9	
	8		9					
9		3				5		1
				7		6		
		1	3					
				6			1	2
			2	1	6			7

Hard # 981

	6			8				1
					2		9	8
1	5		3					
		1		9				2
			1		6			
4				5		1		
					8		3	5
8	9		7					
5				2			7	

Hard # 982

8		9				1	5	
7				6				
							9	8
			7	2		4		1
6		4		9	1			
9	7							
				7				5
	3	1				8		2

Hard # 983

		8	5					
				6	7	2	1	
	2						5	
		5		7			6	2
		3				7		
8	1			2		4		
	4						9	
	8	9	4	3				
					5	6		

Hard # 984

		3	4				9	
		7						4
1		9		7		2		
				9				1
8			3		6			5
7				2				
		1		3		6		8
3						5		
	7				1	3		

Hard # 985

```
. . 4 | . . . | . . .
5 . 9 | . 2 7 | . . 1
. . . | . . 9 | . . 4
------+-------+------
8 5 . | . 1 . | . . .
. 3 . | 6 . 5 | . 8 .
. . . | . 7 . | . 5 9
------+-------+------
6 . . | 2 . . | . . .
4 . . | 1 8 . | 6 . 5
. . . | . . . | 2 . .
```

Hard # 986

```
. . 6 | 2 . . | . . 7
1 . . | . . . | 9 . 4
5 . . | . 4 . | . . 1
------+-------+------
. . . | 1 7 3 | . . .
. . . | 4 . 9 | . . .
. . 8 | 6 5 . | . . .
------+-------+------
3 . . | . 9 . | . . 8
8 . 4 | . . . | . . 3
9 . . | . 5 1 | . . .
```

Hard # 987

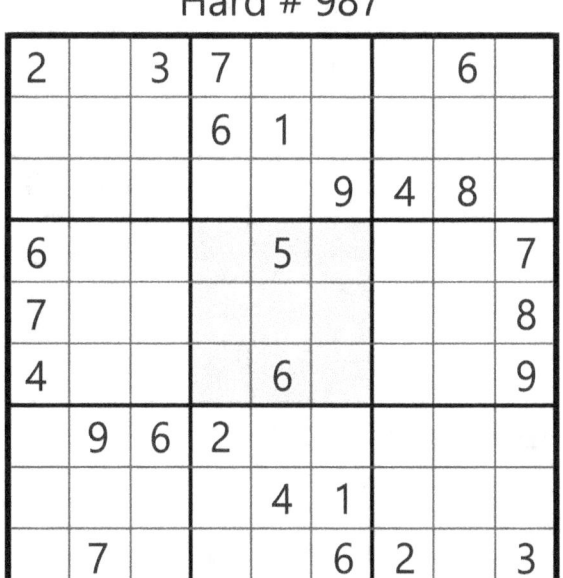

```
2 . 3 | 7 . . | . 6 .
. . . | 6 1 . | . . .
. . . | . 9 4 | 8 . .
------+-------+------
6 . . | . 5 . | . . 7
7 . . | . . . | . . 8
4 . . | . 6 . | . . 9
------+-------+------
. 9 6 | 2 . . | . . .
. . . | . 4 1 | . . .
. 7 . | . . 6 | 2 . 3
```

Hard # 988

```
. . . | . . . | 8 . 1
. . . | 2 . 3 | 6 . .
. 9 . | . . 1 | . . 3
------+-------+------
. 7 . | . 2 4 | 1 . .
. 4 . | . . . | 7 . .
. . 5 | 7 8 . | . 6 .
------+-------+------
7 . . | 4 . . | . 5 .
. . 9 | 6 . 5 | . . .
2 . 3 | . . . | . . .
```

Hard # 989

```
2 . . | . . 6 | . . .
. 7 . | 9 4 . | 5 . .
3 . . | 1 . . | . . .
------+-------+------
5 3 . | 8 . . | . . .
4 . . | . . . | . . 3
. . . | . . 7 | . 1 2
------+-------+------
. . . | . . 8 | . . 7
. 1 . | 3 6 . | . 2 .
. . 2 | . . . | . . 5
```

Hard # 990

```
. 5 . | 9 7 . | . . .
7 . . | . 6 . | . 9 8
. 1 6 | . . . | . . .
------+-------+------
. 6 7 | 8 . . | . . .
. 2 . | . . . | 5 . .
. . . | . . 7 | 4 2 .
------+-------+------
. . . | . . . | 9 4 .
4 9 . | . 2 . | . . 5
. . . | . 1 9 | . 7 .
```

Hard # 991

3			2					
8	9		6			2		4
			5				6	
2		9						
	5	4				8	7	
						9		1
	3			7				
5		2		6			8	9
				8				6

Hard # 992

			8			2	7	
	8		5		4			
			3				9	8
6		9		4				
1								7
			5			9		2
5	1			6				
		1		5			2	
	6	8		7				

Hard # 993

			9	7				
						9	4	
6	1			4	3			
3						7		
5	6						3	1
		8						9
			1	6			2	8
	4	1						
			5	8				

Hard # 994

7			6				3	
1			2					
	3					2	6	9
			4					3
2	1						9	8
5				7				
3	9	5					4	
				4				1
	4			5				6

Hard # 995

				3	1		4	
4			6					
9		7						
5			4		6		7	
	2						8	
	3		9		2			6
						7		8
						7		9
2			6	5				

Hard # 996

			9			4	2	
5			2					
		6				3	7	
	7		2					9
	5		1			9		4
6						4		5
		2	3			8		
			8					4
	9	1		4				

Hard # 997

				7		4		
			1					
			4	8	3	2	6	
7		5				6		1
	3						2	
1		8				5		9
	5	2	9	3	1			
					5			
		3		6				

Hard # 998

		7		3				
		4	6	7			3	9
	6				9	2		
						1		4
			4		8			
5		3						
		5	9				1	
8	7			2	6	4		
				5		9		

Hard # 999

	2							
		5			6		7	
8	1		2					9
2			5	8	4			
1								8
			1	3	7			6
9					5		4	2
	6		3			1		
							6	

Hard # 1000

5	3							
						8	9	
		2					1	3
	4		7		6		3	
		3	8		1	4		
	7		2		3		5	
1	9					5		
	5	6						
							7	6

Hard # 1001

2			1	8		6		
	7					2		
				3	9			
	5	6	4	9		7		
		4		1	8	3	2	
			9	6				
		5					1	
		7		2	5			4

Hard # 1002

	4						3	2
				9	8			6
	1				2			
		5	3					
3	6			1			5	7
				5	9			
			2			6		
7			1	4				
1	8					9		

Hard # 1003

6	5					7	3	
		2		9				1
			3					
			5			1		8
			7		2			
8		4			3			
				4				
2				1		6		
	6	9					5	4

Hard # 1004

	1		4		7	9		2
			6				3	8
		9					4	
	2						6	
	1					4		
8						5		
	2			9				
3	4		7					
9		7	1		6		8	

Hard # 1005

6			8		5			
		4	6	9			8	
		1					9	
8	3			6	7			
			4	3			6	2
	6					7		
	4			7	9	3		
			5		6			1

Hard # 1006

	9				3	5		
8					5			
		3			8		6	9
	1	5					3	
			4		7			
	7					9	2	
5	8		9			3		
			5					4
		2	1				9	

Hard # 1007

2		4			6			
	7				1		6	
	3			7				5
		2			3	1		
6								3
		8	1			4		
5				8			3	
	4		5				9	
			3			5		2

Hard # 1008

	1	8					4	
		7		4	5			
				6	7		5	
8						5	3	
			9		8			
	5	6						1
	2		6	8				
			7	3		9		
	4					7	8	

Hard # 1009

```
2 8 . | . . 9 | . . .
1 . 7 | 2 . . | . . 5
. . 5 | . 7 . | . . .
------+-------+------
. . . | 3 7 . | 1 . .
7 . . | . . . | . . 9
. 4 . | 1 8 . | . . .
------+-------+------
. . . | 9 . . | 3 . .
3 . . | . . 8 | 1 . 2
. . . | 6 . . | . 4 8
```

Hard # 1010

```
. . 1 | . 2 . | 3 . 7
. 5 7 | . . . | . . .
. . . | . 7 . | 1 5 .
------+-------+------
. 4 . | . . . | . 2 .
8 . . | 5 . 2 | . . 3
. 3 . | . . . | . 1 .
------+-------+------
. 6 9 | . 3 . | . . .
. . . | . . . | 7 4 .
4 . 8 | . 1 . | 6 . .
```

Hard # 1011

```
. 6 1 | 8 7 . | . . .
. . . | . . . | . . 1
. . . | 4 . . | 6 8 .
------+-------+------
3 . . | . . 5 | . . 6
. . . | 1 . 9 | . . .
4 . 8 | . . . | . . 9
------+-------+------
. 7 4 | . . 2 | . . .
6 . . | . . . | . . .
. . . | . 5 3 | 7 4 .
```

Hard # 1012

```
4 . . | . 8 . | . . .
1 7 . | . . . | . 3 .
. . . | 6 . 9 | . . .
------+-------+------
7 9 . | . . 3 | . 2 .
. . 1 | . . . | 6 . .
. 4 . | 1 . . | . 7 3
------+-------+------
. . . | 9 . 5 | . . .
. 6 . | . . . | . 8 7
. . . | . 2 . | . . 1
```

Hard # 1013

```
6 . . | . . . | . 1 .
3 . . | 7 . . | 9 . .
. 1 . | . 6 5 | . . 4
------+-------+------
. . . | . . . | 3 . .
. 6 . | 5 . 2 | . 4 .
. . 1 | . . . | . . .
------+-------+------
8 . . | 1 7 . | . 3 .
. . 2 | . . 8 | . . 9
. 4 . | . . . | . . 6
```

Hard # 1014

```
. . . | 9 . . | . . 3
5 . . | 3 . 8 | 4 . .
. 2 . | . . . | . 7 .
------+-------+------
. 5 4 | . 9 . | . 8 .
. . 1 | . . . | 7 . .
. 7 . | . 6 . | 3 4 .
------+-------+------
. 3 . | . . . | . 6 .
. . 8 | 7 . 5 | . . 1
9 . . | . 2 . | . . .
```

Hard # 1015

```
. . . | . . 4 | . . 6
. . 2 | . . . | 5 4 .
. . 5 | . . . | 2 . 3
------+-------+------
. 7 5 | . . . | 9 8 .
1 . . | . . . | . . 2
. 8 9 | . . . | 3 7 .
------+-------+------
8 . 4 | . 5 . | . . .
. 6 7 | . 2 . | . . .
5 . . | 8 . . | . . .
```

Hard # 1016

```
7 . 8 | . 5 . | . . .
. 3 . | . 7 . | 9 . .
5 4 . | 6 . . | . . .
------+-------+------
. 7 . | . . 3 | . . 2
2 . . | . . . | . . 4
3 . . | 8 . . | . 7 .
------+-------+------
. . . | . 5 . | . 8 7
. . 1 | . 8 . | . 9 .
. . . | . 2 . | 6 . 3
```

Hard # 1017

```
. . . | 5 6 . | 3 2 .
3 . . | . 9 . | . . .
. . . | . . 7 | . . 5
------+-------+------
. . . | . 1 3 | 9 . .
. 4 . | . . . | . 5 .
. 2 1 | 6 . . | . . .
------+-------+------
8 . 6 | . . . | . . .
. . . | . 6 . | . . 7
1 9 . | 4 2 . | . . .
```

Hard # 1018

```
. . . | . . 6 | . 3 .
. . 7 | 4 2 5 | . . .
. . 9 | . . . | . . 2
------+-------+------
9 1 . | . . . | 3 . .
. . 3 | 9 . 7 | 2 . .
. . 4 | . . . | . 1 9
------+-------+------
7 . . | . . . | 8 . .
. . . | 8 5 9 | 6 . .
. 4 . | 1 . . | . . .
```

Hard # 1019

```
. . 6 | 9 5 . | . . .
3 5 . | 4 2 . | . . .
8 . . | . . . | . 2 .
------+-------+------
5 . . | . . 7 | . . .
. 7 . | 3 . 1 | . 9 .
. . 4 | . . . | . . 3
------+-------+------
. 8 . | . . . | . . 7
. . . | . 1 3 | . 8 2
. . . | . 6 5 | 1 . .
```

Hard # 1020

```
1 6 . | . 4 . | 5 . .
. . . | 3 1 . | . . .
. . 8 | . . . | 4 1 .
------+-------+------
7 . . | 6 . . | . . .
4 . . | . . . | . . 9
. . . | . 2 . | . . 8
------+-------+------
. 5 2 | . . . | 3 . .
. . . | 9 3 . | . . .
. . 9 | . 8 . | . 7 5
```

Hard # 1021

6		8						
				4	3			5
	5		6	9				
				7		3	5	
	7	2				9	1	
	6	1		5				
				1	8		3	
4			5	3				
						5		2

Hard # 1022

5	8			1				2
4		1						3
			3		8			4
			4	5				
	7						2	
			1	2				
1			5		2			
8						2		6
9				3			4	8

Hard # 1023

			4	8	7	3		
								8
	5			2			6	
2		1	3	8	5			
			1	9	4	5		6
	7			3			9	
4								
	9	6	5	7				

Hard # 1024

2			3				6	7
	8	7						
3				5				
	9			4	6	5		
		4				9		
		8	9	2			1	
				9				5
						2	8	
4	3				2			6

Hard # 1025

			2			4	3	
7			5					
			7				6	8
		9					7	5
2								1
3	8					6		
5	4			6				
							5	2
	1	7			4			

Hard # 1026

			2	9	7	8		5
	7			4			3	
			5					
					6		5	4
		6				7		
4	8			3				
					9			
		5			3			7
9		4		6	7		1	

Hard # 1027

	4	5				2		7
		9		4	1			
3				8				
7								
9		1	4		2	8		5
								3
				2				4
			8	6		5		
1		6				9	3	

Hard # 1028

7						2		4
		3			6		9	
4	8		1					
		4			2			
	1		9		3		8	
			6			5		
				4			3	6
	7		5			4		
6		1						8

Hard # 1029

	2			8				
8	3				7			
	5					2	3	
	8	4	1					9
			3		9			
2					8	1	7	
	1	5					9	
			7				1	6
				9			4	

Hard # 1030

						7		
4	1				5			
7					3		8	
		8			4	6	5	
		6				1		
	3	1	5			9		
	9		3					1
		8					3	7
	4							

Hard # 1031

			6					
		4					3	
6	5	2		8				7
	1		9			4		
		3		1		5		
		5			7		1	
4				5		7	6	8
	9					2		
					6			

Hard # 1032

	6							3
7				6				
3	1			9	8			
	9		5			7		6
6		3			1		4	
			7	2			5	9
			9					7
4						8		

Hard # 1033

			6			9		7
			4		3	5		
				5	8			
3				8	2			5
	1						9	
6			9	5				1
		3	8					
		8	5		6			
4		2			1			

Hard # 1034

						7		6
8			4		1			
		3					8	1
5			8		9			3
	1					5		
3			6		2			7
4	5					6		
			5		8			9
7		9						

Hard # 1035

		5	3	4		9		
		8				4		2
1				6		3		
			3	6				7
3				1	8			
	8		7					4
7		4				1		
	9		6	4		8		

Hard # 1036

		7	6			3		
		4	1					
						8	7	5
				1			3	7
	7		3		9		6	
6	8			4				
2	3	9						
						5	9	
		8				4	7	

Hard # 1037

	3		9			2		
			4					
8						9	5	
	1		5					8
5	8					2	1	
2			7			4		
7	5							9
			3					
		2			9	3		

Hard # 1038

			8	9				
8		9				2	3	
		1		2				
	1		2	5				6
4								8
5			8	7			2	
			3			7		
	3	4				6		1
			1	6				

Hard # 1039

	6	8						
			7			4		1
		9		4				3
			1	3			2	
			5		7			
	4			8	9			
8				5		6		
2		5			6			
						2	3	

Hard # 1040

	8	6	4					
9								
	5		3				7	9
		5	1					
	6	3				1	8	
					8	2		
2	9				5		6	
								2
					1	8	9	

Hard # 1041

2			6	7		1		8
	8						6	
6				4				
9		5					8	
			4		3			
	1					2		7
				3				2
	4					7		
7		2		5	6			9

Hard # 1042

				8				9
			7			1		8
	9		4	6				
4						6		
1	3			4			9	2
		2						7
			1	4		2		
3		5			2			
9				7				

Hard # 1043

	2				5			
7	3							6
1			6	2	4			3
					2	9		
	8						7	
		3	1					
2				8	4	3		7
3							5	4
			5				1	

Hard # 1044

1			2		3		5	8
9			8					
	8					6		1
				6	9			
		8				4		
			7	3				
7		5					3	
			1					4
2	4		3		8			6

Hard # 1045

	8							
			5	2			9	
6	5		9		3			4
			7			3	5	
5								8
	7	8			2			
2			6		5		1	3
	3			1	4			
							6	

Hard # 1046

2		7						
		9		4				
	6		3		7		2	1
1			4		3			
		8				7		
			6		5			2
7	5		1		8	3		
			5			6		
						1		8

Hard # 1047

			7			2		
			2	6		7		5
	1					3		
			8		3	9	5	
3								8
	2	9	1		7			
	9					4		
6		5		7	2			
	7				8			

Hard # 1048

			7	3				6
						4	5	
		8		5		2		
6				9	1			8
			7		2			
9			8	6				7
		1		8		9		
	5	6						
7			3	1				

Hard # 1049

9				2	3			
4		1						
				1		8	5	
	4	2		8	7			
	7					2		
			2	6		4	3	
7	2		9					
						8		6
		5	7					2

Hard # 1050

						7	5	
	7				9			
3	2			6				9
		9	2	4		5		
5								3
	2		9	7	8			
6				3			7	8
			1			3		
	8	3						

Solution

Solution # 1
```
8 3 7 9 6 5 4 2 1
2 4 6 1 3 7 5 8 9
9 5 1 4 8 2 6 3 7
3 7 9 2 5 8 1 4 6
4 1 5 3 9 6 2 7 8
6 2 8 7 4 1 9 5 3
1 8 4 5 7 9 3 6 2
5 6 2 8 1 3 7 9 4
7 9 3 6 2 4 8 1 5
```

Solution # 2
```
6 7 8 9 5 4 2 3 1
3 4 1 8 7 2 9 5 6
2 5 9 1 6 3 4 7 8
5 9 4 7 8 1 3 6 2
1 3 7 2 4 6 8 9 5
8 2 6 5 3 9 1 4 7
4 8 3 6 2 7 5 1 9
7 1 5 3 9 8 6 2 4
9 6 2 4 1 5 7 8 3
```

Solution # 3
```
4 3 6 7 9 5 1 8 2
9 5 8 2 1 6 3 7 4
1 2 7 3 4 8 9 6 5
2 9 1 6 8 3 4 5 7
6 4 3 5 7 9 8 2 1
8 7 5 1 2 4 6 3 9
7 6 2 9 3 1 5 4 8
3 1 4 8 5 7 2 9 6
5 8 9 4 6 2 7 1 3
```

Solution # 4
```
1 4 6 9 2 7 3 5 8
5 3 2 8 6 4 7 9 1
7 9 8 3 5 1 4 6 2
4 8 1 2 7 6 5 3 9
2 6 3 4 9 5 8 1 7
9 5 7 1 3 8 6 2 4
3 1 5 7 8 2 9 4 6
8 2 9 6 4 3 1 7 5
6 7 4 5 1 9 2 8 3
```

Solution # 5
```
5 7 8 3 6 1 2 4 9
2 9 6 7 8 4 5 3 1
1 3 4 9 2 5 7 6 8
7 5 1 4 3 8 9 2 6
4 6 3 2 1 9 8 7 5
8 2 9 5 7 6 3 1 4
9 8 7 1 4 3 6 5 2
6 1 2 8 5 7 4 9 3
3 4 5 6 9 2 1 8 7
```

Solution # 6
```
5 6 3 2 1 8 7 4 9
7 2 8 3 9 4 6 1 5
9 4 1 6 5 7 8 2 3
8 1 9 5 4 3 2 6 7
4 3 6 1 7 2 9 5 8
2 5 7 8 6 9 1 3 4
3 7 4 9 2 6 5 8 1
6 8 5 7 3 1 4 9 2
1 9 2 4 8 5 3 7 6
```

Solution # 7
```
9 1 6 7 2 8 4 5 3
4 2 8 3 1 5 9 7 6
5 7 3 4 6 9 2 1 8
6 5 4 2 9 7 3 8 1
1 8 2 6 5 3 7 4 9
7 3 9 1 8 4 6 2 5
2 6 5 9 7 1 8 3 4
3 9 1 8 4 2 5 6 7
8 4 7 5 3 6 1 9 2
```

Solution # 8
```
6 3 4 7 5 8 2 1 9
7 9 8 3 1 2 4 5 6
2 1 5 4 9 6 8 7 3
3 8 1 2 6 5 7 9 4
5 2 6 9 4 7 1 3 8
4 7 9 1 8 3 6 2 5
1 4 3 8 2 9 5 6 7
9 6 2 5 7 4 3 8 1
8 5 7 6 3 1 9 4 2
```

Solution # 9
```
1 8 5 4 9 6 7 3 2
7 9 6 3 2 8 4 5 1
4 2 3 7 1 5 8 6 9
3 1 4 9 6 7 2 8 5
8 6 9 5 4 2 1 7 3
5 7 2 1 8 3 9 4 6
2 5 8 6 7 1 3 9 4
9 3 1 8 5 4 6 2 7
6 4 7 2 3 9 5 1 8
```

Solution # 10
```
2 3 4 6 7 9 1 8 5
5 7 6 2 1 8 9 4 3
9 8 1 3 5 4 7 2 6
8 2 5 4 9 1 3 6 7
3 4 7 5 8 6 2 9 1
6 1 9 7 2 3 8 5 4
1 9 3 8 4 5 6 7 2
4 6 2 9 3 7 5 1 8
7 5 8 1 6 2 4 3 9
```

Solution # 11
```
9 5 1 6 8 2 4 7 3
7 3 2 4 9 5 8 6 1
6 4 8 7 1 3 5 9 2
2 6 5 1 3 8 7 4 9
1 8 4 2 7 9 6 3 5
3 9 7 5 4 6 2 1 8
5 2 3 9 6 4 1 8 7
4 1 9 8 2 7 3 5 6
8 7 6 3 5 1 9 2 4
```

Solution # 12
```
7 8 2 5 9 6 4 1 3
4 3 5 8 1 2 9 6 7
1 6 9 4 3 7 2 8 5
5 9 1 6 2 8 3 7 4
6 7 8 9 4 3 5 2 1
3 2 4 1 7 5 6 9 8
2 5 7 3 6 1 8 4 9
9 1 3 2 8 4 7 5 6
8 4 6 7 5 9 1 3 2
```

Solution # 13
```
8 9 4 2 7 1 5 3 6
6 2 5 9 4 3 7 8 1
3 1 7 6 8 5 2 4 9
5 4 3 1 2 9 6 7 8
2 7 6 4 3 8 9 1 5
1 8 9 7 5 6 4 2 3
9 6 2 3 1 7 8 5 4
7 5 1 8 6 4 3 9 2
4 3 8 5 9 2 1 6 7
```

Solution # 14
```
8 5 2 6 7 3 1 4 9
4 6 1 2 8 9 3 5 7
9 3 7 4 5 1 2 8 6
3 2 4 8 9 5 7 6 1
1 7 6 3 2 4 5 9 8
5 8 9 7 1 6 4 2 3
6 9 5 1 4 7 8 3 2
2 1 3 5 6 8 9 7 4
7 4 8 9 3 2 6 1 5
```

Solution # 15
```
1 3 5 4 2 7 9 6 8
9 7 2 5 6 8 4 3 1
8 4 6 1 9 3 2 7 5
6 5 4 2 8 9 7 1 3
7 8 1 3 4 6 5 2 9
3 2 9 7 1 5 8 4 6
2 9 8 6 7 1 3 5 4
5 6 7 9 3 4 1 8 2
4 1 3 8 5 2 6 9 7
```

Solution # 16
```
6 2 5 1 4 7 9 8 3
3 7 9 2 6 8 4 5 1
1 4 8 9 5 3 6 7 2
4 8 2 3 9 6 5 1 7
7 9 6 5 1 2 8 3 4
5 3 1 8 7 4 2 9 6
8 6 3 7 2 9 1 4 5
9 5 4 6 3 1 7 2 8
2 1 7 4 8 5 3 6 9
```

Solution # 17
```
1 2 3 8 7 9 6 5 4
5 7 8 4 6 2 1 9 3
9 6 4 5 1 3 2 7 8
6 5 9 7 3 8 4 2 1
2 4 7 1 5 6 3 8 9
8 3 1 2 9 4 5 6 7
3 9 5 6 4 7 8 1 2
7 8 6 3 2 1 9 4 5
4 1 2 9 8 5 7 3 6
```

Solution # 18
```
4 5 1 7 8 2 9 6 3
7 3 9 4 6 5 2 8 1
6 2 8 1 9 3 4 7 5
5 9 3 8 2 1 7 4 6
1 7 6 3 4 9 5 2 8
2 8 4 5 7 6 1 3 9
3 6 2 9 5 4 8 1 7
8 4 5 6 1 7 3 9 2
9 1 7 2 3 8 6 5 4
```

Solution # 19
```
8 4 1 5 2 3 6 9 7
3 7 6 8 1 9 4 5 2
9 2 5 4 7 6 1 8 3
1 9 7 6 5 2 8 3 4
2 6 4 7 3 8 5 1 9
5 8 3 1 9 4 2 7 6
4 3 9 2 8 1 7 6 5
6 5 8 9 4 7 3 2 1
7 1 2 3 6 5 9 4 8
```

Solution # 20
```
9 8 1 6 4 7 5 2 3
4 7 5 3 2 1 6 8 9
2 6 3 9 8 5 1 7 4
6 4 2 5 1 8 9 3 7
7 5 8 2 3 9 4 1 6
3 1 9 4 7 6 2 5 8
1 3 4 7 9 2 8 6 5
8 9 6 1 5 3 7 4 2
5 2 7 8 6 4 3 9 1
```

Solution # 21
```
6 9 1 4 8 7 3 5 2
3 4 5 2 1 9 8 7 6
8 7 2 6 5 3 4 1 9
2 6 4 7 9 8 1 3 5
1 3 8 5 2 4 9 6 7
7 5 9 1 3 6 2 4 8
4 8 7 9 6 1 5 2 3
9 2 6 3 4 5 7 8 1
5 1 3 8 7 2 6 9 4
```

Solution # 22
```
4 3 8 5 1 6 9 2 7
5 1 2 8 7 9 4 6 3
6 9 7 4 2 3 1 5 8
8 4 5 2 6 1 3 7 9
2 6 3 9 4 7 5 8 1
1 7 9 3 8 5 2 4 6
3 5 4 6 9 8 7 1 2
7 2 6 1 3 4 8 9 5
9 8 1 7 5 2 6 3 4
```

Solution # 23
```
9 3 4 6 5 1 7 2 8
2 1 5 9 8 7 3 6 4
7 6 8 4 3 2 1 5 9
6 5 1 8 7 4 2 9 3
8 2 9 1 6 3 5 4 7
4 7 3 2 9 5 8 1 6
3 4 7 5 1 9 6 8 2
1 9 6 7 2 8 4 3 5
5 8 2 3 4 6 9 7 1
```

Solution # 24
```
5 6 7 4 8 3 9 2 1
3 2 4 1 9 6 8 7 5
1 9 8 7 2 5 4 6 3
6 8 3 2 5 9 7 1 4
2 4 9 3 7 1 6 5 8
7 1 5 6 4 8 3 9 2
8 5 2 9 3 7 1 4 6
4 7 1 8 6 2 5 3 9
9 3 6 5 1 4 2 8 7
```

Solution # 25
```
5 9 6 2 3 8 1 7 4
4 2 1 5 9 7 6 3 8
8 7 3 4 1 6 2 5 9
2 8 9 7 6 3 5 4 1
1 3 4 8 2 5 7 9 6
6 5 7 1 4 9 3 8 2
3 4 8 6 5 1 9 2 7
9 1 2 3 7 4 8 6 5
7 6 5 9 8 2 4 1 3
```

Solution # 26
```
3 1 2 8 9 5 6 4 7
7 9 8 6 4 3 1 2 5
5 4 6 7 1 2 8 3 9
1 7 3 4 2 8 5 9 6
9 8 5 1 3 6 4 7 2
2 6 4 9 5 7 3 1 8
6 3 7 2 8 4 9 5 1
4 2 1 5 6 9 7 8 3
8 5 9 3 7 1 2 6 4
```

Solution # 27
```
4 9 8 7 1 3 2 5 6
1 2 7 4 5 6 3 9 8
5 3 6 2 8 9 1 4 7
3 8 4 1 9 2 7 6 5
9 6 2 5 3 7 8 1 4
7 1 5 6 4 8 9 2 3
8 4 3 9 6 1 5 7 2
6 7 1 3 2 5 4 8 9
2 5 9 8 7 4 6 3 1
```

Solution # 28
```
4 5 2 3 7 8 6 1 9
9 3 1 2 6 4 5 7 8
6 8 7 5 9 1 4 2 3
5 7 3 6 4 2 8 9 1
2 4 6 1 8 9 3 5 7
1 9 8 7 3 5 2 4 6
3 6 4 9 2 7 1 8 5
8 1 9 4 5 6 7 3 2
7 2 5 8 1 3 9 6 4
```

Solution # 29
```
1 3 9 6 2 5 7 8 4
8 6 7 9 3 4 2 5 1
2 5 4 7 8 1 3 6 9
5 9 2 3 7 6 1 4 8
3 4 6 8 1 2 5 9 7
7 1 8 4 5 9 6 2 3
4 8 1 2 6 3 9 7 5
6 7 3 5 9 8 4 1 2
9 2 5 1 4 7 8 3 6
```

Solution # 30
```
1 7 6 4 9 8 2 5 3
4 2 3 7 5 6 8 9 1
8 9 5 2 1 3 4 7 6
7 8 2 9 3 1 6 4 5
6 1 4 8 7 5 3 2 9
3 5 9 6 4 2 1 8 7
9 6 7 1 2 4 5 3 8
5 4 8 3 6 9 7 1 2
2 3 1 5 8 7 9 6 4
```

Solution # 31
```
6 9 5 7 4 2 8 1 3
2 7 1 9 3 8 4 5 6
8 4 3 6 5 1 2 9 7
7 5 2 1 8 4 6 3 9
9 3 4 5 2 6 1 7 8
1 8 6 3 9 7 5 4 2
5 6 8 4 7 9 3 2 1
4 1 9 2 6 3 7 8 5
3 2 7 8 1 5 9 6 4
```

Solution # 32
```
5 8 4 7 3 6 1 2 9
9 2 3 5 1 8 4 6 7
7 1 6 9 2 4 8 5 3
2 6 7 4 8 9 5 3 1
8 4 1 3 7 5 2 9 6
3 9 5 2 6 1 7 4 8
6 7 9 1 4 2 3 8 5
1 5 2 8 9 3 6 7 4
4 3 8 6 5 7 9 1 2
```

Solution # 33
```
3 7 9 6 1 5 2 4 8
4 1 5 8 2 7 9 3 6
8 6 2 4 9 3 7 5 1
9 4 7 5 3 6 8 1 2
6 5 8 2 4 1 3 7 9
2 3 1 7 8 9 4 6 5
5 2 3 9 6 4 1 8 7
7 9 4 1 5 8 6 2 3
1 8 6 3 7 2 5 9 4
```

Solution # 34
```
3 1 5 8 7 4 9 6 2
7 4 6 2 9 5 1 8 3
2 8 9 6 3 1 4 5 7
9 6 3 1 2 8 5 7 4
4 5 2 3 6 7 8 9 1
8 7 1 5 4 9 3 2 6
6 9 4 7 8 3 2 1 5
5 2 8 4 1 6 7 3 9
1 3 7 9 5 2 6 4 8
```

Solution # 35
```
5 2 3 1 6 4 7 8 9
7 4 6 5 9 8 3 2 1
1 8 9 3 2 7 5 4 6
8 3 7 6 4 5 1 9 2
9 1 4 7 3 2 8 6 5
2 6 5 8 1 9 4 7 3
4 5 2 9 8 1 6 3 7
3 7 8 2 5 6 9 1 4
6 9 1 4 7 3 2 5 8
```

Solution # 36
```
2 4 6 3 7 9 1 8 5
7 9 8 5 1 4 2 3 6
5 3 1 6 2 8 9 7 4
3 6 5 2 4 7 8 9 1
8 1 2 9 6 3 4 5 7
9 7 4 8 5 1 6 2 3
4 5 9 7 8 6 3 1 2
6 8 7 1 3 2 5 4 9
1 2 3 4 9 5 7 6 8
```

Solution # 37
```
8 7 3 6 4 1 2 9 5
4 9 5 2 3 8 1 6 7
2 6 1 7 5 9 4 3 8
1 5 4 3 7 2 9 8 6
6 3 8 4 9 5 7 2 1
9 2 7 1 8 6 5 4 3
7 4 2 5 6 3 8 1 9
5 8 6 9 1 4 3 7 2
3 1 9 8 2 7 6 5 4
```

Solution # 38
```
8 6 9 3 1 4 2 5 7
3 1 4 5 2 7 9 8 6
2 7 5 8 9 6 4 3 1
5 4 6 9 8 3 7 1 2
9 8 1 2 7 5 3 6 4
7 2 3 4 6 1 5 9 8
4 9 2 6 5 8 1 7 3
6 3 7 1 4 9 8 2 5
1 5 8 7 3 2 6 4 9
```

Solution # 39
```
1 4 3 2 9 6 7 8 5
8 9 5 3 7 4 1 6 2
7 2 6 1 5 8 4 3 9
3 6 8 5 4 1 9 2 7
5 1 2 7 3 9 8 4 6
9 7 4 6 8 2 5 1 3
2 5 7 4 1 3 6 9 8
6 8 1 9 2 7 3 5 4
4 3 9 8 6 5 2 7 1
```

Solution # 40
```
1 2 4 7 8 3 9 5 6
3 9 6 4 1 5 2 8 7
5 7 8 9 2 6 4 3 1
4 6 7 8 9 1 3 2 5
2 8 5 6 3 4 7 1 9
9 1 3 5 7 2 6 4 8
6 5 1 3 4 9 8 7 2
8 4 9 2 5 7 1 6 3
7 3 2 1 6 8 5 9 4
```

Solution # 41
```
4 8 5 9 7 1 6 2 3
3 7 2 6 5 8 9 1 4
9 1 6 2 4 3 7 8 5
8 6 3 4 1 9 2 5 7
5 2 1 7 3 6 4 9 8
7 4 9 5 8 2 1 3 6
2 9 4 3 6 5 8 7 1
6 5 8 1 9 7 3 4 2
1 3 7 8 2 4 5 6 9
```

Solution # 42
```
8 2 3 1 7 6 5 4 9
1 4 9 3 2 5 6 7 8
7 5 6 8 9 4 3 2 1
6 7 2 5 4 8 9 1 3
5 1 4 7 3 9 2 8 6
9 3 8 6 1 2 7 5 4
4 9 7 2 6 1 8 3 5
2 8 1 9 5 3 4 6 7
3 6 5 4 8 7 1 9 2
```

Solution # 43
```
4 5 6 1 7 8 3 9 2
1 9 7 4 3 2 8 6 5
2 3 8 6 9 5 7 1 4
6 8 4 3 5 9 1 2 7
9 7 3 2 6 1 5 4 8
5 1 2 8 4 7 6 3 9
3 6 5 7 2 4 9 8 1
7 4 1 9 8 3 2 5 6
8 2 9 5 1 6 4 7 3
```

Solution # 44
```
5 8 2 9 1 3 6 7 4
6 4 1 7 5 2 3 9 8
7 9 3 6 4 8 1 5 2
8 5 6 1 9 7 4 2 3
1 2 9 5 3 4 8 6 7
4 3 7 2 8 6 9 1 5
3 6 5 4 2 1 7 8 9
9 1 4 8 7 5 2 3 6
2 7 8 3 6 9 5 4 1
```

Solution # 45
```
5 1 8 4 6 2 3 9 7
3 4 6 9 8 7 2 5 1
2 9 7 3 1 5 6 4 8
6 8 4 2 3 9 1 7 5
9 5 1 8 7 6 4 2 3
7 3 2 5 4 1 8 6 9
4 7 5 1 2 8 9 3 6
8 2 9 6 5 3 7 1 4
1 6 3 7 9 4 5 8 2
```

Solution # 46
```
8 4 2 1 9 3 7 6 5
1 7 5 8 6 4 3 9 2
3 6 9 5 7 2 1 4 8
7 5 6 2 4 8 9 3 1
4 3 8 9 1 7 2 5 6
2 9 1 6 3 5 4 8 7
5 2 7 4 8 9 6 1 3
6 8 4 3 2 1 5 7 9
9 1 3 7 5 6 8 2 4
```

Solution # 47
```
5 7 3 9 6 8 4 1 2
6 4 9 7 2 1 5 3 8
8 1 2 4 3 5 9 6 7
3 9 6 1 8 4 7 2 5
7 2 4 6 5 9 1 8 3
1 5 8 2 7 3 6 9 4
2 3 1 5 9 7 8 4 6
9 6 7 8 4 2 3 5 1
4 8 5 3 1 6 2 7 9
```

Solution # 48
```
5 4 2 9 3 7 6 8 1
8 9 6 2 1 5 4 3 7
7 3 1 4 6 8 2 9 5
1 5 8 3 2 4 9 7 6
2 6 3 1 7 9 8 5 4
4 7 9 8 5 6 3 1 2
9 2 5 6 8 1 7 4 3
3 1 4 7 9 2 5 6 8
6 8 7 5 4 3 1 2 9
```

Solution # 49
```
9 5 4 2 6 3 7 1 8
1 3 6 8 7 9 2 4 5
7 2 8 5 4 1 6 9 3
2 8 7 4 9 6 5 3 1
5 6 3 7 1 8 4 2 9
4 9 1 3 5 2 8 7 6
8 1 5 9 2 7 3 6 4
3 7 9 6 8 4 1 5 2
6 4 2 1 3 5 9 8 7
```

Solution # 50
```
9 2 1 4 3 8 6 5 7
8 5 3 7 6 2 4 9 1
4 7 6 5 9 1 2 3 8
5 9 8 3 2 6 7 1 4
1 3 4 8 5 7 9 2 6
7 6 2 9 1 4 3 8 5
2 1 5 6 7 3 8 4 9
3 4 7 1 8 9 5 6 2
6 8 9 2 4 5 1 7 3
```

Solution # 51
```
9 6 1 5 8 7 3 4 2
3 5 2 9 4 6 7 1 8
7 4 8 1 2 3 5 9 6
4 8 9 7 5 1 6 2 3
5 7 6 2 3 4 9 8 1
2 1 3 6 9 8 4 5 7
6 3 5 8 1 9 2 7 4
8 9 7 4 6 2 1 3 5
1 2 4 3 7 5 8 6 9
```

Solution # 52
```
9 4 1 2 6 5 8 3 7
5 8 2 3 9 7 1 6 4
3 7 6 4 1 8 9 2 5
4 9 7 6 3 1 2 5 8
2 6 5 8 4 9 3 7 1
8 1 3 5 7 2 4 9 6
6 3 9 7 8 4 5 1 2
1 2 8 9 5 6 7 4 3
7 5 4 1 2 3 6 8 9
```

Solution # 53
```
3 8 5 9 1 4 6 2 7
2 9 6 3 5 7 8 4 1
1 7 4 6 2 8 5 3 9
6 3 9 4 8 5 1 7 2
4 2 8 7 6 1 3 9 5
5 1 7 2 9 3 4 6 8
9 5 2 8 3 6 7 1 4
7 6 1 5 4 9 2 8 3
8 4 3 1 7 2 9 5 6
```

Solution # 54
```
5 9 6 1 8 7 4 2 3
3 4 1 5 2 9 7 8 6
7 8 2 3 4 6 9 5 1
4 6 9 2 1 3 8 7 5
2 3 5 4 7 8 1 6 9
1 7 8 6 9 5 3 4 2
9 1 4 7 6 2 5 3 8
6 5 7 8 3 1 2 9 4
8 2 3 9 5 4 6 1 7
```

Solution # 55
```
4 6 8 1 7 9 5 3 2
3 9 7 2 5 6 1 4 8
2 1 5 8 4 3 7 6 9
1 8 3 5 2 4 9 7 6
9 5 6 3 1 7 8 2 4
7 4 2 9 6 8 3 5 1
6 3 1 7 9 2 4 8 5
8 2 9 4 3 5 6 1 7
5 7 4 6 8 1 2 9 3
```

Solution # 56
```
6 5 4 2 3 7 8 9 1
3 1 9 6 8 5 4 2 7
2 8 7 9 4 1 3 5 6
4 2 3 8 6 9 1 7 5
9 6 5 1 7 3 2 8 4
1 7 8 4 5 2 6 3 9
8 4 2 5 9 6 7 1 3
5 3 1 7 2 4 9 6 8
7 9 6 3 1 8 5 4 2
```

Solution # 57
```
7 3 9 6 5 2 4 8 1
5 1 2 4 3 8 6 7 9
8 4 6 1 7 9 2 3 5
6 5 4 9 8 7 3 1 2
1 2 7 3 4 6 9 5 8
9 8 3 2 1 5 7 6 4
4 6 8 5 9 3 1 2 7
2 7 1 8 6 4 5 9 3
3 9 5 7 2 1 8 4 6
```

Solution # 58
```
8 4 7 5 2 9 1 6 3
6 1 2 7 3 4 5 9 8
5 3 9 8 1 6 2 4 7
2 8 4 6 7 1 9 3 5
3 7 1 9 4 5 8 2 6
9 5 6 2 8 3 7 1 4
7 6 8 3 9 2 4 5 1
4 9 3 1 5 8 6 7 2
1 2 5 4 6 7 3 8 9
```

Solution # 59
```
8 6 5 2 3 1 4 7 9
3 1 4 7 6 9 2 8 5
7 9 2 5 4 8 3 1 6
9 2 6 4 7 3 1 5 8
1 4 7 6 8 5 9 2 3
5 3 8 1 9 2 6 4 7
6 7 3 8 1 4 5 9 2
2 8 1 9 5 6 7 3 4
4 5 9 3 2 7 8 6 1
```

Solution # 60
```
1 3 9 2 5 8 6 7 4
8 4 7 9 1 6 3 2 5
5 2 6 3 7 4 8 9 1
2 8 5 4 6 7 1 3 9
6 9 3 5 2 1 7 4 8
7 1 4 8 9 3 5 6 2
9 5 8 7 3 2 4 1 6
3 6 2 1 4 5 9 8 7
4 7 1 6 8 9 2 5 3
```

Solution # 61
```
7 4 5 9 2 8 3 6 1
8 2 3 1 6 7 5 9 4
6 1 9 4 5 3 8 2 7
9 8 2 5 4 6 1 7 3
3 5 1 7 8 2 6 4 9
4 6 7 3 9 1 2 8 5
1 9 8 6 3 4 7 5 2
2 3 4 8 7 5 9 1 6
5 7 6 2 1 9 4 3 8
```

Solution # 62
```
4 1 8 5 6 7 9 2 3
5 6 9 3 1 2 4 8 7
7 2 3 9 4 8 6 5 1
6 9 7 4 8 1 5 3 2
2 8 5 6 7 3 1 9 4
1 3 4 2 9 5 7 6 8
3 4 2 7 5 6 8 1 9
8 7 6 1 2 9 3 4 5
9 5 1 8 3 4 2 7 6
```

Solution # 63
```
2 1 3 9 8 6 4 5 7
8 5 6 7 4 2 3 1 9
9 4 7 1 3 5 6 2 8
4 7 8 3 2 1 9 6 5
5 6 2 4 9 8 7 3 1
1 3 9 6 5 7 8 4 2
6 2 5 8 7 4 1 9 3
7 9 4 2 1 3 5 8 6
3 8 1 5 6 9 2 7 4
```

Solution # 64
```
6 1 3 4 9 2 5 8 7
8 9 7 3 5 6 4 1 2
5 4 2 1 8 7 3 9 6
7 8 1 2 4 5 9 6 3
4 2 6 8 3 9 1 7 5
3 5 9 7 6 1 8 2 4
1 7 5 9 2 4 6 3 8
9 3 4 6 7 8 2 5 1
2 6 8 5 1 3 7 4 9
```

Solution # 65
```
5 3 8 1 9 4 2 6 7
9 6 7 5 2 3 8 4 1
1 2 4 6 8 7 3 9 5
7 9 2 8 4 1 5 3 6
4 8 3 7 5 6 1 2 9
6 1 5 2 3 9 7 8 4
8 7 9 3 6 5 4 1 2
2 5 6 4 1 8 9 7 3
3 4 1 9 7 2 6 5 8
```

Solution # 66
```
8 4 6 1 9 2 3 7 5
1 2 3 5 6 7 8 4 9
5 9 7 3 4 8 2 6 1
7 6 5 2 1 4 9 8 3
4 3 1 8 5 9 6 2 7
9 8 2 6 7 3 5 1 4
3 1 4 9 8 6 7 5 2
2 5 8 7 3 1 4 9 6
6 7 9 4 2 5 1 3 8
```

Solution # 67
```
7 1 4 3 2 6 9 8 5
2 8 9 7 5 4 3 1 6
3 5 6 1 9 8 7 4 2
1 2 5 8 7 9 4 6 3
9 4 3 5 6 1 8 2 7
8 6 7 4 3 2 1 5 9
4 3 2 9 1 5 6 7 8
6 9 1 2 8 7 5 3 4
5 7 8 6 4 3 2 9 1
```

Solution # 68
```
3 1 8 4 9 5 6 2 7
7 5 2 1 6 3 4 9 8
4 9 6 2 8 7 3 1 5
9 2 3 7 4 1 5 8 6
8 7 1 6 5 2 9 3 4
5 6 4 9 3 8 1 7 2
2 3 5 8 1 4 7 6 9
1 8 9 5 7 6 2 4 3
6 4 7 3 2 9 8 5 1
```

Solution # 69
```
4 8 3 2 1 5 9 6 7
5 1 6 4 7 9 3 8 2
2 7 9 6 3 8 5 1 4
6 3 5 1 4 7 8 2 9
1 4 2 9 8 3 7 5 6
7 9 8 5 2 6 4 3 1
9 5 7 3 6 2 1 4 8
3 6 1 8 9 4 2 7 5
8 2 4 7 5 1 6 9 3
```

Solution # 70
```
8 7 9 5 6 4 2 1 3
4 3 1 8 7 2 5 6 9
2 5 6 3 9 1 4 7 8
9 6 7 2 8 5 3 4 1
3 8 2 4 1 9 7 5 6
5 1 4 6 3 7 9 8 2
1 4 3 9 5 8 6 2 7
6 2 8 7 4 3 1 9 5
7 9 5 1 2 6 8 3 4
```

Solution # 71
```
2 8 9 7 3 1 5 6 4
6 1 4 9 5 8 7 3 2
7 3 5 6 2 4 1 9 8
9 7 2 8 6 5 3 4 1
1 5 6 2 4 3 9 8 7
8 4 3 1 7 9 2 5 6
3 2 7 5 8 6 4 1 9
5 6 1 4 9 2 8 7 3
4 9 8 3 1 7 6 2 5
```

Solution # 72
```
6 3 5 2 8 7 1 9 4
2 7 1 3 4 9 6 8 5
4 8 9 5 6 1 7 2 3
1 9 3 7 5 4 8 6 2
7 6 2 9 3 8 4 5 1
8 5 4 1 2 6 9 3 7
5 4 7 8 9 2 3 1 6
3 1 8 6 7 5 2 4 9
9 2 6 4 1 3 5 7 8
```

Solution # 73
```
8 3 7 6 5 1 9 2 4
5 9 6 2 4 3 7 8 1
1 4 2 8 7 9 5 3 6
9 7 5 4 1 2 3 6 8
2 1 8 3 6 7 4 9 5
3 6 4 5 9 8 2 1 7
4 5 3 9 8 6 1 7 2
6 2 1 7 3 4 8 5 9
7 8 9 1 2 5 6 4 3
```

Solution # 74
```
3 8 2 1 9 4 7 5 6
9 5 1 7 6 8 2 4 3
6 4 7 5 2 3 9 8 1
5 1 4 8 3 2 6 7 9
2 3 9 6 5 7 4 1 8
7 6 8 4 1 9 3 2 5
8 7 5 3 4 6 1 9 2
4 9 3 2 8 1 5 6 7
1 2 6 9 7 5 8 3 4
```

Solution # 75
```
1 4 9 5 8 2 3 6 7
5 8 3 6 9 7 2 4 1
7 6 2 4 1 3 5 8 9
6 3 4 2 5 9 1 7 8
8 7 5 3 6 1 9 2 4
2 9 1 8 7 4 6 5 3
4 5 7 9 3 6 8 1 2
9 2 8 1 4 5 7 3 6
3 1 6 7 2 8 4 9 5
```

Solution # 76
```
2 4 8 5 1 3 7 6 9
5 6 1 4 7 9 8 2 3
3 7 9 2 8 6 1 4 5
4 1 5 3 2 7 9 8 6
8 3 7 9 6 1 2 5 4
9 2 6 8 4 5 3 1 7
6 8 3 7 5 2 4 9 1
1 9 4 6 3 8 5 7 2
7 5 2 1 9 4 6 3 8
```

Solution # 77
```
3 6 8 7 5 9 4 2 1
5 4 1 8 2 6 9 3 7
9 2 7 4 1 3 8 6 5
7 8 3 6 9 1 2 5 4
2 9 5 3 8 4 1 7 6
6 1 4 5 7 2 3 9 8
1 7 6 9 3 8 5 4 2
8 5 9 2 4 7 6 1 3
4 3 2 1 6 5 7 8 9
```

Solution # 78
```
2 9 7 6 8 4 3 1 5
6 5 8 2 3 1 4 9 7
3 1 4 5 9 7 2 6 8
5 7 3 4 1 9 6 8 2
8 4 2 7 5 6 9 3 1
9 6 1 8 2 3 5 7 4
7 2 6 3 4 8 1 5 9
1 8 5 9 6 2 7 4 3
4 3 9 1 7 5 8 2 6
```

Solution # 79
```
9 8 7 4 3 1 5 2 6
6 5 4 8 9 2 7 1 3
3 2 1 5 6 7 4 9 8
4 1 6 2 8 3 9 7 5
8 3 5 7 4 9 2 6 1
2 7 9 1 5 6 8 3 4
7 4 2 3 1 5 6 8 9
5 9 3 6 7 8 1 4 2
1 6 8 9 2 4 3 5 7
```

Solution # 80
```
6 4 7 1 3 8 9 2 5
8 3 9 4 5 2 7 6 1
5 2 1 6 9 7 4 3 8
1 6 3 9 2 4 8 5 7
2 8 4 7 1 5 3 9 6
9 7 5 8 6 3 1 4 2
3 1 2 5 8 9 6 7 4
7 5 8 3 4 6 2 1 9
4 9 6 2 7 1 5 8 3
```

Solution # 81
```
4 9 1 5 6 2 7 3 8
5 7 6 3 4 8 9 1 2
8 2 3 9 1 7 6 4 5
9 6 8 2 7 1 4 5 3
2 5 4 6 3 9 1 8 7
1 3 7 8 5 4 2 9 6
3 4 9 7 8 6 5 2 1
6 1 5 4 2 3 8 7 9
7 8 2 1 9 5 3 6 4
```

Solution # 82
```
4 5 2 9 3 8 7 1 6
7 9 8 5 1 6 3 4 2
3 1 6 2 7 4 5 9 8
9 8 3 4 5 2 1 6 7
2 4 5 7 6 1 9 8 3
6 7 1 3 8 9 4 2 5
1 3 4 6 2 7 8 5 9
8 2 7 1 9 5 6 3 4
5 6 9 8 4 3 2 7 1
```

Solution # 83
```
1 3 2 7 8 9 5 4 6
5 9 7 2 4 6 3 8 1
6 8 4 3 5 1 9 2 7
4 2 9 8 6 5 7 1 3
7 5 3 1 2 4 6 9 8
8 6 1 9 3 7 2 5 4
3 1 8 5 7 2 4 6 9
2 7 6 4 9 8 1 3 5
9 4 5 6 1 3 8 7 2
```

Solution # 84
```
7 3 8 9 5 6 4 2 1
5 2 9 3 1 4 6 8 7
1 4 6 8 7 2 5 3 9
8 7 1 6 3 5 2 9 4
3 6 2 1 4 9 8 7 5
9 5 4 2 8 7 3 1 6
6 1 3 4 9 8 7 5 2
4 9 7 5 2 3 1 6 8
2 8 5 7 6 1 9 4 3
```

Solution # 85
```
3 9 2 7 1 6 5 4 8
1 8 7 4 5 3 2 9 6
6 4 5 8 2 9 7 1 3
9 2 4 5 7 8 6 3 1
7 1 3 6 9 4 8 2 5
5 6 8 2 3 1 9 7 4
8 3 1 9 6 2 4 5 7
2 7 6 1 4 5 3 8 9
4 5 9 3 8 7 1 6 2
```

Solution # 86
```
6 3 2 5 9 1 4 8 7
9 7 1 4 8 3 5 2 6
5 8 4 2 7 6 3 9 1
3 6 8 7 1 2 9 4 5
1 4 9 6 3 5 8 7 2
2 5 7 9 4 8 1 6 3
8 2 3 1 6 9 7 5 4
7 9 5 3 2 4 6 1 8
4 1 6 8 5 7 2 3 9
```

Solution # 87
```
7 3 6 2 4 1 5 9 8
1 5 2 6 9 8 7 4 3
9 8 4 3 7 5 1 2 6
3 7 9 4 8 2 6 1 5
4 2 5 1 3 6 9 8 7
6 1 8 9 5 7 4 3 2
5 4 7 8 2 9 3 6 1
2 6 3 5 1 4 8 7 9
8 9 1 7 6 3 2 5 4
```

Solution # 88
```
3 5 8 4 9 6 2 7 1
7 9 6 3 2 1 8 5 4
1 4 2 5 8 7 6 9 3
2 6 4 9 5 8 3 1 7
9 8 3 1 7 2 4 6 5
5 7 1 6 3 4 9 8 2
6 1 5 2 4 9 7 3 8
8 2 9 7 1 3 5 4 6
4 3 7 8 6 5 1 2 9
```

Solution # 89
```
6 8 5 9 7 1 3 2 4
7 2 4 3 8 6 1 5 9
3 1 9 5 2 4 8 6 7
8 6 7 4 5 3 9 1 2
4 9 2 6 1 8 5 7 3
5 3 1 2 9 7 4 8 6
1 4 3 7 6 5 2 9 8
2 5 6 8 4 9 7 3 1
9 7 8 1 3 2 6 4 5
```

Solution # 90
```
2 4 3 5 8 6 9 7 1
5 9 7 3 2 1 6 4 8
1 8 6 4 7 9 3 2 5
3 5 8 7 6 2 4 1 9
9 2 4 1 5 3 8 6 7
7 6 1 8 9 4 2 5 3
8 3 5 6 4 7 1 9 2
4 7 9 2 1 8 5 3 6
6 1 2 9 3 5 7 8 4
```

Solution # 91
```
7 5 9 1 2 4 3 6 8
3 8 4 9 7 6 2 5 1
2 6 1 5 8 3 7 4 9
9 7 3 4 6 5 1 8 2
1 4 6 2 3 8 5 9 7
8 2 5 7 1 9 4 3 6
4 3 7 8 9 1 6 2 5
6 1 8 3 5 2 9 7 4
5 9 2 6 4 7 8 1 3
```

Solution # 92
```
5 1 8 4 7 3 2 9 6
6 4 7 1 2 9 3 8 5
9 2 3 8 5 6 7 4 1
8 3 9 2 6 4 5 1 7
4 5 2 7 3 1 8 6 9
7 6 1 5 9 8 4 2 3
2 9 4 3 1 5 6 7 8
3 7 6 9 8 2 1 5 4
1 8 5 6 4 7 9 3 2
```

Solution # 93
```
4 7 9 5 3 6 1 2 8
2 3 5 1 8 4 9 7 6
8 1 6 7 2 9 4 5 3
6 4 2 9 7 8 3 1 5
9 8 3 2 1 5 6 4 7
1 5 7 4 6 3 8 9 2
7 2 8 6 4 1 5 3 9
5 6 1 3 9 2 7 8 4
3 9 4 8 5 7 2 6 1
```

Solution # 94
```
9 7 5 4 1 8 2 6 3
1 8 2 5 6 3 9 7 4
4 6 3 7 2 9 1 5 8
6 9 4 1 7 5 3 8 2
3 1 7 6 8 2 5 4 9
5 2 8 3 9 4 6 1 7
2 4 9 8 5 6 7 3 1
7 3 6 2 4 1 8 9 5
8 5 1 9 3 7 4 2 6
```

Solution # 95
```
7 2 6 3 4 9 5 1 8
3 4 1 5 8 7 2 9 6
9 5 8 1 2 6 7 4 3
4 1 9 2 5 3 8 6 7
5 7 2 9 6 8 4 3 1
6 8 3 4 7 1 9 2 5
2 6 4 7 1 5 3 8 9
1 9 5 8 3 4 6 7 2
8 3 7 6 9 2 1 5 4
```

Solution # 96
```
1 4 6 9 5 3 2 7 8
9 2 8 7 1 6 3 4 5
7 3 5 4 8 2 6 1 9
2 7 4 5 6 1 9 8 3
6 5 9 3 8 4 1 2 7
3 8 1 2 9 7 5 6 4
4 1 2 3 7 9 8 5 6
8 6 3 1 4 5 7 9 2
5 9 7 6 2 8 4 3 1
```

Solution # 97
```
3 2 5 9 4 7 8 6 1
4 9 7 8 6 1 2 3 5
1 6 8 3 5 2 7 4 9
6 4 1 5 7 3 9 2 8
5 7 9 2 8 6 4 1 3
2 8 3 1 9 4 5 7 6
7 5 6 4 1 8 3 9 2
9 1 2 7 3 5 6 8 4
8 3 4 6 2 9 1 5 7
```

Solution # 98
```
2 7 8 3 4 6 5 1 9
6 3 4 9 1 5 2 7 8
9 5 1 8 7 2 3 4 6
1 4 5 6 3 8 9 2 7
7 6 2 5 9 1 8 3 4
3 8 9 7 2 4 1 6 5
8 9 3 2 6 7 4 5 1
5 1 7 4 8 3 6 9 2
4 2 6 1 5 9 7 8 3
```

Solution # 99
```
2 4 7 9 3 8 5 1 6
9 8 5 6 2 1 4 7 3
1 6 3 5 4 7 2 8 9
4 2 1 3 7 9 8 6 5
3 7 8 1 6 5 9 2 4
6 5 9 4 8 2 1 3 7
8 1 4 7 5 3 6 9 2
5 3 2 8 9 6 7 4 1
7 9 6 2 1 4 3 5 8
```

Solution # 100
```
2 5 7 4 1 3 8 6 9
1 6 9 7 8 2 4 3 5
3 4 8 6 5 9 2 7 1
8 7 1 5 3 6 9 2 4
6 9 4 8 2 7 5 1 3
5 3 2 1 9 4 6 8 7
4 1 5 2 7 8 3 9 6
7 2 3 9 6 5 1 4 8
9 8 6 3 4 1 7 5 2
```

Solution # 101
```
8 9 3 4 7 1 2 5 6
2 6 4 8 5 3 7 1 9
5 1 7 2 9 6 3 4 8
6 8 2 7 1 4 5 9 3
3 5 1 6 2 9 4 8 7
4 7 9 5 3 8 1 6 2
9 3 5 1 6 2 8 7 4
1 4 6 3 8 7 9 2 5
7 2 8 9 4 5 6 3 1
```

Solution # 102
```
8 9 4 3 1 2 5 6 7
5 6 1 8 7 4 2 3 9
2 7 3 9 6 5 1 8 4
1 8 2 5 9 6 7 4 3
4 5 9 7 8 3 6 2 1
6 3 7 2 4 1 9 5 8
9 4 6 1 5 8 3 7 2
7 2 5 4 3 9 8 1 6
3 1 8 6 2 7 4 9 5
```

Solution # 103
```
9 5 8 6 4 7 3 1 2
3 2 7 8 1 9 5 6 4
4 1 6 5 3 2 8 7 9
7 8 1 3 9 4 2 5 6
5 3 9 7 2 6 1 4 8
2 6 4 1 5 8 9 3 7
8 4 3 9 7 5 6 2 1
1 9 2 4 6 3 7 8 5
6 7 5 2 8 1 4 9 3
```

Solution # 104
```
9 4 6 3 1 8 7 2 5
5 7 1 4 2 6 3 8 9
8 2 3 7 5 9 6 4 1
4 5 7 2 6 3 9 1 8
1 9 2 8 4 7 5 3 6
6 3 8 5 9 1 4 7 2
7 6 4 1 8 5 2 9 3
3 1 9 6 7 2 8 5 4
2 8 5 9 3 4 1 6 7
```

Solution # 105
```
9 7 6 2 3 5 4 1 8
4 3 1 8 7 9 6 5 2
8 2 5 4 1 6 3 7 9
6 4 7 5 9 3 2 8 1
2 5 8 7 4 1 9 6 3
1 9 3 6 2 8 5 4 7
5 1 4 3 8 2 7 9 6
7 8 2 9 6 4 1 3 5
3 6 9 1 5 7 8 2 4
```

Solution # 106
```
9 2 5 1 8 4 7 6 3
7 4 6 9 3 2 5 1 8
3 1 8 5 7 6 4 2 9
2 6 4 3 9 5 1 8 7
8 7 9 2 4 1 3 5 6
5 3 1 8 6 7 9 4 2
1 8 2 7 5 3 6 9 4
4 9 7 6 1 8 2 3 5
6 5 3 4 2 9 8 7 1
```

Solution # 107
```
1 4 8 2 9 3 5 7 6
6 2 7 1 4 5 8 3 9
3 5 9 7 8 6 4 2 1
4 1 3 8 2 7 6 9 5
9 6 2 5 3 4 1 8 7
7 8 5 6 1 9 2 4 3
2 7 4 9 5 1 3 6 8
8 9 1 3 6 2 7 5 4
5 3 6 4 7 8 9 1 2
```

Solution # 108
```
4 6 1 3 7 9 2 8 5
3 8 9 2 5 1 4 6 7
5 7 2 6 4 8 3 1 9
9 1 5 4 3 2 6 7 8
2 4 8 9 6 7 5 3 1
7 3 6 1 8 5 9 2 4
1 5 4 7 2 3 8 9 6
6 9 3 8 1 4 7 5 2
8 2 7 5 9 6 1 4 3
```

Solution # 109
```
1 6 5 7 3 9 2 4 8
8 7 9 2 1 4 3 6 5
2 3 4 5 6 8 9 7 1
6 4 7 8 9 2 1 5 3
3 9 2 1 7 5 4 8 6
5 8 1 3 4 6 7 2 9
7 1 8 6 2 3 5 9 4
4 5 3 9 8 7 6 1 2
9 2 6 4 5 1 8 3 7
```

Solution # 110
```
3 5 7 8 6 1 4 9 2
9 2 1 5 7 4 3 8 6
8 6 4 2 3 9 1 7 5
1 8 2 6 5 7 9 3 4
7 4 6 3 9 8 5 2 1
5 3 9 4 1 2 8 6 7
4 9 3 7 2 5 6 1 8
6 7 8 1 4 3 2 5 9
2 1 5 9 8 6 7 4 3
```

Solution # 111
```
7 9 6 5 3 2 1 8 4
1 5 2 8 4 7 3 6 9
8 4 3 6 9 1 7 5 2
3 1 9 2 5 6 8 4 7
6 2 8 3 7 4 5 9 1
5 7 4 9 1 8 6 2 3
2 3 7 4 6 5 9 1 8
4 6 1 7 8 9 2 3 5
9 8 5 1 2 3 4 7 6
```

Solution # 112
```
4 7 3 5 2 1 9 6 8
1 2 8 6 3 9 5 4 7
5 9 6 4 7 8 2 3 1
9 6 1 8 5 4 3 7 2
7 3 4 1 9 2 6 8 5
2 8 5 7 6 3 4 1 9
3 1 7 9 4 5 8 2 6
8 4 9 2 1 6 7 5 3
6 5 2 3 8 7 1 9 4
```

Solution # 113
```
7 9 2 4 8 5 3 6 1
8 1 5 9 3 6 7 4 2
4 6 3 1 7 2 5 8 9
2 3 8 6 9 7 4 1 5
9 4 7 8 5 1 6 2 3
6 5 1 2 4 3 8 9 7
3 8 4 7 2 9 1 5 6
5 2 6 3 1 8 9 7 4
1 7 9 5 6 4 2 3 8
```

Solution # 114
```
4 5 8 6 3 7 1 2 9
2 6 7 1 9 8 3 4 5
9 1 3 5 4 2 7 8 6
5 2 9 7 6 4 8 3 1
6 3 4 8 2 1 5 9 7
8 7 1 3 5 9 4 6 2
7 9 5 4 8 6 2 1 3
1 8 2 9 7 3 6 5 4
3 4 6 2 1 5 9 7 8
```

Solution # 115
```
2 8 5 3 7 6 1 4 9
6 9 7 5 1 4 8 2 3
4 3 1 2 8 9 6 7 5
1 2 9 6 4 3 7 5 8
3 7 8 1 5 2 4 9 6
5 4 6 7 9 8 2 3 1
9 5 2 4 6 1 3 8 7
7 1 3 8 2 5 9 6 4
8 6 4 9 3 7 5 1 2
```

Solution # 116
```
5 2 8 4 3 1 7 9 6
3 4 9 5 6 7 2 8 1
1 7 6 9 2 8 5 4 3
9 5 7 1 4 3 8 6 2
4 6 1 8 5 2 9 3 7
8 3 2 6 7 9 4 1 5
2 1 4 3 8 5 6 7 9
6 9 5 7 1 4 3 2 8
7 8 3 2 9 6 1 5 4
```

Solution # 117
```
9 6 3 4 8 5 7 1 2
1 8 4 9 7 2 3 6 5
7 5 2 6 1 3 4 8 9
8 2 6 7 9 1 5 3 4
4 3 7 5 6 8 9 2 1
5 1 9 2 3 4 8 7 6
6 7 1 3 5 9 2 4 8
3 4 5 8 2 6 1 9 7
2 9 8 1 4 7 6 5 3
```

Solution # 118
```
4 3 7 8 2 6 9 5 1
9 2 8 4 1 5 7 3 6
1 6 5 3 7 9 8 4 2
3 1 2 5 9 4 6 8 7
7 9 4 2 6 8 5 1 3
8 5 6 7 3 1 4 2 9
5 7 1 9 4 2 3 6 8
2 4 9 6 8 3 1 7 5
6 8 3 1 5 7 2 9 4
```

Solution # 119
```
3 4 8 7 6 2 1 5 9
9 2 7 1 5 4 6 8 3
6 5 1 9 8 3 7 2 4
1 7 6 4 3 5 2 9 8
5 9 2 6 7 8 3 4 1
4 8 3 2 1 9 5 7 6
8 6 9 5 2 1 4 3 7
7 3 5 8 4 6 9 1 2
2 1 4 3 9 7 8 6 5
```

Solution # 120
```
8 4 2 5 6 1 9 7 3
3 1 9 4 7 2 6 8 5
7 5 6 3 8 9 4 2 1
6 9 3 2 4 7 1 5 8
5 2 7 6 1 8 3 4 9
1 8 4 9 3 5 2 6 7
9 7 5 1 2 4 8 3 6
2 3 1 8 5 6 7 9 4
4 6 8 7 9 3 5 1 2
```

Solution # 121
```
1 8 6 5 3 9 2 4 7
4 3 7 6 2 8 5 9 1
2 5 9 1 7 4 8 6 3
6 9 5 4 8 1 7 3 2
3 7 1 2 5 6 4 8 9
8 4 2 7 9 3 6 1 5
9 2 3 8 4 7 1 5 6
5 1 4 9 6 2 3 7 8
7 6 8 3 1 5 9 2 4
```

Solution # 122
```
9 4 2 6 8 3 5 1 7
6 3 7 1 4 5 2 8 9
8 1 5 2 9 7 4 6 3
1 6 3 5 2 9 8 7 4
7 9 8 4 3 1 6 5 2
5 2 4 7 6 8 9 3 1
3 5 9 8 7 4 1 2 6
4 8 6 3 1 2 7 9 5
2 7 1 9 5 6 3 4 8
```

Solution # 123
```
6 7 1 5 3 8 4 2 9
9 8 3 1 2 4 6 7 5
2 5 4 6 9 7 8 1 3
7 4 8 2 6 5 3 9 1
1 6 2 3 8 9 7 5 4
5 3 9 7 4 1 2 6 8
3 9 6 8 1 2 5 4 7
4 2 5 9 7 3 1 8 6
8 1 7 4 5 6 9 3 2
```

Solution # 124
```
5 7 6 9 1 3 4 8 2
8 2 3 4 6 7 9 5 1
4 9 1 8 5 2 7 6 3
9 3 2 7 8 4 6 1 5
1 6 4 3 2 5 8 9 7
7 8 5 6 9 1 3 2 4
6 1 7 5 3 9 2 4 8
3 5 9 2 4 8 1 7 6
2 4 8 1 7 6 5 3 9
```

Solution # 125
```
2 1 4 6 3 8 9 5 7
9 5 8 4 7 2 6 1 3
7 6 3 9 5 1 4 2 8
8 9 6 1 2 7 3 4 5
1 3 5 8 9 4 2 7 6
4 7 2 5 6 3 1 8 9
5 2 9 7 1 6 8 3 4
3 8 7 2 4 9 5 6 1
6 4 1 3 8 5 7 9 2
```

Solution # 126
```
8 5 1 3 6 4 2 7 9
4 6 7 9 8 2 1 5 3
9 3 2 5 1 7 6 4 8
1 9 6 2 7 8 5 3 4
3 4 5 6 9 1 8 2 7
2 7 8 4 3 5 9 1 6
6 1 4 7 2 9 3 8 5
5 2 9 8 4 3 7 6 1
7 8 3 1 5 6 4 9 2
```

Solution # 127
```
5 8 1 9 4 3 7 2 6
4 6 9 2 7 5 1 3 8
2 7 3 1 8 6 4 9 5
3 4 7 5 2 1 8 6 9
1 9 2 7 6 8 5 4 3
8 5 6 3 9 4 2 1 7
7 2 8 6 1 9 3 5 4
9 3 4 8 5 2 6 7 1
6 1 5 4 3 7 9 8 2
```

Solution # 128
```
7 5 2 9 1 8 3 6 4
8 3 1 5 4 6 2 9 7
6 9 4 7 3 2 8 5 1
2 7 3 4 6 1 5 8 9
4 6 9 8 7 5 1 2 3
5 1 8 3 2 9 4 7 6
1 8 5 6 9 4 7 3 2
3 4 6 2 5 7 9 1 8
9 4 7 2 5 3 6 1 8
```

Solution # 129
```
5 7 9 3 8 4 6 1 2
2 3 6 5 1 7 4 8 9
4 1 8 2 9 6 3 5 7
7 2 1 6 4 3 8 9 5
9 4 3 8 5 1 7 2 6
8 6 5 9 7 2 1 3 4
6 9 4 1 3 5 2 7 8
3 5 7 4 2 8 9 6 1
1 8 2 7 6 9 5 4 3
```

Solution # 130
```
2 1 4 7 3 5 8 9 6
7 8 9 6 2 1 3 4 5
5 3 6 9 8 4 1 2 7
3 9 5 8 1 6 2 7 4
4 6 8 2 7 3 5 1 9
1 7 2 4 5 9 6 8 3
8 4 3 5 9 2 7 6 1
9 2 1 3 6 7 4 5 8
6 5 7 1 4 8 9 3 2
```

Solution # 131
```
5 8 4 9 2 1 6 3 7
9 1 7 5 3 6 2 8 4
6 3 2 7 4 8 1 9 5
2 7 9 6 1 4 8 5 3
4 5 8 2 9 3 7 6 1
3 6 1 8 7 5 4 2 9
8 4 6 3 5 7 9 1 2
7 9 5 1 6 2 3 4 8
1 2 3 4 8 9 5 7 6
```

Solution # 132
```
6 3 9 2 1 4 8 7 5
8 7 1 9 6 5 4 3 2
5 4 2 3 7 8 6 1 9
1 2 6 4 8 7 9 5 3
3 8 4 5 9 1 2 6 7
7 9 5 6 2 3 1 8 4
4 6 7 8 5 9 3 2 1
9 1 8 7 3 2 5 4 6
2 5 3 1 4 6 7 9 8
```

Solution # 133
```
7 5 2 3 6 1 9 4 8
4 6 1 8 9 5 3 7 2
3 8 9 2 7 4 5 6 1
2 4 8 9 5 6 1 3 7
5 9 7 1 2 3 6 8 4
6 1 3 4 8 7 2 9 5
1 7 4 5 3 9 8 2 6
9 2 6 7 1 8 4 5 3
8 3 5 6 4 2 7 1 9
```

Solution # 134
```
4 2 5 9 6 3 1 7 8
7 1 3 4 8 2 6 5 9
8 9 6 1 5 7 4 2 3
5 6 8 2 4 1 9 3 7
2 3 9 6 7 8 5 4 1
1 4 7 5 3 9 2 8 6
6 7 2 8 9 4 3 1 5
9 8 4 3 1 5 7 6 2
3 5 1 7 2 6 8 9 4
```

Solution # 135
```
5 4 3 2 6 9 8 1 7
1 9 7 3 5 8 4 2 6
6 8 2 7 1 4 5 3 9
8 1 5 4 9 6 2 7 3
9 2 4 8 7 3 1 6 5
7 3 6 5 2 1 9 4 8
2 7 9 1 3 5 6 8 4
3 5 8 6 4 2 7 9 1
4 6 1 9 8 7 3 5 2
```

Solution # 136
```
9 4 1 5 7 6 2 3 8
7 8 3 9 2 4 5 6 1
2 6 5 8 1 3 7 4 9
8 5 2 7 3 9 4 1 6
4 7 9 1 6 5 8 2 3
1 3 6 2 4 8 9 7 5
6 9 4 3 5 2 1 8 7
3 1 8 4 9 7 6 5 2
5 2 7 6 8 1 3 9 4
```

Solution # 137
```
2 7 5 3 4 8 1 9 6
1 3 4 9 7 6 2 8 5
8 9 6 1 5 2 4 3 7
4 2 8 7 1 5 3 6 9
3 6 1 8 9 4 7 5 2
7 5 9 6 2 3 8 1 4
5 1 2 4 3 9 6 7 8
9 8 7 2 6 1 5 4 3
6 4 3 5 8 7 9 2 1
```

Solution # 138
```
7 2 3 5 6 9 4 1 8
8 5 1 2 7 4 9 6 3
4 9 6 8 1 3 5 2 7
5 6 2 7 3 1 8 4 9
1 7 9 4 8 6 2 3 5
3 8 4 9 2 5 1 7 6
2 1 8 3 5 7 6 9 4
9 3 5 6 4 2 7 8 1
6 4 7 1 9 8 3 5 2
```

Solution # 139
```
4 2 7 9 1 8 6 5 3
3 9 6 5 4 7 2 8 1
5 1 8 6 3 2 9 4 7
8 5 9 1 6 4 3 7 2
1 7 2 3 8 9 4 6 5
6 4 3 2 7 5 1 9 8
7 6 4 8 2 3 5 1 9
9 3 1 7 5 6 8 2 4
2 8 5 4 9 1 7 3 6
```

Solution # 140
```
1 5 3 7 9 6 4 8 2
9 8 2 4 5 1 3 6 7
4 6 7 3 8 2 5 9 1
6 4 9 5 2 7 1 3 8
7 3 8 9 1 4 2 5 6
5 2 1 6 3 8 7 4 9
8 7 6 2 4 3 9 1 5
3 1 5 8 7 9 6 2 4
2 9 4 1 6 5 8 7 3
```

Solution # 141

```
2 4 9 7 5 1 6 3 8
8 6 7 4 2 3 9 5 1
1 3 5 8 6 9 2 4 7
6 5 2 9 3 7 8 1 4
9 7 1 6 4 8 5 2 3
4 8 3 5 1 2 7 9 6
3 9 8 2 7 4 1 6 5
7 1 6 3 9 5 4 8 2
5 2 4 1 8 6 3 7 9
```

Solution # 142

```
9 4 6 7 2 1 3 8 5
8 2 1 9 5 3 7 6 4
3 7 5 6 4 8 2 9 1
1 3 2 4 6 9 8 5 7
6 5 9 2 8 7 1 4 3
4 8 7 1 3 5 9 2 6
7 6 8 5 1 2 4 3 9
5 9 3 8 7 4 6 1 2
2 1 4 3 9 6 5 7 8
```

Solution # 143

```
3 5 2 1 6 7 4 9 8
9 8 6 5 4 3 7 2 1
1 4 7 8 9 2 5 6 3
6 2 4 3 7 5 8 1 9
8 1 5 4 2 9 6 3 7
7 3 9 6 1 8 2 4 5
4 7 8 2 3 1 9 5 6
2 9 3 7 5 6 1 8 4
5 6 1 9 8 4 3 7 2
```

Solution # 144

```
7 6 9 2 5 8 1 4 3
5 8 3 9 4 1 7 2 6
4 1 2 7 6 3 5 8 9
2 3 8 6 9 5 4 7 1
9 4 1 8 2 7 3 6 5
6 7 5 3 1 4 2 9 8
1 2 6 5 7 9 8 3 4
3 9 4 1 8 2 6 5 7
8 5 7 4 3 6 9 1 2
```

Solution # 145

```
9 4 1 5 6 2 7 3 8
2 5 6 3 7 8 9 4 1
8 3 7 4 1 9 2 5 6
5 7 9 2 4 6 8 1 3
4 1 3 8 5 7 6 9 2
6 2 8 9 3 1 4 7 5
3 9 5 6 8 4 1 2 7
7 8 4 1 2 3 5 6 9
1 6 2 7 9 5 3 8 4
```

Solution # 146

```
6 3 8 7 2 4 1 5 9
9 7 5 3 1 8 4 2 6
4 2 1 9 5 6 3 8 7
1 6 2 5 3 7 8 9 4
3 9 7 8 4 2 5 6 1
8 5 4 6 9 1 2 7 3
2 8 3 4 6 9 7 1 5
7 4 6 1 8 5 9 3 2
5 1 9 2 7 3 6 4 8
```

Solution # 147

```
3 5 6 9 7 1 8 2 4
8 4 9 3 2 6 5 7 1
1 2 7 5 8 4 9 6 3
2 6 5 7 4 8 3 1 9
9 7 3 1 5 2 4 8 6
4 1 8 6 3 9 2 5 7
6 3 1 8 9 5 7 4 2
5 9 2 4 1 7 6 3 8
7 8 4 2 6 3 1 9 5
```

Solution # 148

```
2 3 9 8 6 7 1 5 4
6 5 4 1 9 3 7 2 8
1 7 8 5 4 2 6 3 9
7 6 1 4 3 5 8 9 2
4 8 5 6 2 9 3 1 7
9 2 3 7 8 1 5 4 6
5 9 2 3 7 6 4 8 1
3 4 7 2 1 8 9 6 5
8 1 6 9 5 4 2 7 3
```

Solution # 149

```
5 6 7 2 1 8 4 9 3
2 3 9 6 4 7 8 1 5
4 8 1 9 3 5 2 6 7
3 5 4 7 8 9 1 2 6
6 1 8 4 2 3 5 7 9
9 7 2 1 5 6 3 4 8
8 4 6 3 9 2 7 5 1
1 9 3 5 7 4 6 8 2
7 2 5 8 6 1 9 3 4
```

Solution # 150

```
5 7 1 3 9 6 8 4 2
6 2 8 7 5 4 9 3 1
4 3 9 8 2 1 6 5 7
9 5 2 4 1 8 3 7 6
7 4 3 2 6 5 1 8 9
8 1 6 9 3 7 5 2 4
2 6 4 1 8 3 7 9 5
3 9 5 6 7 2 4 1 8
1 8 7 5 4 9 2 6 3
```

Solution # 151

```
3 1 5 7 6 4 8 9 2
2 9 6 5 8 3 1 7 4
7 8 4 9 2 1 5 6 3
6 2 1 4 3 9 7 5 8
4 3 9 8 5 7 6 2 1
5 7 8 6 1 2 4 3 9
8 5 2 3 4 6 9 1 7
1 6 7 2 9 8 3 4 5
9 4 3 1 7 5 2 8 6
```

Solution # 152

```
6 4 8 3 9 1 5 7 2
3 9 7 6 5 2 4 8 1
1 5 2 4 7 8 6 9 3
5 7 9 2 1 4 8 3 6
4 1 3 8 6 7 9 2 5
8 2 6 9 3 5 1 4 7
2 6 4 5 8 3 7 1 9
7 8 5 1 2 9 3 6 4
9 3 1 7 4 6 2 5 8
```

Solution # 153

```
8 3 1 4 7 9 2 6 5
7 2 9 5 3 6 4 8 1
6 5 4 1 8 2 9 7 3
4 7 5 2 6 1 8 3 9
2 1 8 3 9 5 6 4 7
9 6 3 8 4 7 1 5 2
5 4 2 6 1 3 7 9 8
3 8 7 9 2 4 5 1 6
1 9 6 7 5 8 3 2 4
```

Solution # 154

```
6 3 8 5 9 4 1 7 2
9 7 1 6 2 8 4 5 3
5 4 2 7 3 1 9 8 6
3 2 9 4 1 7 8 6 5
4 6 7 3 8 5 2 1 9
1 8 5 9 6 2 3 4 7
2 1 6 8 7 9 5 3 4
8 5 3 2 4 6 7 9 1
7 9 4 1 5 3 6 2 8
```

Solution # 155

```
8 5 1 3 6 4 2 7 9
7 3 2 5 9 1 4 6 8
9 4 6 8 7 2 1 5 3
2 8 3 7 1 9 5 4 6
4 1 9 6 5 3 8 2 7
5 6 7 4 2 8 3 9 1
6 2 5 1 8 7 9 3 4
3 9 8 2 4 6 7 1 5
1 7 4 9 3 5 6 8 2
```

Solution # 156

```
6 9 3 5 7 8 2 1 4
5 7 1 4 2 3 8 9 6
4 2 8 1 9 6 5 7 3
2 3 5 7 6 4 9 8 1
1 6 7 2 8 9 4 3 5
9 8 4 3 5 1 7 6 2
3 5 2 8 1 7 6 4 9
8 4 6 9 3 5 1 2 7
7 1 9 6 4 2 3 5 8
```

Solution # 157

```
9 8 3 4 7 1 5 6 2
4 2 6 3 8 5 9 1 7
5 7 1 9 2 6 8 4 3
2 6 8 5 1 3 4 7 9
7 1 5 2 4 9 6 3 8
3 9 4 7 6 8 1 2 5
6 5 2 1 9 7 3 8 4
1 3 7 8 5 4 2 9 6
8 4 9 6 3 2 7 5 1
```

Solution # 158

```
5 8 4 7 1 6 3 2 9
6 9 1 3 8 2 4 5 7
3 2 7 5 4 9 6 8 1
8 1 6 4 5 3 7 9 2
9 4 5 1 2 7 8 6 3
2 7 3 9 6 8 1 4 5
1 6 8 2 7 5 9 3 4
7 5 9 8 3 4 2 1 6
4 3 2 6 9 1 5 7 8
```

Solution # 159

```
4 2 7 3 8 1 5 6 9
9 3 5 6 4 7 2 8 1
1 6 8 5 9 2 3 7 4
2 5 9 8 7 3 1 4 6
8 4 3 1 5 6 9 2 7
6 7 1 4 2 9 8 3 5
3 8 6 9 1 4 7 5 2
5 9 2 7 6 8 4 1 3
7 1 4 2 3 5 6 9 8
```

Solution # 160

```
6 2 4 7 5 3 1 8 9
3 8 5 1 4 9 2 7 6
1 7 9 2 6 8 3 5 4
7 9 1 4 3 5 8 6 2
4 6 8 9 1 2 5 3 7
2 5 3 6 8 7 4 9 1
8 4 6 5 7 1 9 2 3
9 3 7 8 2 4 6 1 5
5 1 2 3 9 6 7 4 8
```

Solution # 161

```
8 9 4 7 5 3 2 6 1
7 5 1 8 6 2 3 4 9
6 2 3 4 9 1 5 8 7
2 3 7 9 8 6 1 5 4
4 1 8 3 7 5 6 9 2
5 6 9 1 2 4 8 7 3
3 4 6 5 1 7 9 2 8
9 7 2 6 3 8 4 1 5
1 8 5 2 4 9 7 3 6
```

Solution # 162

```
4 9 3 5 2 8 6 7 1
2 6 8 4 7 1 9 3 5
1 7 5 3 9 6 4 2 8
5 1 9 6 3 7 8 4 2
7 8 6 1 4 2 5 9 3
3 4 2 8 5 9 1 6 7
8 3 4 7 6 5 2 1 9
6 2 1 9 8 3 7 5 4
9 5 7 2 1 4 3 8 6
```

Solution # 163

```
9 2 4 3 8 6 5 1 7
8 7 5 4 9 1 3 6 2
1 6 3 7 2 5 8 9 4
2 9 6 8 5 3 4 7 1
5 3 7 9 1 4 2 8 6
4 1 8 6 7 2 9 3 5
6 5 9 1 4 8 7 2 3
3 8 2 5 6 7 1 4 9
7 4 1 2 3 9 6 5 8
```

Solution # 164

```
4 1 5 2 6 8 7 3 9
9 8 6 3 4 7 1 2 5
7 3 2 1 5 9 6 8 4
1 7 4 9 2 3 8 5 6
5 9 3 8 1 6 2 4 7
6 2 8 4 7 5 9 1 3
3 4 1 7 9 2 5 6 8
2 5 7 6 8 4 3 9 1
8 6 9 5 3 1 4 7 2
```

Solution # 165

```
2 6 4 9 1 7 8 3 5
5 7 9 4 3 8 2 1 6
8 1 3 5 6 2 7 4 9
1 3 7 6 2 9 4 5 8
4 2 8 3 7 5 9 6 1
9 5 6 1 8 4 3 2 7
6 9 2 7 4 1 5 8 3
3 4 5 8 9 6 1 7 2
7 8 1 2 5 3 6 9 4
```

Solution # 166

```
5 1 2 3 9 4 8 7 6
4 6 7 8 1 5 9 3 2
8 3 9 7 6 2 4 5 1
7 8 5 2 4 3 6 1 9
6 2 1 5 8 9 7 4 3
9 4 3 6 7 1 2 8 5
1 5 6 4 2 7 3 9 8
3 7 8 9 5 6 1 2 4
2 9 4 1 3 8 5 6 7
```

Solution # 167

```
1 8 4 5 6 3 9 2 7
6 5 7 2 1 9 8 3 4
2 3 9 4 8 7 5 6 1
4 6 1 8 3 5 2 7 9
9 7 5 1 4 2 3 8 6
8 2 3 9 7 6 4 1 5
7 1 8 3 9 4 6 5 2
3 9 2 6 5 1 7 4 8
5 4 6 7 2 8 1 9 3
```

Solution # 168

```
6 1 9 2 5 7 3 8 4
7 4 3 8 6 9 2 1 5
5 2 8 4 1 3 9 6 7
9 8 5 1 3 6 7 4 2
1 7 4 9 2 8 5 3 6
3 6 2 5 7 4 8 9 1
2 9 6 3 4 5 1 7 8
4 3 1 7 8 2 6 5 9
8 5 7 6 9 1 4 2 3
```

Solution # 169

```
5 7 1 3 8 6 2 4 9
8 4 3 1 2 9 6 5 7
2 6 9 5 7 4 1 3 8
3 9 5 2 6 8 4 7 1
7 2 8 9 4 1 3 6 5
6 1 4 7 3 5 8 9 2
9 8 6 4 5 2 7 1 3
4 5 7 8 1 3 9 2 6
1 3 2 6 9 7 5 8 4
```

Solution # 170

```
5 6 4 3 2 9 8 1 7
3 8 7 1 5 4 9 2 6
9 1 2 6 7 8 4 3 5
2 9 8 5 4 3 6 7 1
7 5 1 8 6 2 3 9 4
4 3 6 7 9 1 5 8 2
6 4 3 2 8 7 1 5 9
8 2 5 9 1 6 7 4 3
1 7 9 4 3 5 2 6 8
```

Solution # 171

```
6 2 7 4 9 5 1 8 3
4 3 9 7 8 1 5 2 6
1 5 8 6 3 2 9 4 7
5 8 4 2 6 3 7 9 1
2 1 3 9 5 7 8 6 4
7 9 6 8 1 4 3 5 2
8 4 1 3 2 9 6 7 5
3 6 2 5 7 8 4 1 9
9 7 5 1 4 6 2 3 8
```

Solution # 172

```
2 4 9 8 6 1 5 7 3
8 3 1 2 7 5 4 9 6
7 5 6 9 4 3 1 8 2
9 8 3 4 1 2 7 6 5
4 1 7 5 8 6 3 2 9
5 6 2 3 9 7 8 4 1
3 7 5 6 2 8 9 1 4
1 2 4 7 5 9 6 3 8
6 9 8 1 3 4 2 5 7
```

Solution # 173

```
4 9 8 1 6 2 5 7 3
1 5 2 4 7 3 6 8 9
6 3 7 8 5 9 2 1 4
2 4 6 9 8 5 7 3 1
8 1 9 7 3 6 4 2 5
5 7 3 2 4 1 9 6 8
7 2 5 3 9 8 1 4 6
3 6 4 5 1 7 8 9 2
9 8 1 6 2 4 3 5 7
```

Solution # 174

```
2 6 8 3 5 4 9 1 7
5 7 9 1 6 8 4 2 3
4 1 3 2 7 9 8 5 6
9 5 2 4 1 3 7 6 8
3 4 6 7 8 2 5 9 1
1 8 7 5 9 6 3 4 2
8 2 4 6 3 5 1 7 9
6 3 1 9 4 7 2 8 5
7 9 5 8 2 1 6 3 4
```

Solution # 175

```
5 7 2 4 8 6 1 3 9
4 6 3 9 1 2 5 8 7
1 9 8 3 5 7 2 4 6
3 5 7 2 9 8 4 6 1
8 2 6 1 3 4 9 7 5
9 4 1 7 6 5 3 2 8
7 3 5 6 2 9 8 1 4
6 1 9 8 4 3 7 5 2
2 8 4 5 7 1 6 9 3
```

Solution # 176
```
2 4 5 7 1 6 9 3 8
9 1 7 3 8 4 6 5 2
6 8 3 2 9 5 7 4 1
7 2 9 6 4 1 3 8 5
8 3 6 5 7 2 4 1 9
1 5 4 9 3 8 2 7 6
4 6 1 8 2 3 5 9 7
5 7 8 4 6 9 1 2 3
3 9 2 1 5 7 8 6 4
```

Solution # 177
```
3 2 1 6 4 7 9 5 8
8 9 4 3 5 1 2 6 7
7 5 6 8 2 9 4 3 1
2 6 7 1 8 4 3 9 5
9 1 5 2 7 3 6 8 4
4 8 3 5 9 6 1 7 2
1 7 9 4 3 8 5 2 6
5 4 8 9 6 2 7 1 3
6 3 2 7 1 5 8 4 9
```

Solution # 178
```
3 1 9 6 4 2 5 8 7
8 4 6 3 5 7 2 1 9
7 2 5 8 1 9 4 3 6
4 6 2 5 8 3 9 7 1
9 3 1 2 7 4 6 5 8
5 7 8 1 9 6 3 2 4
6 5 7 4 3 8 1 9 2
1 8 4 9 2 5 7 6 3
2 9 3 7 6 1 8 4 5
```

Solution # 179
```
8 6 3 4 1 2 9 5 7
2 4 9 5 7 8 3 1 6
5 7 1 3 9 6 4 8 2
9 3 7 2 4 5 1 6 8
6 2 5 8 3 1 7 9 4
1 8 4 9 6 7 2 3 5
4 5 6 1 2 3 8 7 9
3 9 8 7 5 4 6 2 1
7 1 2 6 8 9 5 4 3
```

Solution # 180
```
8 1 6 9 7 4 5 3 2
7 4 3 5 1 2 6 9 8
5 2 9 8 6 3 4 1 7
1 9 5 6 2 8 7 4 3
2 6 4 3 9 7 8 5 1
3 7 8 1 4 5 2 6 9
4 5 1 7 8 9 3 2 6
6 8 2 4 3 1 9 7 5
9 3 7 2 5 6 1 8 4
```

Solution # 181
```
9 6 5 7 8 2 1 4 3
4 8 3 1 9 6 2 5 7
7 2 1 5 3 4 6 8 9
1 5 9 8 6 3 7 2 4
6 7 2 4 5 9 8 3 1
3 4 8 2 7 1 9 6 5
5 1 4 9 2 8 3 7 6
2 9 6 3 4 7 5 1 8
8 3 7 6 1 5 4 9 2
```

Solution # 182
```
6 4 8 3 9 5 7 1 2
7 3 9 8 1 2 6 4 5
5 2 1 6 7 4 9 3 8
3 6 4 1 5 7 8 2 9
9 1 2 4 6 8 5 7 3
8 5 7 9 2 3 4 6 1
4 8 5 2 3 6 1 9 7
1 7 3 5 4 9 2 8 6
2 9 6 7 8 1 3 5 4
```

Solution # 183
```
2 3 1 9 6 5 4 8 7
5 6 8 4 1 7 9 2 3
4 7 9 2 8 3 1 5 6
6 9 2 5 4 8 3 7 1
1 5 7 3 2 6 8 9 4
3 8 4 1 7 9 5 6 2
9 4 6 7 5 1 2 3 8
8 1 3 6 9 2 7 4 5
7 2 5 8 3 4 6 1 9
```

Solution # 184
```
1 8 4 6 9 5 2 7 3
5 2 3 4 7 8 1 6 9
9 7 6 3 1 2 8 4 5
3 5 9 2 8 7 6 1 4
8 6 2 9 4 1 5 3 7
4 1 7 5 3 6 9 2 8
2 9 5 7 6 4 3 8 1
6 4 1 8 5 3 7 9 2
7 3 8 1 2 9 4 5 6
```

Solution # 185
```
9 4 6 2 7 5 3 1 8
7 1 8 3 6 4 2 5 9
2 5 3 9 8 1 4 7 6
4 2 1 6 5 9 8 3 7
3 9 5 8 2 7 6 4 1
6 8 7 4 1 3 9 2 5
5 3 4 1 9 8 7 6 2
1 6 9 7 3 2 5 8 4
8 7 2 5 4 6 1 9 3
```

Solution # 186
```
2 5 3 1 4 6 7 9 8
6 9 4 5 7 8 1 3 2
1 8 7 9 3 2 6 4 5
9 2 6 3 8 1 5 7 4
3 4 8 6 5 7 2 1 9
7 1 5 4 2 9 8 6 3
4 7 1 2 9 5 3 8 6
5 6 9 8 1 3 4 2 7
8 3 2 7 6 4 9 5 1
```

Solution # 187
```
9 4 1 5 7 2 3 8 6
7 6 5 8 4 3 9 1 2
8 3 2 6 1 9 4 5 7
4 1 8 2 3 7 6 9 5
5 7 6 9 8 4 2 3 1
2 9 3 1 6 5 7 4 8
6 2 9 3 5 1 8 7 4
1 8 4 7 9 6 5 2 3
3 5 7 4 2 8 1 6 9
```

Solution # 188
```
1 4 9 3 6 5 7 2 8
6 5 8 9 7 2 3 1 4
2 7 3 8 4 1 5 9 6
7 9 5 2 1 4 6 8 3
3 6 4 7 8 9 2 5 1
8 2 1 5 3 6 4 7 9
4 3 2 1 9 7 8 6 5
5 1 6 4 2 8 9 3 7
9 8 7 6 5 3 1 4 2
```

Solution # 189
```
3 6 4 9 2 1 8 7 5
9 5 8 6 3 7 4 2 1
2 7 1 4 8 5 9 6 3
4 1 2 3 6 9 7 5 8
7 9 5 1 4 8 2 3 6
6 8 3 5 7 2 1 9 4
8 4 9 7 5 3 6 1 2
5 2 7 8 1 6 3 4 9
1 3 6 2 9 4 5 8 7
```

Solution # 190
```
5 8 9 6 1 4 3 7 2
3 6 7 9 5 2 4 8 1
2 1 4 8 7 3 5 6 9
6 4 5 3 2 9 8 1 7
8 9 1 4 6 7 2 3 5
7 2 3 1 8 5 6 9 4
1 5 6 7 4 8 9 2 3
9 7 2 5 3 6 1 4 8
4 3 8 2 9 1 7 5 6
```

Solution # 191
```
9 7 8 3 2 6 5 1 4
3 2 6 4 5 1 9 8 7
5 4 1 9 7 8 2 3 6
8 1 3 2 4 5 6 7 9
7 5 4 6 9 3 8 2 1
2 6 9 8 1 7 4 5 3
6 3 2 7 8 9 1 4 5
1 8 7 5 6 4 3 9 2
4 9 5 1 3 2 7 6 8
```

Solution # 192
```
8 3 2 4 6 7 1 9 5
4 9 7 5 8 1 6 2 3
6 1 5 9 3 2 8 4 7
7 2 1 8 4 3 9 5 6
3 5 8 6 2 9 4 7 1
9 6 4 7 1 5 2 3 8
1 7 9 2 5 6 3 8 4
2 4 3 1 7 8 5 6 9
5 8 6 3 9 4 7 1 2
```

Solution # 193
```
9 4 3 6 2 8 1 5 7
5 8 6 9 1 7 3 2 4
7 1 2 4 5 3 6 9 8
1 6 9 7 3 4 2 8 5
3 2 4 5 8 6 9 7 1
8 7 5 1 9 2 4 3 6
2 9 7 8 6 1 5 4 3
6 3 8 2 4 5 7 1 9
4 5 1 3 7 9 8 6 2
```

Solution # 194
```
6 2 4 8 1 3 9 5 7
8 5 7 2 6 9 4 1 3
1 3 9 4 5 7 2 6 8
7 4 8 9 3 6 1 2 5
5 6 2 1 7 4 3 8 9
3 9 1 5 8 2 6 7 4
9 7 3 6 2 8 5 4 1
4 1 6 7 9 5 8 3 2
2 8 5 3 4 1 7 9 6
```

Solution # 195
```
2 7 5 8 4 1 3 6 9
8 6 3 7 2 9 5 4 1
4 1 9 6 3 5 8 7 2
6 3 8 2 5 7 9 1 4
1 5 7 4 9 6 2 3 8
9 4 2 1 8 3 6 5 7
3 2 1 9 6 4 7 8 5
5 8 4 3 7 2 1 9 6
7 9 6 5 1 8 4 2 3
```

Solution # 196
```
5 1 4 6 2 3 7 8 9
3 2 7 8 5 9 1 6 4
8 9 6 1 7 4 5 3 2
6 5 1 2 9 7 3 4 8
9 3 8 5 4 1 6 2 7
7 4 2 3 6 8 9 1 5
1 6 5 9 8 2 4 7 3
2 7 3 4 1 5 8 9 6
4 8 9 7 3 6 2 5 1
```

Solution # 197
```
2 3 1 6 4 8 5 9 7
8 9 7 1 5 2 4 6 3
6 4 5 9 7 3 8 2 1
5 6 2 3 8 4 1 7 9
7 8 9 2 1 6 3 4 5
3 1 4 7 9 5 6 8 2
4 2 3 5 6 7 9 1 8
1 5 8 4 2 9 7 3 6
9 7 6 8 3 1 2 5 4
```

Solution # 198
```
2 3 9 7 4 6 8 5 1
8 7 6 2 5 1 4 9 3
5 1 4 9 3 8 6 2 7
7 6 1 4 2 5 9 3 8
3 4 8 1 7 9 5 6 2
9 5 2 6 8 3 1 7 4
4 2 5 8 9 7 3 1 6
6 9 7 3 1 4 2 8 5
1 8 3 5 6 2 7 4 9
```

Solution # 199
```
8 3 2 4 6 7 5 1 9
4 6 1 5 9 3 2 7 8
5 9 7 8 2 1 4 6 3
2 5 6 3 4 9 7 8 1
3 7 4 2 1 8 6 9 5
1 8 9 7 5 6 3 2 4
6 1 5 9 7 4 8 3 2
9 4 3 6 8 2 1 5 7
7 2 8 1 3 5 9 4 6
```

Solution # 200
```
5 6 8 7 2 4 1 3 9
1 3 7 8 5 9 4 6 2
2 4 9 1 3 6 5 8 7
4 8 6 9 7 2 3 1 5
7 5 3 6 8 1 9 2 4
9 1 2 3 4 5 8 7 6
3 9 5 2 1 7 6 4 8
6 2 1 4 9 8 7 5 3
8 7 4 5 6 3 2 9 1
```

Solution # 201
```
8 2 5 4 3 7 6 9 1
6 7 3 5 9 1 4 8 2
4 1 9 8 2 6 3 5 7
1 5 7 6 8 3 2 4 9
2 9 8 7 5 4 1 3 6
3 4 6 9 1 2 8 7 5
9 3 2 1 7 8 5 6 4
5 6 1 3 4 9 7 2 8
7 8 4 2 6 5 9 1 3
```

Solution # 202
```
4 1 7 3 8 6 2 5 9
8 2 9 4 5 7 3 1 6
6 5 3 1 9 2 8 4 7
7 8 2 6 3 1 5 9 4
1 3 5 9 2 4 6 7 8
9 4 6 5 7 8 1 2 3
5 7 8 2 6 9 4 3 1
2 6 4 7 1 3 9 8 5
3 9 1 8 4 5 7 6 2
```

Solution # 203
```
2 1 9 8 5 7 4 3 6
5 3 8 9 6 4 2 7 1
4 6 7 2 1 3 5 8 9
6 8 3 7 4 5 1 9 2
9 2 4 1 3 8 6 5 7
1 7 5 6 2 9 3 4 8
7 4 6 3 9 2 8 1 5
8 5 1 4 7 6 9 2 3
3 9 2 5 8 1 7 6 4
```

Solution # 204
```
4 5 3 7 2 1 8 9 6
2 1 6 9 5 8 7 3 4
9 8 7 6 3 4 2 1 5
3 4 9 8 6 7 1 5 2
6 2 8 1 9 5 4 7 3
1 7 5 3 4 2 6 8 9
5 6 1 2 8 9 3 4 7
8 3 4 5 7 6 9 2 1
7 9 2 4 1 3 5 6 8
```

Solution # 205
```
4 5 3 6 1 7 8 2 9
9 1 8 3 2 4 6 7 5
6 2 7 9 5 8 4 3 1
5 7 9 4 6 3 1 8 2
8 3 1 7 9 2 5 4 6
2 4 6 1 8 5 3 9 7
1 6 4 8 7 9 2 5 3
3 9 5 2 4 6 7 1 8
7 8 2 5 3 1 9 6 4
```

Solution # 206
```
8 4 6 3 1 5 9 2 7
2 5 3 9 4 7 6 8 1
9 1 7 8 2 6 5 4 3
1 8 2 7 6 3 4 5 9
3 9 5 4 8 2 1 7 6
7 6 4 1 5 9 8 3 2
4 7 9 6 3 8 2 1 5
5 3 1 2 9 4 7 6 8
6 2 8 5 7 1 3 9 4
```

Solution # 207
```
9 3 1 4 8 5 7 6 2
2 4 7 3 1 6 8 9 5
6 5 8 2 9 7 4 3 1
8 2 6 7 4 9 1 5 3
1 7 4 5 2 3 6 8 9
5 9 3 8 6 1 2 4 7
3 1 5 6 7 8 9 2 4
4 6 9 1 5 2 3 7 8
7 8 2 9 3 4 5 1 6
```

Solution # 208
```
8 3 9 7 5 1 4 6 2
2 1 4 6 9 8 7 3 5
5 7 6 4 3 2 1 8 9
4 9 1 2 6 3 8 5 7
6 8 7 5 4 9 3 2 1
3 5 2 1 8 7 6 9 4
7 2 8 3 1 5 9 4 6
9 6 5 8 7 4 2 1 3
1 4 3 9 2 6 5 7 8
```

Solution # 209
```
1 4 2 7 3 5 6 8 9
5 6 9 8 4 1 2 7 3
8 3 7 2 9 6 4 1 5
7 5 6 4 2 8 9 3 1
2 9 3 1 5 7 8 6 4
4 1 8 3 6 9 5 2 7
3 7 4 5 8 2 1 9 6
9 2 5 6 1 3 7 4 8
6 8 1 9 7 4 3 5 2
```

Solution # 210
```
2 1 3 8 5 6 9 4 7
6 4 8 1 9 7 2 5 3
9 5 7 3 2 4 1 6 8
5 3 4 6 8 1 7 9 2
1 7 6 2 4 9 8 3 5
8 2 9 7 3 5 6 1 4
4 9 1 5 7 8 3 2 6
7 6 2 4 1 3 5 8 9
3 8 5 9 6 2 4 7 1
```

Solution # 211
```
1 5 6 4 2 7 3 9 8
3 2 7 8 9 6 1 4 5
9 4 8 1 5 3 6 7 2
7 1 3 9 6 5 2 8 4
4 6 2 7 1 8 5 3 9
8 9 5 3 4 2 7 1 6
5 3 1 6 8 4 9 2 7
6 7 4 2 3 9 8 5 1
2 8 9 5 7 1 4 6 3
```

Solution # 212
```
8 7 4 1 3 5 2 6 9
6 9 1 4 2 8 5 3 7
2 3 5 6 9 7 8 4 1
9 2 8 3 5 1 4 7 6
3 1 6 2 7 4 9 8 5
4 5 7 9 8 6 1 2 3
5 8 2 7 1 3 6 9 4
7 6 9 5 4 2 3 1 8
1 4 3 8 6 9 7 5 2
```

Solution # 213
```
8 9 5 4 2 3 1 7 6
4 3 1 6 9 7 2 5 8
7 6 2 1 5 8 9 3 4
1 7 4 2 3 9 8 6 5
5 2 6 8 4 1 3 9 7
9 8 3 5 7 6 4 2 1
6 5 8 3 1 2 7 4 9
3 4 7 9 8 5 6 1 2
2 1 9 7 6 4 5 8 3
```

Solution # 214
```
7 2 5 8 4 9 1 3 6
4 6 9 2 3 1 7 8 5
3 8 1 5 6 7 4 9 2
5 4 3 7 9 8 6 2 1
9 1 6 4 2 3 8 5 7
8 7 2 1 5 6 3 4 9
2 9 7 6 8 4 5 1 3
6 3 8 9 1 5 2 7 4
1 5 4 3 7 2 9 6 8
```

Solution # 215
```
3 9 2 4 8 7 5 1 6
4 5 7 1 6 2 8 3 9
6 8 1 5 9 3 2 4 7
5 2 9 3 7 1 4 6 8
8 7 3 9 4 6 1 5 2
1 4 6 2 5 8 7 9 3
2 3 5 7 1 9 6 8 4
9 1 8 6 2 4 3 7 5
7 6 4 8 3 5 9 2 1
```

Solution # 216
```
6 4 3 7 2 9 8 5 1
5 9 7 3 1 8 6 2 4
1 2 8 6 5 4 9 7 3
9 3 6 8 7 2 4 1 5
7 5 4 9 6 1 3 8 2
2 8 1 4 3 5 7 6 9
4 6 2 1 8 3 5 9 7
8 1 9 5 4 7 2 3 6
3 7 5 2 9 6 1 4 8
```

Solution # 217
```
2 3 7 6 8 4 9 1 5
5 4 1 7 9 3 8 2 6
6 9 8 1 2 5 7 4 3
7 5 4 8 3 1 2 6 9
3 2 9 4 7 6 1 5 8
1 8 6 2 5 9 4 3 7
4 1 5 9 6 8 3 7 2
8 7 3 5 1 2 6 9 4
9 6 2 3 4 7 5 8 1
```

Solution # 218
```
2 5 1 7 4 3 6 8 9
4 9 7 6 8 5 2 1 3
8 6 3 2 1 9 7 4 5
6 2 8 4 9 7 5 3 1
3 1 4 8 5 6 9 2 7
5 7 9 3 2 1 8 6 4
1 3 6 9 7 2 4 5 8
7 4 2 5 3 8 1 9 6
9 8 5 1 6 4 3 7 2
```

Solution # 219
```
5 7 2 4 8 9 6 3 1
9 1 8 2 3 6 5 7 4
3 6 4 7 5 1 8 9 2
2 9 7 1 4 8 3 5 6
8 4 1 3 6 5 9 2 7
6 5 3 9 2 7 4 1 8
7 8 6 5 9 2 1 4 3
4 2 5 8 1 3 7 6 9
1 3 9 6 7 4 2 8 5
```

Solution # 220
```
6 2 8 7 1 4 3 5 9
1 7 9 2 5 3 8 6 4
5 4 3 6 9 8 1 2 7
3 6 2 8 7 5 4 9 1
9 8 4 3 2 1 6 7 5
7 1 5 9 4 6 2 3 8
4 9 6 5 8 2 7 1 3
8 3 7 1 6 9 5 4 2
2 5 1 4 3 7 9 8 6
```

Solution # 221
```
9 5 1 4 2 8 6 3 7
4 8 7 9 3 6 2 1 5
3 6 2 7 5 1 8 9 4
1 2 5 3 7 4 9 8 6
8 9 4 6 1 2 7 5 3
7 3 6 8 9 5 1 4 2
6 1 8 5 4 7 3 2 9
5 7 3 2 8 9 4 6 1
2 4 9 1 6 3 5 7 8
```

Solution # 222
```
4 2 7 8 9 3 1 5 6
1 6 8 2 4 5 3 9 7
3 9 5 1 6 7 4 8 2
6 3 1 5 7 4 8 2 9
5 8 2 9 3 1 7 6 4
7 4 9 6 2 8 5 1 3
9 5 3 4 1 6 2 7 8
2 1 4 7 8 9 6 3 5
8 7 6 3 5 2 9 4 1
```

Solution # 223
```
8 3 7 2 1 9 4 5 6
1 9 6 5 4 3 2 7 8
5 4 2 8 7 6 9 3 1
7 1 8 9 6 5 3 4 2
9 5 3 1 2 4 8 6 7
6 2 4 3 8 7 5 1 9
3 7 5 6 9 2 1 8 4
4 8 9 7 3 1 6 2 5
2 6 1 4 5 8 7 9 3
```

Solution # 224
```
9 1 5 2 3 7 6 4 8
4 3 8 5 9 6 1 2 7
7 6 2 1 4 8 3 5 9
5 8 7 3 6 1 4 9 2
1 4 3 9 7 2 5 8 6
6 2 9 4 8 5 7 1 3
8 7 1 6 2 4 9 3 5
3 5 6 8 1 9 2 7 4
2 9 4 7 5 3 8 6 1
```

Solution # 225
```
8 1 3 9 4 5 6 2 7
6 2 9 7 1 3 5 8 4
5 7 4 8 2 6 1 9 3
4 5 8 3 6 9 2 7 1
9 6 2 4 7 1 3 5 8
7 3 1 5 8 2 4 6 9
3 4 7 6 5 8 9 1 2
1 9 6 2 3 7 8 4 5
2 8 5 1 9 4 7 3 6
```

Solution # 226
```
3 8 9 7 6 1 4 2 5
2 7 5 8 9 4 3 1 6
6 4 1 2 3 5 7 8 9
7 9 8 3 4 6 2 5 1
1 2 3 5 7 8 9 6 4
4 5 6 1 2 9 8 7 3
8 6 4 9 5 7 1 3 2
9 3 7 6 1 2 5 4 8
5 1 2 4 8 3 6 9 7
```

Solution # 227
```
7 3 6 9 4 8 5 1 2
4 1 5 7 3 2 8 6 9
2 8 9 6 1 5 7 3 4
8 4 7 5 2 1 3 9 6
3 9 2 8 6 7 4 5 1
5 6 1 3 9 4 2 8 7
6 5 8 2 7 9 1 4 3
9 2 4 1 8 3 6 7 5
1 7 3 4 5 6 9 2 8
```

Solution # 228
```
4 7 9 6 2 1 3 8 5
6 8 3 9 5 7 1 2 4
5 2 1 3 8 4 9 6 7
3 9 2 1 7 6 5 4 8
1 5 4 8 9 2 6 7 3
7 6 8 4 3 5 2 9 1
8 1 7 5 6 9 4 3 2
9 3 5 2 4 8 7 1 6
2 4 6 7 1 3 8 5 9
```

Solution # 229
```
3 2 8 4 7 1 5 9 6
5 1 6 9 2 8 3 4 7
9 7 4 6 5 3 8 1 2
1 9 7 3 8 2 4 6 5
2 4 3 1 6 5 9 7 8
8 6 5 7 4 9 1 2 3
7 5 2 8 9 4 6 3 1
6 3 9 5 1 7 2 8 4
4 8 1 2 3 6 7 5 9
```

Solution # 230
```
9 4 6 2 3 1 5 8 7
7 8 3 4 6 5 1 9 2
1 2 5 7 9 8 4 3 6
4 1 7 8 2 6 3 5 9
3 6 8 1 5 9 7 2 4
5 9 2 3 7 4 6 1 8
2 3 4 5 8 7 9 6 1
6 5 1 9 4 2 8 7 3
8 7 9 6 1 3 2 4 5
```

Solution # 231
```
5 7 9 6 4 2 8 3 1
8 4 1 5 9 3 7 2 6
2 3 6 1 7 8 5 9 4
6 9 5 7 8 4 2 1 3
7 2 4 3 1 5 9 6 8
1 8 3 9 2 6 4 7 5
4 5 7 2 6 1 3 8 9
3 6 2 8 5 9 1 4 7
9 1 8 4 3 7 6 5 2
```

Solution # 232
```
7 9 8 3 6 5 1 2 4
6 5 2 9 4 1 3 7 8
3 4 1 8 2 7 6 5 9
8 1 3 7 9 6 5 4 2
4 6 7 2 5 8 9 3 1
9 2 5 4 1 3 7 8 6
5 8 4 6 3 9 2 1 7
2 3 6 1 7 4 8 9 5
1 7 9 5 8 2 4 6 3
```

Solution # 233
```
1 3 5 8 2 6 4 7 9
9 6 7 3 4 1 5 2 8
4 8 2 9 7 5 3 6 1
8 7 9 5 1 4 6 3 2
3 5 1 6 8 2 7 9 4
2 4 6 7 9 3 1 8 5
7 9 3 4 5 8 2 1 6
5 2 8 1 6 7 9 4 3
6 1 4 2 3 9 8 5 7
```

Solution # 234
```
3 9 1 2 8 7 4 6 5
8 6 7 3 5 4 1 9 2
4 5 2 6 9 1 7 3 8
7 3 8 9 4 2 6 5 1
1 4 5 8 3 6 9 2 7
9 2 6 1 7 5 3 8 4
6 1 4 5 2 3 8 7 9
2 7 9 4 6 8 5 1 3
5 8 3 7 1 9 2 4 6
```

Solution # 235
```
2 3 1 9 6 4 5 7 8
4 6 8 5 3 7 9 2 1
9 7 5 8 2 1 6 4 3
1 2 3 6 7 5 4 8 9
6 9 4 2 1 8 7 3 5
5 8 7 3 4 9 2 1 6
8 4 2 1 5 6 3 9 7
3 1 6 7 9 2 8 5 4
7 5 9 4 8 3 1 6 2
```

Solution # 236
```
1 8 2 6 3 7 4 9 5
4 7 5 9 8 1 2 3 6
6 9 3 5 2 4 1 7 8
9 5 6 3 4 2 8 1 7
8 3 7 1 6 5 9 4 2
2 4 1 7 9 8 5 6 3
7 1 9 2 5 3 6 8 4
5 6 8 4 7 9 3 2 1
3 2 4 8 1 6 7 5 9
```

Solution # 237
```
2 6 4 7 3 5 8 1 9
1 5 8 4 9 6 2 3 7
3 7 9 2 1 8 5 6 4
8 4 6 9 2 3 1 7 5
7 2 1 5 6 4 3 9 8
5 9 3 8 7 1 6 4 2
4 1 7 6 8 2 9 5 3
9 3 2 1 5 7 4 8 6
6 8 5 3 4 9 7 2 1
```

Solution # 238
```
6 3 1 4 7 2 5 9 8
4 5 2 8 9 1 6 3 7
8 7 9 6 3 5 4 1 2
2 4 3 5 6 7 1 8 9
7 1 5 2 8 9 3 4 6
9 6 8 1 4 3 7 2 5
1 2 7 9 5 4 8 6 3
3 8 4 7 2 6 9 5 1
5 9 6 3 1 8 2 7 4
```

Solution # 239
```
8 5 1 6 9 3 2 4 7
9 6 3 2 4 7 1 5 8
2 4 7 5 8 1 6 9 3
6 1 9 4 3 2 8 7 5
5 3 8 7 6 9 4 2 1
4 7 2 1 5 8 3 6 9
3 8 6 9 7 4 5 1 2
7 2 4 3 1 5 9 8 6
1 9 5 8 2 6 7 3 4
```

Solution # 240
```
5 3 1 8 2 6 4 7 9
6 2 7 3 4 9 8 5 1
8 9 4 1 5 7 3 2 6
3 1 5 4 8 2 6 9 7
2 4 6 7 9 1 5 8 3
7 8 9 5 6 3 2 1 4
1 7 2 6 3 5 9 4 8
9 6 8 2 7 4 1 3 5
4 5 3 9 1 8 7 6 2
```

Solution # 241
```
6 4 1 3 2 8 7 5 9
3 7 5 6 9 4 1 2 8
2 8 9 1 5 7 6 4 3
5 1 3 7 8 6 2 9 4
4 2 7 9 1 3 8 6 5
9 6 8 2 4 5 3 7 1
7 9 4 8 3 2 5 1 6
1 3 2 5 6 9 4 8 7
8 5 6 4 7 1 9 3 2
```

Solution # 242
```
7 4 6 3 8 2 5 1 9
2 5 1 7 9 6 4 3 8
8 9 3 4 1 5 2 7 6
3 8 2 9 6 1 7 5 4
9 7 5 8 2 4 1 6 3
1 6 4 5 3 7 8 9 2
4 2 9 1 7 3 6 8 5
6 1 8 2 5 9 3 4 7
5 3 7 6 4 8 9 2 1
```

Solution # 243
```
2 8 5 4 7 6 9 3 1
9 6 1 2 5 3 8 4 7
7 3 4 9 1 8 5 6 2
8 2 7 6 9 1 4 5 3
4 5 2 3 6 9 1 7 8
3 1 6 7 8 5 2 9 4
5 4 7 8 9 1 6 2 3
1 9 3 6 4 2 7 8 5
6 2 8 5 3 7 4 1 9
```

Solution # 244
```
5 3 9 4 7 1 8 6 2
7 2 8 3 5 6 4 9 1
6 4 1 2 8 9 5 7 3
9 6 4 1 3 5 7 2 8
1 7 5 9 2 8 3 4 6
3 8 2 6 4 7 1 5 9
4 1 3 7 9 2 6 8 5
2 5 6 8 1 4 9 3 7
8 9 7 5 6 3 2 1 4
```

Solution # 245
```
2 5 8 3 1 4 9 6 7
3 4 6 9 2 7 1 8 5
7 9 1 6 5 8 3 4 2
8 3 2 4 7 9 6 5 1
5 6 4 2 8 1 7 3 9
1 7 9 5 6 3 4 2 8
4 1 7 8 3 2 5 9 6
6 8 3 7 9 5 2 1 4
9 2 5 1 4 6 8 7 3
```

Solution # 246
```
1 9 4 | 3 2 6 | 7 5 8
2 6 7 | 8 5 9 | 3 1 4
5 8 3 | 4 1 7 | 6 9 2
------+-------+------
3 4 9 | 7 8 1 | 5 2 6
7 2 6 | 5 3 4 | 9 8 1
8 5 1 | 9 6 2 | 4 7 3
------+-------+------
9 7 8 | 2 4 3 | 1 6 5
6 3 5 | 1 9 8 | 2 4 7
4 1 2 | 6 7 5 | 8 3 9
```

Solution # 247
```
6 5 1 | 9 8 7 | 2 3 4
2 9 4 | 5 6 3 | 1 7 8
3 7 8 | 1 2 4 | 9 6 5
------+-------+------
7 1 2 | 6 9 8 | 5 4 3
9 8 5 | 3 4 2 | 6 1 7
4 6 3 | 7 5 1 | 8 9 2
------+-------+------
8 3 9 | 2 7 6 | 4 5 1
5 4 7 | 8 1 9 | 3 2 6
1 2 6 | 4 3 5 | 7 8 9
```

Solution # 248
```
7 5 1 | 6 8 4 | 3 9 2
4 3 9 | 1 2 7 | 6 8 5
8 2 6 | 9 5 3 | 7 1 4
------+-------+------
6 8 5 | 4 7 2 | 1 3 9
2 9 3 | 8 1 6 | 5 4 7
1 4 7 | 3 9 5 | 2 6 8
------+-------+------
3 6 2 | 7 4 9 | 8 5 1
5 1 8 | 2 6 9 | 4 7 3
9 7 4 | 5 3 1 | 8 2 6
```

Solution # 249
```
9 4 1 | 6 5 2 | 7 8 3
2 7 8 | 4 3 1 | 5 6 9
6 5 3 | 7 9 8 | 1 4 2
------+-------+------
4 2 7 | 8 1 9 | 6 3 5
3 8 5 | 2 7 6 | 4 9 1
1 6 9 | 5 4 3 | 2 7 8
------+-------+------
8 9 2 | 1 6 4 | 3 5 7
5 1 4 | 3 8 7 | 9 2 6
7 3 6 | 9 2 5 | 8 1 4
```

Solution # 250
```
2 7 6 | 3 1 8 | 9 5 4
9 3 5 | 6 4 2 | 8 7 1
4 8 1 | 5 9 7 | 3 6 2
------+-------+------
1 9 3 | 7 2 4 | 6 8 5
6 4 7 | 8 3 5 | 2 1 9
5 2 8 | 9 6 1 | 4 3 7
------+-------+------
3 6 4 | 1 7 9 | 5 2 8
7 5 2 | 4 8 6 | 1 9 3
8 1 9 | 2 5 3 | 7 4 6
```

Solution # 251
```
8 5 1 | 7 4 2 | 3 9 6
4 7 3 | 9 6 1 | 2 8 5
6 2 9 | 3 5 8 | 7 4 1
------+-------+------
9 3 6 | 1 8 7 | 4 5 2
7 1 5 | 2 9 4 | 8 6 3
2 4 8 | 6 3 5 | 9 1 7
------+-------+------
5 9 7 | 4 1 3 | 6 2 8
1 6 2 | 8 7 9 | 5 3 4
3 8 4 | 5 2 6 | 1 7 9
```

Solution # 252
```
3 7 1 | 2 9 5 | 4 8 6
5 9 4 | 8 6 7 | 3 1 2
8 2 6 | 4 3 1 | 7 9 5
------+-------+------
9 4 3 | 6 1 8 | 5 2 7
7 8 2 | 9 5 4 | 6 3 1
1 6 5 | 7 2 3 | 9 4 8
------+-------+------
2 1 7 | 5 4 9 | 8 6 3
4 3 8 | 1 7 6 | 2 5 9
6 5 9 | 3 8 2 | 1 7 4
```

Solution # 253
```
2 9 8 | 6 7 5 | 3 1 4
5 1 4 | 3 2 9 | 8 6 7
7 3 6 | 4 1 8 | 9 2 5
------+-------+------
8 7 9 | 1 5 3 | 2 4 6
6 5 3 | 9 4 2 | 1 7 8
1 4 2 | 7 8 6 | 5 3 9
------+-------+------
4 8 5 | 2 3 7 | 6 9 1
3 6 1 | 5 9 4 | 7 8 2
9 2 7 | 8 6 1 | 4 5 3
```

Solution # 254
```
1 9 6 | 4 8 2 | 3 7 5
3 8 4 | 7 5 6 | 2 1 9
7 5 2 | 1 9 3 | 6 4 8
------+-------+------
6 7 3 | 5 4 8 | 9 2 1
8 4 9 | 2 7 1 | 5 6 3
5 2 1 | 6 3 9 | 4 8 7
------+-------+------
4 1 5 | 3 6 7 | 8 9 2
9 6 7 | 8 2 5 | 1 3 4
2 3 8 | 9 1 4 | 7 5 6
```

Solution # 255
```
6 2 4 | 8 3 9 | 1 5 7
9 8 3 | 7 1 5 | 2 6 4
1 7 5 | 4 2 6 | 8 9 3
------+-------+------
8 4 2 | 1 9 7 | 6 3 5
7 5 1 | 6 8 3 | 4 2 9
3 6 9 | 5 4 2 | 7 1 8
------+-------+------
4 9 8 | 3 6 1 | 5 7 2
5 3 6 | 2 7 8 | 9 4 1
2 1 7 | 9 5 4 | 3 8 6
```

Solution # 256
```
3 5 1 | 7 4 6 | 2 9 8
2 4 8 | 9 5 3 | 6 7 1
6 9 7 | 8 2 1 | 4 3 5
------+-------+------
4 7 9 | 1 8 2 | 3 5 6
1 2 6 | 5 3 9 | 7 8 4
8 3 5 | 6 7 4 | 9 1 2
------+-------+------
5 8 2 | 3 6 7 | 1 4 9
7 1 4 | 2 9 5 | 8 6 3
9 6 3 | 4 1 8 | 5 2 7
```

Solution # 257
```
7 2 8 | 6 9 5 | 3 4 1
4 5 1 | 3 2 7 | 9 6 8
6 9 3 | 1 4 8 | 2 7 5
------+-------+------
5 8 6 | 4 7 3 | 1 2 9
1 7 2 | 8 5 9 | 6 3 4
9 3 4 | 2 6 1 | 8 5 7
------+-------+------
2 4 7 | 9 1 6 | 5 8 3
3 6 9 | 5 8 4 | 7 1 2
8 1 5 | 7 3 2 | 4 9 6
```

Solution # 258
```
5 7 4 | 9 6 1 | 3 8 2
2 1 6 | 5 3 8 | 9 7 4
8 3 9 | 4 7 2 | 1 6 5
------+-------+------
9 8 1 | 6 4 3 | 2 5 7
7 4 2 | 8 9 5 | 6 3 1
6 5 3 | 2 1 7 | 4 9 8
------+-------+------
1 6 5 | 7 2 9 | 8 4 3
4 2 8 | 3 5 6 | 7 1 9
3 9 7 | 1 8 4 | 5 2 6
```

Solution # 259
```
3 7 5 | 8 2 4 | 9 1 6
4 8 6 | 3 9 1 | 2 7 5
1 2 9 | 5 7 6 | 8 3 4
------+-------+------
2 6 1 | 9 4 7 | 3 5 8
7 4 3 | 2 8 5 | 6 9 1
9 5 8 | 1 6 3 | 4 2 7
------+-------+------
6 1 2 | 7 3 8 | 5 4 9
8 9 7 | 4 5 2 | 1 6 3
5 3 4 | 6 1 9 | 7 8 2
```

Solution # 260
```
4 9 5 | 8 7 1 | 3 6 2
3 2 8 | 9 6 5 | 1 4 7
1 6 7 | 4 3 2 | 9 8 5
------+-------+------
6 5 3 | 1 4 9 | 2 7 8
2 4 1 | 7 8 3 | 6 5 9
7 8 9 | 5 2 6 | 4 3 1
------+-------+------
8 7 6 | 2 1 4 | 5 9 3
9 1 4 | 3 5 8 | 7 2 6
5 3 2 | 6 9 7 | 8 1 4
```

Solution # 261
```
8 4 3 | 7 6 1 | 2 5 9
6 9 1 | 8 5 2 | 4 7 3
7 2 5 | 9 4 3 | 6 8 1
------+-------+------
9 3 7 | 4 2 6 | 8 1 5
5 6 4 | 3 1 8 | 7 9 2
2 1 8 | 5 7 9 | 3 6 4
------+-------+------
1 8 6 | 2 9 4 | 5 3 7
3 7 2 | 1 8 5 | 9 4 6
4 5 9 | 6 3 7 | 1 2 8
```

Solution # 262
```
1 3 7 | 5 4 2 | 8 6 9
5 6 9 | 7 8 3 | 2 1 4
4 8 2 | 9 6 1 | 5 3 7
------+-------+------
6 9 4 | 2 1 7 | 3 8 5
3 5 1 | 8 9 6 | 7 4 2
7 2 8 | 4 3 5 | 6 9 1
------+-------+------
2 1 6 | 3 5 9 | 4 7 8
8 7 3 | 1 2 4 | 9 5 6
9 4 5 | 6 7 8 | 1 2 3
```

Solution # 263
```
5 6 2 | 3 1 4 | 7 9 8
9 4 8 | 7 2 5 | 1 6 3
3 1 7 | 6 9 8 | 2 4 5
------+-------+------
6 3 4 | 8 5 2 | 9 1 7
2 7 9 | 1 3 6 | 8 5 4
1 8 5 | 4 7 9 | 6 3 2
------+-------+------
4 2 6 | 9 8 3 | 5 7 1
8 9 1 | 5 4 7 | 3 2 6
7 5 3 | 2 6 1 | 4 8 9
```

Solution # 264
```
3 8 5 | 2 7 1 | 6 4 9
4 6 2 | 5 9 3 | 8 7 1
7 1 9 | 6 4 8 | 5 2 3
------+-------+------
2 3 6 | 7 1 5 | 4 9 8
8 4 1 | 9 2 6 | 3 5 7
9 5 7 | 3 8 4 | 2 1 6
------+-------+------
5 2 3 | 1 6 9 | 7 8 4
1 7 8 | 4 3 2 | 9 6 5
6 9 4 | 8 5 7 | 1 3 2
```

Solution # 265
```
8 5 1 | 9 2 7 | 3 6 4
6 9 4 | 3 1 8 | 2 5 7
2 7 3 | 6 4 5 | 9 1 8
------+-------+------
3 2 6 | 1 7 4 | 5 8 9
7 4 9 | 5 8 3 | 6 2 1
5 1 8 | 2 6 9 | 4 7 3
------+-------+------
1 8 5 | 4 3 2 | 7 9 6
9 3 7 | 8 5 6 | 1 4 2
4 6 2 | 7 9 1 | 8 3 5
```

Solution # 266
```
5 9 2 | 7 8 6 | 1 3 4
1 4 8 | 2 3 9 | 6 5 7
7 3 6 | 5 4 1 | 8 2 9
------+-------+------
2 8 3 | 1 6 7 | 4 9 5
4 1 7 | 9 5 8 | 3 6 2
6 5 9 | 3 2 4 | 7 8 1
------+-------+------
3 7 1 | 8 9 5 | 2 4 6
9 2 4 | 6 1 3 | 5 7 8
8 6 5 | 4 7 2 | 9 1 3
```

Solution # 267
```
9 4 6 | 3 5 7 | 1 8 2
3 5 8 | 1 6 2 | 4 9 7
2 1 7 | 8 4 9 | 5 6 3
------+-------+------
1 9 3 | 5 2 6 | 8 7 4
6 8 5 | 4 7 1 | 3 2 9
7 2 4 | 9 8 3 | 6 5 1
------+-------+------
8 3 9 | 7 1 5 | 2 4 6
5 7 2 | 6 3 4 | 9 1 8
4 6 1 | 2 9 8 | 7 3 5
```

Solution # 268
```
5 8 3 | 9 2 7 | 4 1 6
6 7 9 | 3 4 1 | 8 5 2
4 2 1 | 6 5 8 | 9 3 7
------+-------+------
2 3 5 | 4 1 9 | 7 6 8
9 6 8 | 2 7 5 | 1 4 3
1 4 7 | 8 3 6 | 2 9 5
------+-------+------
7 5 4 | 1 6 2 | 3 8 9
8 1 6 | 7 9 3 | 5 2 4
3 9 2 | 5 8 4 | 6 7 1
```

Solution # 269
```
1 6 8 | 5 4 2 | 7 9 3
4 7 5 | 3 1 9 | 2 6 8
3 9 2 | 6 7 8 | 5 1 4
------+-------+------
2 5 9 | 1 6 4 | 3 8 7
8 1 3 | 2 9 7 | 4 5 6
6 4 7 | 8 5 3 | 1 2 9
------+-------+------
5 8 1 | 4 3 6 | 9 7 2
7 3 6 | 9 2 5 | 8 4 1
9 2 4 | 7 8 1 | 6 3 5
```

Solution # 270
```
5 4 6 | 1 2 8 | 9 7 3
9 1 8 | 3 6 7 | 4 2 5
3 2 7 | 5 9 4 | 8 1 6
------+-------+------
8 3 4 | 7 5 1 | 2 6 9
2 6 5 | 8 4 9 | 1 3 7
7 9 1 | 2 3 6 | 5 8 4
------+-------+------
6 5 2 | 9 8 3 | 7 4 1
4 7 9 | 6 1 2 | 3 5 8
1 8 3 | 4 7 5 | 6 9 2
```

Solution # 271
```
8 4 1 | 6 3 9 | 2 7 5
5 6 2 | 4 7 1 | 9 3 8
9 7 3 | 8 5 2 | 4 1 6
------+-------+------
2 9 6 | 5 1 8 | 3 4 7
3 1 8 | 2 4 7 | 5 6 9
7 5 4 | 3 9 6 | 1 8 2
------+-------+------
6 3 5 | 9 8 4 | 7 2 1
4 2 7 | 1 6 5 | 8 9 3
1 8 9 | 7 2 3 | 6 5 4
```

Solution # 272
```
5 2 8 | 7 1 4 | 6 3 9
3 7 6 | 5 8 9 | 1 2 4
4 9 1 | 6 2 3 | 7 8 5
------+-------+------
6 4 3 | 2 7 1 | 9 5 8
2 1 5 | 8 9 6 | 3 4 7
7 8 9 | 4 3 5 | 2 1 6
------+-------+------
8 5 2 | 3 6 7 | 4 9 1
1 6 4 | 9 5 2 | 8 7 3
9 3 7 | 1 4 8 | 5 6 2
```

Solution # 273
```
1 4 7 | 6 2 9 | 8 3 5
5 6 2 | 4 3 8 | 1 7 9
8 9 3 | 7 5 1 | 4 6 2
------+-------+------
9 3 1 | 2 8 7 | 6 5 4
2 7 8 | 5 4 6 | 3 9 1
6 5 4 | 1 9 3 | 2 8 7
------+-------+------
4 2 6 | 8 7 5 | 9 1 3
3 1 5 | 9 6 2 | 7 4 8
7 8 9 | 3 1 4 | 5 2 6
```

Solution # 274
```
3 6 4 | 1 8 5 | 7 2 9
8 7 1 | 4 2 9 | 5 3 6
9 5 2 | 6 3 7 | 4 1 8
------+-------+------
7 4 8 | 9 5 3 | 1 6 2
1 2 3 | 7 6 4 | 9 8 5
6 9 5 | 8 1 2 | 3 7 4
------+-------+------
4 1 6 | 3 9 8 | 2 5 7
2 3 9 | 5 7 6 | 8 4 1
5 8 7 | 2 4 1 | 6 9 3
```

Solution # 275
```
1 8 6 | 3 9 7 | 4 2 5
9 7 3 | 5 4 2 | 8 6 1
5 2 4 | 6 1 8 | 3 7 9
------+-------+------
6 1 8 | 4 5 3 | 2 9 7
2 4 5 | 7 6 9 | 1 3 8
3 9 7 | 2 8 1 | 5 4 6
------+-------+------
4 6 9 | 1 3 5 | 7 8 2
8 5 2 | 9 7 4 | 6 1 3
7 3 1 | 8 2 6 | 9 5 4
```

Solution # 276
```
2 5 9 | 1 3 6 | 7 4 8
6 4 3 | 2 7 8 | 5 9 1
8 1 7 | 4 9 5 | 2 3 6
------+-------+------
5 8 2 | 7 6 3 | 4 1 9
3 9 4 | 5 1 2 | 6 8 7
1 7 6 | 8 4 9 | 3 5 2
------+-------+------
4 6 5 | 9 2 1 | 8 7 3
9 3 8 | 6 5 7 | 1 2 4
7 2 1 | 3 8 4 | 9 6 5
```

Solution # 277
```
6 5 3 | 2 7 4 | 8 9 1
7 9 2 | 6 1 8 | 4 3 5
8 4 1 | 3 5 9 | 7 6 2
------+-------+------
3 7 4 | 9 2 5 | 1 8 6
9 6 8 | 4 3 1 | 2 5 7
2 1 5 | 7 8 6 | 9 4 3
------+-------+------
5 8 6 | 1 4 7 | 3 2 9
1 2 9 | 8 6 3 | 5 7 4
4 3 7 | 5 9 2 | 6 1 8
```

Solution # 278
```
2 1 9 | 5 8 7 | 4 6 3
3 8 7 | 2 6 4 | 1 5 9
6 4 5 | 3 9 1 | 2 8 7
------+-------+------
1 7 6 | 8 4 3 | 9 2 5
8 9 4 | 6 2 5 | 7 3 1
5 3 2 | 1 7 9 | 8 4 6
------+-------+------
4 2 1 | 9 3 6 | 5 7 8
7 5 3 | 4 1 8 | 6 9 2
9 6 8 | 7 5 2 | 3 1 4
```

Solution # 279
```
3 2 4 | 7 9 1 | 5 8 6
9 5 8 | 6 3 2 | 4 7 1
6 7 1 | 4 8 5 | 2 9 3
------+-------+------
1 4 2 | 9 5 6 | 8 3 7
8 6 9 | 2 7 3 | 1 5 4
7 3 5 | 8 1 4 | 9 6 2
------+-------+------
2 1 3 | 5 6 9 | 7 4 8
5 8 6 | 1 4 7 | 3 2 9
4 9 7 | 3 2 8 | 6 1 5
```

Solution # 280
```
6 3 9 | 4 1 5 | 8 2 7
7 4 2 | 8 3 9 | 1 6 5
8 1 5 | 7 2 6 | 3 4 9
------+-------+------
4 2 7 | 1 8 3 | 9 5 6
9 6 8 | 2 5 7 | 4 1 3
3 5 1 | 9 6 4 | 7 8 2
------+-------+------
1 7 4 | 5 9 2 | 6 3 8
5 8 3 | 6 7 1 | 2 9 4
2 9 6 | 3 4 8 | 5 7 1
```

Solution # 281
```
6 1 2 3 9 8 7 4 5
7 8 9 4 5 6 3 1 2
3 4 5 2 1 7 6 9 8
1 3 7 8 6 4 2 5 9
9 2 6 5 3 1 8 7 4
8 5 4 7 2 9 1 6 3
4 7 1 9 8 3 5 2 6
2 6 3 1 4 5 9 8 7
5 9 8 6 7 2 4 3 1
```

Solution # 282
```
2 3 6 7 4 1 8 9 5
4 7 8 9 5 3 6 2 1
1 5 9 8 2 6 7 3 4
5 2 3 4 1 8 9 6 7
8 1 7 6 9 5 2 4 3
6 9 4 2 3 7 5 1 8
7 4 5 1 6 9 3 8 2
9 8 2 3 7 4 1 5 6
3 6 1 5 8 2 4 7 9
```

Solution # 283
```
7 8 6 4 9 3 2 5 1
4 5 9 8 1 2 6 3 7
3 1 2 7 6 5 9 8 4
8 2 4 1 5 9 3 7 6
9 6 1 3 2 7 8 4 5
5 7 3 6 4 8 1 9 2
1 4 5 9 8 6 7 2 3
6 3 8 2 7 4 5 1 9
2 9 7 5 3 1 4 6 8
```

Solution # 284
```
4 1 2 5 6 7 3 9 8
6 3 7 9 8 1 4 5 2
9 5 8 3 2 4 7 1 6
3 9 1 6 4 8 5 2 7
2 6 5 1 7 3 8 4 9
7 8 4 2 5 9 1 6 3
8 2 6 7 1 5 9 3 4
5 4 9 8 3 6 2 7 1
1 7 3 4 9 2 6 8 5
```

Solution # 285
```
4 8 2 5 7 3 6 1 9
1 3 5 8 6 9 4 2 7
9 7 6 1 4 2 5 3 8
3 6 1 7 9 4 2 8 5
8 4 7 2 5 1 9 6 3
5 2 9 6 3 8 7 4 1
7 1 4 9 8 6 3 5 2
2 9 3 4 1 5 8 7 6
6 5 8 3 2 7 1 9 4
```

Solution # 286
```
8 4 5 2 9 7 3 6 1
2 1 3 5 4 6 8 9 7
6 7 9 1 3 8 4 2 5
4 2 6 3 8 5 1 7 9
1 3 7 4 6 9 5 8 2
9 5 8 7 1 2 6 4 3
3 9 2 6 5 4 7 1 8
5 8 4 9 7 1 2 3 6
7 6 1 8 2 3 9 5 4
```

Solution # 287
```
3 7 9 5 4 6 1 2 8
6 4 8 2 1 7 5 9 3
5 2 1 8 9 3 4 6 7
2 8 7 6 5 1 9 3 4
1 9 6 4 3 8 7 5 2
4 5 3 7 2 9 8 1 6
9 6 2 1 7 4 3 8 5
8 3 4 9 6 5 2 7 1
7 1 5 3 8 2 6 4 9
```

Solution # 288
```
4 8 7 3 2 5 1 9 6
6 3 2 9 8 1 4 7 5
9 5 1 4 6 7 2 3 8
5 6 8 2 1 3 7 4 9
2 4 3 7 9 6 5 8 1
7 1 9 5 4 8 6 2 3
8 2 4 1 5 9 3 6 7
3 9 5 6 7 2 8 1 4
1 7 6 8 3 4 9 5 2
```

Solution # 289
```
3 9 8 2 6 1 4 7 5
6 4 2 5 7 3 1 9 8
5 1 7 9 4 8 2 3 6
9 3 1 6 2 7 8 5 4
4 8 6 3 9 5 7 2 1
2 7 5 8 1 4 3 6 9
8 2 3 1 5 6 9 4 7
1 6 4 7 3 9 5 8 2
7 5 9 4 8 2 6 1 3
```

Solution # 290
```
4 1 7 2 5 6 8 9 3
5 3 8 7 9 1 6 4 2
9 6 2 8 3 4 5 1 7
6 4 3 9 8 2 1 7 5
7 8 9 3 1 5 4 2 6
1 2 5 4 6 7 3 8 9
8 9 4 6 2 3 7 5 1
3 7 1 5 4 9 2 6 8
2 5 6 1 7 8 9 3 4
```

Solution # 291
```
8 9 3 5 4 7 6 1 2
2 5 1 3 9 6 7 8 4
7 4 6 8 2 1 3 5 9
5 6 4 7 1 8 2 9 3
1 2 9 6 3 4 5 7 8
3 7 8 9 5 2 4 6 1
4 8 2 1 7 5 9 3 6
6 3 5 2 8 9 1 4 7
9 1 7 4 6 3 8 2 5
```

Solution # 292
```
3 6 9 5 4 8 1 7 2
4 2 8 1 9 7 5 6 3
5 1 7 6 2 3 4 9 8
6 9 2 4 8 1 7 3 5
7 5 4 9 3 2 8 1 6
1 8 3 7 6 5 2 4 9
9 3 1 8 5 4 6 2 7
8 4 6 2 7 9 3 5 1
2 7 5 3 1 6 9 8 4
```

Solution # 293
```
5 7 6 4 8 9 2 3 1
2 3 4 5 1 7 9 8 6
1 9 8 3 6 2 4 5 7
4 8 3 2 7 6 5 1 9
7 1 5 9 4 3 8 6 2
6 2 9 8 5 1 3 7 4
8 4 1 6 2 5 7 9 3
3 5 7 1 9 4 6 2 8
9 6 2 7 3 8 1 4 5
```

Solution # 294
```
7 4 2 8 6 3 1 5 9
6 8 5 1 7 9 2 4 3
3 1 9 4 2 5 6 8 7
2 3 4 6 9 1 5 7 8
1 6 8 5 3 7 4 9 2
9 5 7 2 8 4 3 1 6
8 7 1 3 4 2 9 6 5
5 9 3 7 1 6 8 2 4
4 2 6 9 5 8 7 3 1
```

Solution # 295
```
7 4 1 6 5 3 8 9 2
9 6 8 4 7 2 5 1 3
2 5 3 1 8 9 7 4 6
8 3 6 5 9 4 1 2 7
4 1 7 3 2 6 9 5 8
5 9 2 8 1 7 6 3 4
6 7 4 9 3 5 2 8 1
1 2 9 7 4 8 3 6 5
3 8 5 2 6 1 4 7 9
```

Solution # 296
```
6 9 3 7 2 1 4 5 8
1 8 7 4 9 5 6 2 3
4 5 2 8 6 3 9 7 1
3 2 6 5 4 8 7 1 9
8 4 9 1 7 6 2 3 5
7 1 5 2 3 9 8 6 4
2 6 1 3 8 4 5 9 7
5 7 4 9 1 2 3 8 6
9 3 8 6 5 7 1 4 2
```

Solution # 297
```
5 3 9 2 1 8 7 4 6
6 1 2 4 9 7 8 3 5
8 7 4 5 3 6 2 1 9
1 9 5 7 8 4 6 2 3
3 6 8 1 2 9 5 7 4
2 4 7 3 6 5 9 8 1
7 8 3 6 5 1 4 9 2
4 2 6 9 7 3 1 5 8
9 5 1 8 4 2 3 6 7
```

Solution # 298
```
7 6 8 4 2 5 9 3 1
5 1 4 3 7 9 6 2 8
2 9 3 6 8 1 4 7 5
4 2 5 9 1 3 7 8 6
9 8 7 5 6 2 3 1 4
1 3 6 8 4 7 2 5 9
6 4 2 7 5 8 1 9 3
3 5 1 2 9 4 8 6 7
8 7 9 1 3 6 5 4 2
```

Solution # 299
```
5 3 1 4 9 6 7 8 2
8 6 4 3 2 7 9 1 5
2 9 7 8 5 1 6 4 3
3 1 5 6 4 8 2 7 9
9 7 6 1 3 2 8 5 4
4 8 2 9 7 5 1 3 6
1 4 8 2 6 3 5 9 7
7 2 9 5 8 4 3 6 1
6 5 3 7 1 9 4 2 8
```

Solution # 300
```
8 5 2 1 6 9 3 7 4
4 3 9 2 8 7 1 5 6
6 7 1 3 4 5 2 8 9
9 2 6 7 5 8 4 3 1
3 8 7 4 1 6 9 2 5
5 1 4 9 2 3 8 6 7
7 4 8 6 3 1 5 9 2
2 9 5 8 7 4 6 1 3
1 6 3 5 9 2 7 4 8
```

Solution # 301
```
1 9 6 2 5 8 3 7 4
2 5 7 4 1 3 6 9 8
4 8 3 7 9 6 2 5 1
9 6 5 1 8 4 7 3 2
7 3 4 6 2 9 1 8 5
8 1 2 3 7 5 4 6 9
3 4 8 9 6 1 5 2 7
5 7 1 8 3 2 9 4 6
6 2 9 5 4 7 8 1 3
```

Solution # 302
```
9 1 5 3 2 8 4 7 6
6 2 3 1 4 7 5 9 8
8 4 7 5 9 6 3 2 1
7 9 4 8 6 5 2 1 3
5 3 8 9 1 2 6 4 7
1 6 2 4 7 3 8 5 9
3 7 9 2 8 4 1 6 5
4 8 1 6 5 9 7 3 2
2 5 6 7 3 1 9 8 4
```

Solution # 303
```
4 5 1 7 2 6 8 3 9
9 8 2 1 3 4 7 5 6
6 7 3 9 8 5 1 4 2
5 3 9 4 1 8 2 6 7
7 4 8 6 9 2 5 1 3
2 1 6 5 7 3 9 8 4
1 6 7 8 4 9 3 2 5
3 9 4 2 5 1 6 7 8
8 2 5 3 6 7 4 9 1
```

Solution # 304
```
4 9 8 6 5 7 1 2 3
2 1 5 9 3 4 8 7 6
3 6 7 1 8 2 4 9 5
5 2 6 3 9 8 7 4 1
7 8 1 4 2 5 6 3 9
9 4 3 7 6 1 5 8 2
1 3 2 8 7 6 9 5 4
6 7 9 5 4 3 2 1 8
8 5 4 2 1 9 3 6 7
```

Solution # 305
```
8 9 5 6 4 1 2 3 7
4 1 3 7 8 2 9 6 5
7 2 6 9 3 5 4 1 8
2 4 7 5 1 3 8 9 6
6 3 8 2 9 4 7 5 1
1 5 9 8 6 7 3 2 4
9 7 4 1 2 6 5 8 3
3 8 1 4 5 9 6 7 2
5 6 2 3 7 8 1 4 9
```

Solution # 306
```
7 6 3 2 9 8 1 4 5
4 9 5 1 3 6 2 8 7
1 8 2 4 7 5 3 6 9
3 5 4 9 8 1 7 2 6
8 2 9 3 6 7 4 5 1
6 1 7 5 4 2 9 3 8
9 3 6 7 5 4 8 1 2
5 7 1 8 2 9 6 4 3
2 4 8 6 1 9 5 7 3
```

Solution # 307
```
9 8 6 7 4 2 1 5 3
2 5 3 6 8 1 4 9 7
4 7 1 5 3 9 6 8 2
7 6 8 9 5 4 3 2 1
5 2 4 1 7 3 8 6 9
3 1 9 2 6 8 7 4 5
1 4 5 8 2 7 9 3 6
8 9 2 3 1 6 5 7 4
6 3 7 4 9 5 2 1 8
```

Solution # 308
```
9 6 2 3 7 5 4 1 8
5 4 3 2 1 8 6 9 7
7 1 8 9 4 6 3 5 2
4 9 7 1 5 3 8 2 6
3 2 5 8 6 7 9 4 1
6 8 1 4 2 9 7 3 5
8 7 4 5 3 1 2 6 9
2 5 6 7 9 4 1 8 3
1 3 9 6 8 2 5 7 4
```

Solution # 309
```
3 5 8 2 4 1 9 6 7
2 6 4 5 7 9 1 8 3
9 1 7 3 6 8 2 5 4
8 2 6 7 1 3 5 4 9
5 4 3 9 8 2 6 7 1
1 7 9 4 5 6 8 3 2
6 9 5 1 3 4 7 2 8
7 3 2 8 9 5 4 1 6
4 8 1 6 2 7 3 9 5
```

Solution # 310
```
9 1 4 7 3 5 2 6 8
7 5 6 2 4 8 3 1 9
3 2 8 9 6 1 5 4 7
2 4 5 3 8 6 7 9 1
1 8 7 5 9 4 6 2 3
6 3 9 1 2 7 8 5 4
8 7 1 6 5 9 4 3 2
5 9 2 4 7 3 1 8 6
4 6 3 8 1 2 9 7 5
```

Solution # 311
```
5 2 1 4 7 9 6 3 8
6 7 3 5 2 8 4 9 1
8 4 9 3 6 1 7 5 2
7 8 4 6 9 5 1 2 3
1 5 2 8 3 7 9 4 6
3 9 6 2 1 4 8 7 5
9 3 7 1 5 6 2 8 4
4 1 5 9 8 2 3 6 7
2 6 8 7 4 3 5 1 9
```

Solution # 312
```
5 2 1 7 8 4 6 9 3
8 7 3 6 9 1 5 2 4
4 6 9 3 5 2 1 7 8
6 8 2 1 7 3 4 5 9
9 1 7 2 4 5 8 3 6
3 5 4 8 6 9 2 1 7
7 3 5 4 1 8 9 6 2
1 4 6 9 2 7 3 8 5
2 9 8 5 3 6 7 4 1
```

Solution # 313
```
8 6 4 7 9 1 5 3 2
9 3 2 6 5 8 4 7 1
5 1 7 3 2 4 8 6 9
6 2 1 5 3 9 7 4 8
3 4 8 1 6 7 2 9 5
7 9 5 4 8 2 3 1 6
1 8 6 2 4 3 9 5 7
2 5 3 9 7 6 1 8 4
4 7 9 8 1 5 6 2 3
```

Solution # 314
```
5 4 6 8 2 7 3 9 1
2 8 9 3 4 1 7 6 5
7 3 1 5 6 9 2 4 8
6 7 4 2 1 5 9 8 3
9 5 8 7 3 4 6 1 2
1 2 3 6 9 8 5 7 4
8 9 7 1 5 2 4 3 6
4 6 5 9 8 3 1 2 7
3 1 2 4 7 6 8 5 9
```

Solution # 315
```
8 7 3 6 4 5 9 2 1
1 5 9 3 8 2 6 4 7
4 6 2 9 7 1 5 3 8
6 9 5 2 1 4 8 7 3
2 1 8 7 6 3 4 5 9
3 4 7 5 9 8 2 1 6
5 8 1 4 3 6 7 9 2
7 3 4 8 2 9 1 6 5
9 2 6 1 5 7 3 8 4
```

Solution # 316
```
4 6 3 8 7 2 1 9 5
1 5 7 9 6 3 4 2 8
8 2 9 4 5 1 6 3 7
2 1 5 3 9 6 8 7 4
9 4 6 1 8 7 3 5 2
7 3 8 2 4 5 9 6 1
6 8 1 7 2 9 5 4 3
3 9 2 5 1 4 7 8 6
5 7 4 6 3 8 2 1 9
```

Solution # 317
```
3 9 7 2 6 8 5 1 4
2 8 5 1 3 4 9 7 6
4 1 6 9 5 7 3 2 8
6 2 1 5 4 3 7 8 9
7 5 8 6 1 9 2 4 3
9 4 3 8 7 2 1 6 5
8 7 9 3 2 6 4 5 1
1 6 2 4 9 5 8 3 7
5 3 4 7 8 1 6 9 2
```

Solution # 318
```
9 1 8 6 3 5 4 7 2
7 5 2 1 4 8 9 6 3
4 3 6 2 7 9 8 5 1
5 2 4 9 6 7 1 3 8
1 7 9 8 5 3 6 2 4
6 8 3 4 1 2 5 9 7
8 6 5 3 2 4 7 1 9
2 4 1 7 9 6 3 8 5
3 9 7 5 8 1 2 4 6
```

Solution # 319
```
9 6 8 5 1 7 2 3 4
7 2 4 6 8 3 1 9 5
3 5 1 2 9 4 8 6 7
2 9 6 1 3 5 4 7 8
1 3 7 9 4 8 5 2 6
4 8 5 7 6 2 3 1 9
8 4 2 3 7 9 6 5 1
5 1 9 4 2 6 7 8 3
6 7 3 8 5 1 9 4 2
```

Solution # 320
```
2 4 1 8 6 7 9 3 5
5 7 3 1 4 9 2 8 6
8 6 9 2 5 3 1 7 4
9 2 4 5 7 1 3 6 8
3 1 5 6 2 8 4 9 7
7 8 6 3 9 4 5 2 1
4 9 8 7 1 2 6 5 3
6 3 2 4 8 5 7 1 9
1 5 7 9 3 6 8 4 2
```

Solution # 321
```
6 3 7 4 2 8 5 9 1
5 8 2 1 9 6 4 7 3
1 4 9 3 7 5 2 6 8
3 6 4 5 8 9 7 1 2
9 7 8 2 4 1 3 5 6
2 5 1 6 3 7 8 4 9
4 9 5 8 6 2 1 3 7
8 1 6 7 5 3 9 2 4
7 2 3 9 1 4 6 8 5
```

Solution # 322
```
9 8 2 6 1 7 5 4 3
3 5 6 2 4 9 7 8 1
7 1 4 8 5 3 6 9 2
4 9 5 1 2 6 8 3 7
2 6 8 3 7 5 4 1 9
1 7 3 4 9 8 2 6 5
6 3 7 9 8 2 1 5 4
8 2 1 5 3 4 9 7 6
5 4 9 7 6 1 3 2 8
```

Solution # 323
```
4 6 8 5 2 7 3 1 9
1 7 5 9 6 3 4 2 8
2 9 3 1 8 4 6 5 7
9 4 1 7 5 2 8 3 6
5 8 2 6 3 9 7 4 1
6 3 7 4 1 8 5 9 2
7 1 9 8 4 5 2 6 3
8 2 4 3 9 6 1 7 5
3 5 6 2 7 1 9 8 4
```

Solution # 324
```
6 4 2 3 9 8 5 1 7
8 7 3 5 4 1 6 2 9
9 5 1 2 7 6 8 4 3
2 6 8 4 1 7 9 3 5
4 3 5 9 6 2 7 8 1
1 9 7 8 5 3 4 6 2
3 2 4 7 8 5 1 9 6
5 8 6 1 2 9 3 7 4
7 1 9 6 3 4 2 5 8
```

Solution # 325
```
9 4 3 1 8 5 6 2 7
1 8 2 6 7 3 5 9 4
6 5 7 2 4 9 1 3 8
4 1 5 7 6 2 3 8 9
7 2 8 3 9 1 4 5 6
3 6 9 8 5 4 7 1 2
2 9 1 4 3 6 8 7 5
5 7 4 9 1 8 2 6 3
8 3 6 5 2 7 9 4 1
```

Solution # 326
```
7 6 9 4 8 5 3 1 2
8 1 5 3 2 7 6 4 9
3 4 2 9 1 6 8 5 7
9 8 4 5 6 2 1 7 3
6 5 7 1 3 8 9 2 4
1 2 3 7 9 4 5 6 8
5 3 6 2 4 9 7 8 1
4 9 8 6 7 1 2 3 5
2 7 1 8 5 3 4 9 6
```

Solution # 327
```
1 6 3 4 2 9 5 7 8
4 5 2 7 6 8 9 3 1
7 8 9 1 3 5 2 6 4
9 7 6 5 8 3 1 4 2
5 3 1 2 4 7 8 9 6
2 4 8 9 1 6 3 5 7
6 9 4 8 5 1 7 2 3
3 1 7 6 9 2 4 8 5
8 2 5 3 7 4 6 1 9
```

Solution # 328
```
6 2 8 1 3 5 9 7 4
1 9 7 2 8 4 6 5 3
5 3 4 7 9 6 8 1 2
2 6 5 4 1 3 7 8 9
7 4 1 9 6 8 3 2 5
3 8 9 5 2 7 4 6 1
9 7 6 3 5 1 2 4 8
8 1 3 6 4 2 5 9 7
4 5 2 8 7 9 1 3 6
```

Solution # 329
```
2 8 3 5 4 7 6 9 1
7 4 1 9 6 3 5 2 8
9 6 5 2 1 8 7 4 3
4 3 2 1 5 6 9 8 7
1 9 8 7 2 4 3 5 6
6 5 7 8 3 9 4 1 2
5 2 4 3 7 1 8 6 9
8 7 6 4 9 2 1 3 5
3 1 9 6 8 5 2 7 4
```

Solution # 330
```
9 7 2 1 4 8 6 5 3
3 1 6 5 7 9 4 8 2
4 5 8 6 2 3 1 9 7
6 2 3 4 8 1 5 7 9
1 9 5 3 6 7 8 2 4
7 8 4 2 9 5 3 6 1
8 6 7 9 3 4 2 1 5
5 3 9 8 1 2 7 4 6
2 4 1 7 5 6 9 3 8
```

Solution # 331
```
9 2 6 8 5 1 4 7 3
8 7 5 2 4 3 9 1 6
1 3 4 7 9 6 2 5 8
5 9 8 6 1 2 3 4 7
7 4 2 9 3 5 6 8 1
3 6 1 4 8 7 5 9 2
2 8 7 5 6 4 1 3 9
4 1 9 3 2 8 7 6 5
6 5 3 1 7 9 8 2 4
```

Solution # 332
```
9 5 3 8 7 1 6 4 2
2 6 8 4 5 9 3 7 1
1 7 4 3 2 6 8 5 9
3 4 2 9 1 8 5 6 7
7 8 5 6 4 2 9 1 3
6 1 9 5 3 7 2 8 4
8 9 7 2 6 4 1 3 5
5 2 1 7 8 3 4 9 6
4 3 6 1 9 5 7 2 8
```

Solution # 333
```
8 3 1 6 7 2 5 9 4
7 9 5 4 3 8 2 1 6
6 4 2 9 5 1 7 3 8
1 7 4 3 8 9 6 5 2
9 8 6 2 4 5 3 7 1
2 5 3 7 1 6 4 8 9
4 1 8 5 2 3 9 6 7
3 2 9 8 6 7 1 4 5
5 6 7 1 9 4 8 2 3
```

Solution # 334
```
5 7 9 2 4 8 6 1 3
2 6 1 9 3 5 8 7 4
8 3 4 6 1 7 5 9 2
3 4 8 1 2 6 7 5 9
9 1 7 8 5 3 2 4 6
6 5 2 4 7 9 3 8 1
4 8 3 5 9 2 1 6 7
1 2 5 7 6 4 9 3 8
7 9 6 3 8 1 4 2 5
```

Solution # 335
```
6 1 4 2 5 3 7 8 9
9 5 8 4 6 7 1 3 2
7 2 3 9 1 8 4 6 5
3 4 1 5 8 9 6 2 7
5 8 6 1 7 2 3 9 4
2 7 9 3 4 6 8 5 1
4 6 7 8 2 5 9 1 3
1 9 5 6 3 4 2 7 8
8 3 2 7 9 1 5 4 6
```

Solution # 336
```
8 7 9 6 2 5 1 4 3
4 1 2 8 9 3 6 7 5
6 3 5 7 1 4 9 8 2
7 6 3 9 4 2 8 5 1
5 4 8 3 6 1 7 2 9
9 2 1 5 8 7 4 3 6
3 5 4 1 7 9 2 6 8
2 9 6 4 5 8 3 1 7
1 8 7 2 3 6 5 9 4
```

Solution # 337
```
6 9 3 8 7 4 2 1 5
1 8 4 2 5 6 7 9 3
5 7 2 1 9 3 8 4 6
9 2 6 5 4 1 3 8 7
3 1 7 9 2 8 5 6 4
8 4 5 6 3 7 9 2 1
7 5 1 4 8 2 6 3 9
4 3 8 7 6 9 1 5 2
2 6 9 3 1 5 4 7 8
```

Solution # 338
```
5 8 3 6 2 7 9 1 4
7 1 9 3 4 5 2 8 6
6 2 4 9 8 1 5 3 7
8 4 5 2 1 3 7 6 9
9 3 2 7 6 8 4 5 1
1 6 7 5 9 4 8 2 3
3 7 1 4 5 2 6 9 8
2 9 8 1 7 6 3 4 5
4 5 6 8 3 9 1 7 2
```

Solution # 339
```
5 9 7 8 1 3 6 4 2
8 6 1 5 2 4 7 3 9
2 3 4 6 7 9 1 8 5
7 8 5 2 4 6 3 9 1
4 1 3 9 5 8 2 7 6
6 2 9 7 3 1 4 5 8
3 5 6 4 8 2 9 1 7
9 4 8 1 6 7 5 2 3
1 7 2 3 9 5 8 6 4
```

Solution # 340
```
9 8 6 3 5 4 2 7 1
7 5 1 6 2 9 8 3 4
2 4 3 7 8 1 6 5 9
4 7 2 8 9 5 3 1 6
1 6 5 4 3 7 9 2 8
8 3 9 2 1 6 5 4 7
5 9 8 1 7 2 4 6 3
6 2 7 9 4 3 1 8 5
3 1 4 5 6 8 7 9 2
```

Solution # 341
```
8 6 3 2 7 9 5 1 4
7 4 5 8 3 1 9 2 6
2 1 9 4 6 5 8 3 7
4 8 1 9 5 7 2 6 3
5 3 2 6 8 4 1 7 9
9 7 6 1 2 3 4 8 5
3 9 7 5 1 2 6 4 8
1 5 8 7 4 6 3 9 2
6 2 4 3 9 8 7 5 1
```

Solution # 342
```
3 2 5 6 8 7 4 9 1
7 6 8 4 1 9 2 5 3
4 1 9 3 5 2 6 8 7
9 7 3 2 6 8 5 1 4
1 5 2 9 7 4 8 3 6
8 4 6 1 3 5 9 7 2
2 8 4 7 9 3 1 6 5
5 3 1 8 4 6 7 2 9
6 9 7 5 2 1 3 4 8
```

Solution # 343
```
9 8 1 5 3 4 6 7 2
4 6 2 1 8 7 5 3 9
3 5 7 2 6 9 8 1 4
2 1 8 4 9 5 3 6 7
7 3 4 6 1 8 2 9 5
6 9 5 7 2 3 4 8 1
1 7 6 8 4 2 9 5 3
5 4 9 3 7 6 1 2 8
8 2 3 9 5 1 7 4 6
```

Solution # 344
```
3 6 2 1 4 9 5 8 7
1 7 4 5 8 3 6 2 9
9 8 5 6 2 7 1 4 3
8 9 3 4 1 5 7 6 2
5 2 1 3 7 6 4 9 8
7 4 6 8 9 2 3 5 1
4 3 9 2 5 1 8 7 6
2 1 8 7 6 4 9 3 5
6 5 7 9 3 8 2 1 4
```

Solution # 345
```
3 4 2 6 1 8 5 7 9
7 8 1 9 3 5 2 4 6
6 5 9 4 2 7 3 1 8
4 9 7 2 5 3 6 8 1
1 2 3 8 6 9 4 5 7
5 6 8 7 4 1 9 2 3
2 7 5 3 8 6 1 9 4
9 3 4 1 7 2 8 6 5
8 1 6 5 9 4 7 3 2
```

Solution # 346
```
5 1 7 2 9 4 6 3 8
3 8 2 1 6 5 4 9 7
9 6 4 3 8 7 2 5 1
1 9 8 6 4 3 7 2 5
4 5 3 8 7 2 9 1 6
2 7 6 9 5 1 8 4 3
6 2 5 7 3 9 1 8 4
8 4 1 5 2 6 3 7 9
7 3 9 4 1 8 5 6 2
```

Solution # 347
```
3 1 6 2 4 7 8 5 9
7 5 8 3 9 1 6 2 4
2 9 4 6 5 8 7 3 1
5 2 7 9 6 4 3 1 8
8 3 9 1 2 5 4 6 7
4 6 1 8 7 3 2 9 5
1 7 3 5 8 2 9 4 6
9 8 2 4 1 6 5 7 3
6 4 5 7 3 9 1 8 2
```

Solution # 348
```
9 3 6 4 8 2 1 5 7
5 4 7 3 6 1 2 9 8
2 1 8 9 7 5 3 6 4
6 8 1 7 2 9 5 4 3
3 2 9 6 5 4 7 8 1
4 7 5 8 1 3 9 2 6
7 6 2 1 9 8 4 3 5
1 5 4 2 3 6 8 7 9
8 9 3 5 4 7 6 1 2
```

Solution # 349
```
3 1 7 4 2 5 8 9 6
5 9 8 1 6 7 3 2 4
2 4 6 3 9 8 5 1 7
4 8 5 7 1 2 9 6 3
1 6 9 5 4 3 2 7 8
7 2 3 9 8 6 1 4 5
6 3 4 2 5 9 7 8 1
9 5 1 8 7 4 6 3 2
8 7 2 6 3 1 4 5 9
```

Solution # 350
```
8 5 6 2 4 9 7 3 1
1 2 3 7 6 5 4 8 9
7 4 9 8 1 3 6 5 2
5 9 8 4 3 2 1 6 7
2 3 1 6 5 7 8 9 4
6 7 4 1 9 8 3 2 5
9 1 5 3 8 4 2 7 6
3 6 7 5 2 1 9 4 8
4 8 2 9 7 6 5 1 3
```

Solution # 351
```
1 5 6 9 3 8 2 7 4
2 8 3 4 7 1 6 5 9
7 4 9 6 2 5 3 8 1
4 3 2 7 8 9 5 1 6
6 7 5 1 4 3 9 2 8
8 9 1 5 6 2 7 4 3
3 2 7 8 9 4 1 6 5
5 6 8 3 1 7 4 9 2
9 1 4 2 5 6 8 3 7
```

Solution # 352
```
1 9 3 8 5 7 6 2 4
4 5 6 1 2 9 8 7 3
7 8 2 3 6 4 1 5 9
5 3 8 9 4 1 2 6 7
9 6 7 5 3 2 4 8 1
2 1 4 7 8 6 9 3 5
3 4 5 6 9 8 7 1 2
8 2 1 4 7 3 5 9 6
6 7 9 2 1 5 3 4 8
```

Solution # 353
```
4 2 3 5 8 9 7 6 1
8 1 9 3 6 7 2 5 4
5 6 7 4 1 2 3 9 8
3 4 2 6 7 1 5 8 9
7 9 1 8 5 4 6 2 3
6 8 5 2 9 3 4 1 7
2 7 6 9 4 8 1 3 5
9 5 4 1 3 6 8 7 2
1 3 8 7 2 5 9 4 6
```

Solution # 354
```
7 3 8 4 9 6 5 2 1
4 1 9 2 5 8 3 6 7
5 2 6 1 3 7 4 8 9
9 4 5 6 2 3 7 1 8
3 8 1 5 7 9 6 4 2
6 7 2 8 1 4 9 3 5
8 5 3 7 4 2 1 9 6
2 9 7 3 6 1 8 5 4
1 6 4 9 8 5 2 7 3
```

Solution # 355
```
7 3 5 4 6 9 2 1 8
4 2 6 5 8 1 7 9 3
1 8 9 3 2 7 4 5 6
2 1 7 8 9 4 3 6 5
3 9 8 6 7 5 1 2 4
6 5 4 2 1 3 9 8 7
8 6 1 7 4 2 5 3 9
5 7 2 9 3 8 6 4 1
9 4 3 1 5 6 8 7 2
```

Solution # 356
```
7 2 1 5 8 9 6 4 3
6 5 4 3 7 2 1 8 9
9 3 8 1 4 6 5 2 7
5 6 9 7 2 1 8 3 4
8 1 2 4 9 3 7 5 6
3 4 7 8 6 5 9 1 2
4 7 3 6 5 8 2 9 1
1 9 5 2 3 7 4 6 8
2 8 6 9 1 4 3 7 5
```

Solution # 357
```
4 5 1 6 9 2 8 3 7
2 8 7 5 4 3 9 6 1
3 6 9 7 1 8 4 2 5
6 1 5 3 7 4 2 9 8
9 3 2 1 8 6 7 5 4
8 7 4 2 5 9 6 1 3
7 4 6 9 3 5 1 8 2
1 9 3 8 2 7 5 4 6
5 2 8 4 6 1 3 7 9
```

Solution # 358
```
8 9 5 6 4 7 2 3 1
7 6 3 5 1 2 9 8 4
4 2 1 8 3 9 5 7 6
3 8 4 9 5 6 1 2 7
6 7 9 1 2 3 4 5 8
1 5 2 7 8 4 6 9 3
9 3 8 4 6 5 7 1 2
2 4 7 3 9 1 8 6 5
5 1 6 2 7 8 3 4 9
```

Solution # 359
```
3 7 4 8 2 5 9 1 6
8 6 9 1 7 3 4 2 5
2 1 5 4 6 9 8 7 3
6 9 3 7 8 4 1 5 2
7 4 1 5 3 2 6 8 9
5 2 8 9 1 6 3 4 7
9 3 7 2 4 1 5 6 8
1 5 2 6 9 8 7 3 4
4 8 6 3 5 7 2 9 1
```

Solution # 360
```
5 9 7 2 1 3 6 8 4
2 6 3 5 8 4 9 1 7
4 1 8 9 7 6 3 5 2
8 5 4 6 3 9 2 7 1
9 3 2 1 4 7 5 6 8
1 7 6 8 2 5 4 9 3
7 2 5 3 9 8 1 4 6
6 4 1 7 5 2 8 3 9
3 8 9 4 6 1 7 2 5
```

Solution # 361
```
5 1 6 2 8 9 3 7 4
8 4 7 1 3 6 9 2 5
3 2 9 5 7 4 6 8 1
4 5 2 3 9 7 1 6 8
6 8 3 4 2 1 7 5 9
9 7 1 8 6 5 2 4 3
2 9 5 7 1 8 4 3 6
1 3 4 6 5 2 8 9 7
7 6 8 9 4 3 5 1 2
```

Solution # 362
```
1 9 3 5 6 8 7 4 2
8 7 5 4 1 2 9 3 6
6 4 2 3 9 7 5 1 8
3 8 4 7 5 1 6 2 9
2 5 7 9 4 6 1 8 3
9 6 1 2 8 3 4 7 5
7 2 9 1 3 5 8 6 4
4 1 6 8 2 9 3 5 7
5 3 8 6 7 4 2 9 1
```

Solution # 363
```
5 3 2 6 7 1 8 9 4
1 6 4 5 8 9 7 3 2
7 8 9 2 3 4 6 1 5
4 9 6 8 1 5 2 7 3
2 5 7 4 9 3 1 8 6
3 1 8 7 2 6 4 5 9
6 4 1 9 5 8 3 2 7
9 7 3 1 4 2 5 6 8
8 2 5 3 6 7 9 4 1
```

Solution # 364
```
2 4 7 1 8 5 9 6 3
3 6 8 4 9 7 1 2 5
1 9 5 6 3 2 8 7 4
9 3 6 8 1 4 7 5 2
8 5 4 2 7 9 3 1 6
7 1 2 3 5 6 4 8 9
4 7 3 5 6 1 2 9 8
5 2 9 7 4 8 6 3 1
6 8 1 9 2 3 5 4 7
```

Solution # 365
```
9 7 6 1 4 5 8 3 2
8 2 4 3 9 7 1 5 6
1 3 5 2 8 6 7 9 4
2 4 3 6 1 8 9 7 5
5 8 9 4 7 3 2 6 1
7 6 1 5 2 9 4 8 3
6 1 7 9 3 4 5 2 8
4 5 8 7 6 2 3 1 9
3 9 2 8 5 1 6 4 7
```

Solution # 366
```
4 8 9 7 5 1 2 3 6
2 1 7 6 3 8 5 9 4
3 6 5 9 4 2 7 8 1
1 3 2 8 9 4 6 5 7
7 5 8 3 1 6 4 2 9
9 4 6 2 7 5 3 1 8
8 2 4 1 6 3 9 7 5
6 9 1 5 2 7 8 4 3
5 7 3 4 8 9 1 6 2
```

Solution # 367
```
5 8 6 9 1 2 7 4 3
4 2 9 7 3 5 8 1 6
1 7 3 4 8 6 9 2 5
2 3 4 1 6 9 5 8 7
9 5 7 3 4 8 1 6 2
6 1 8 2 5 7 4 3 9
7 4 5 6 2 1 3 9 8
3 9 2 8 7 4 6 5 1
8 6 1 5 9 3 2 7 4
```

Solution # 368
```
9 5 2 8 3 1 6 4 7
7 8 6 5 9 4 1 3 2
3 1 4 6 7 2 5 9 8
4 6 8 2 1 5 9 7 3
2 7 9 3 6 8 4 1 5
5 3 1 7 4 9 8 2 6
8 4 7 9 2 6 3 5 1
1 2 5 4 8 3 7 6 9
6 9 3 1 5 7 2 8 4
```

Solution # 369
```
6 5 4 7 8 3 9 1 2
2 7 8 5 1 9 6 4 3
1 9 3 6 4 2 7 8 5
5 2 6 4 9 1 8 3 7
3 1 7 8 2 5 4 9 6
4 8 9 3 7 6 2 5 1
7 3 5 9 6 4 1 2 8
8 4 2 1 5 7 3 6 9
9 6 1 2 3 8 5 7 4
```

Solution # 370
```
9 7 2 4 3 8 6 1 5
4 3 5 6 7 1 2 8 9
1 8 6 5 2 9 7 4 3
3 5 4 7 1 2 9 6 8
7 2 8 9 5 6 1 3 4
6 9 1 8 4 3 5 2 7
8 4 7 1 6 5 3 9 2
5 6 3 2 9 4 8 7 1
2 1 9 3 8 7 4 5 6
```

Solution # 371
```
5 7 6 3 2 9 8 1 4
9 4 2 1 5 8 3 7 6
1 3 8 4 7 6 2 9 5
7 2 3 5 6 4 9 8 1
4 8 1 2 9 3 5 6 7
6 9 5 8 1 7 4 3 2
2 5 9 6 3 1 7 4 8
8 1 7 9 4 5 6 2 3
3 6 4 7 8 2 1 5 9
```

Solution # 372
```
2 6 9 3 8 5 1 4 7
4 1 3 9 2 7 6 8 5
8 7 5 4 1 6 9 3 2
5 3 7 8 6 1 2 9 4
9 2 8 5 4 3 7 1 6
6 4 1 7 9 2 8 5 3
3 8 2 6 5 9 4 7 1
7 9 6 1 3 4 5 2 8
1 5 4 2 7 8 3 6 9
```

Solution # 373
```
1 3 9 4 5 7 2 8 6
6 5 2 3 1 8 4 7 9
7 4 8 9 2 6 5 3 1
4 2 5 6 3 1 8 9 7
3 7 1 8 4 9 6 5 2
9 8 6 2 7 5 1 4 3
5 1 3 7 6 4 9 2 8
2 9 4 1 8 3 7 6 5
8 6 7 5 9 2 3 1 4
```

Solution # 374
```
4 6 1 8 9 3 7 5 2
5 9 7 4 2 1 3 8 6
8 2 3 5 6 7 4 1 9
2 1 8 7 4 9 5 6 3
7 4 9 3 5 6 1 2 8
3 5 6 1 8 2 9 4 7
6 7 4 2 3 5 8 9 1
1 8 2 9 7 4 6 3 5
9 3 5 6 1 8 2 7 4
```

Solution # 375
```
3 5 9 4 6 2 8 1 7
8 7 2 1 5 9 3 6 4
1 4 6 8 7 3 2 9 5
4 3 1 6 9 8 5 7 2
7 6 8 5 2 1 9 4 3
9 2 5 7 3 4 6 8 1
5 9 7 2 4 6 1 3 8
6 1 4 3 8 5 7 2 9
2 8 3 9 1 7 4 5 6
```

Solution # 376
```
9 2 8 5 1 3 6 7 4
4 3 5 6 9 7 2 8 1
7 1 6 4 2 8 3 9 5
2 6 3 1 8 4 7 5 9
8 9 4 2 7 5 1 3 6
5 7 1 9 3 6 4 2 8
1 5 7 3 4 9 8 6 2
6 8 2 7 5 1 9 4 3
3 4 9 8 6 2 5 1 7
```

Solution # 377
```
6 3 1 8 5 9 4 2 7
5 2 9 6 4 7 8 1 3
7 4 8 1 3 2 5 6 9
3 1 5 2 8 4 9 7 6
4 8 6 7 9 1 2 3 5
2 9 7 3 6 5 1 4 8
9 7 2 5 1 3 6 8 4
1 6 4 9 7 8 3 5 2
8 5 3 4 2 6 7 9 1
```

Solution # 378
```
5 6 3 2 9 7 4 1 8
9 2 7 1 8 4 3 6 5
1 8 4 3 5 6 7 2 9
3 7 6 4 2 5 8 9 1
8 4 9 6 7 1 5 3 2
2 5 1 9 3 8 6 4 7
7 3 2 5 6 9 1 8 4
4 9 5 8 1 3 2 7 6
6 1 8 7 4 2 9 5 3
```

Solution # 379
```
2 6 4 5 9 7 1 3 8
3 9 5 6 8 1 4 7 2
1 7 8 2 4 3 6 5 9
9 8 2 3 5 6 7 1 4
6 3 1 9 7 4 8 2 5
4 5 7 8 1 2 3 9 6
7 1 9 4 6 5 2 8 3
8 4 3 1 2 9 5 6 7
5 2 6 7 3 8 9 4 1
```

Solution # 380
```
4 7 2 6 9 1 3 5 8
9 5 6 2 3 8 4 1 7
1 3 8 4 5 7 6 2 9
5 4 3 8 1 6 9 7 2
7 6 1 5 2 9 8 3 4
8 2 9 7 4 3 1 6 5
2 8 4 1 6 5 7 9 3
3 1 5 9 7 4 2 8 6
6 9 7 3 8 2 5 4 1
```

Solution # 381
```
1 6 9 7 8 2 4 5 3
7 2 5 3 6 4 1 8 9
8 4 3 5 9 1 7 6 2
6 8 1 2 4 9 3 7 5
3 5 4 1 7 8 2 9 6
9 7 2 6 5 3 8 4 1
4 1 8 9 2 6 5 3 7
5 3 6 8 1 7 9 2 4
2 9 7 4 3 5 6 1 8
```

Solution # 382
```
8 1 7 2 5 9 6 3 4
2 3 6 4 7 8 5 1 9
9 4 5 1 3 6 2 7 8
3 6 1 7 9 2 4 8 5
7 2 9 5 8 4 1 6 3
5 8 4 3 6 1 9 2 7
4 7 3 6 1 5 8 9 2
6 9 2 8 4 3 7 5 1
1 5 8 9 2 7 3 4 6
```

Solution # 383
```
5 4 7 6 9 8 2 3 1
2 8 1 5 7 3 4 6 9
3 6 9 4 2 1 8 7 5
1 9 6 8 3 7 5 2 4
7 2 4 1 6 5 3 9 8
8 5 3 9 4 2 7 1 6
6 1 2 7 5 4 9 8 3
9 3 5 2 8 6 1 4 7
4 7 8 3 1 9 6 5 2
```

Solution # 384
```
9 6 8 2 5 1 7 3 4
7 2 4 8 3 9 5 1 6
3 1 5 7 6 4 9 2 8
5 8 1 9 7 6 2 4 3
4 7 2 5 1 3 8 6 9
6 3 9 4 8 2 1 5 7
2 5 6 3 9 7 4 8 1
8 9 3 1 4 5 6 7 2
1 4 7 6 2 8 3 9 5
```

Solution # 385
```
1 2 9 5 3 4 6 8 7
5 6 8 1 2 7 9 4 3
7 3 4 9 8 6 1 5 2
9 7 5 2 4 8 3 6 1
6 8 2 7 1 3 5 9 4
3 4 1 6 5 9 2 7 8
4 1 3 8 9 5 7 2 6
2 5 6 4 7 1 8 3 9
8 9 7 3 6 2 4 1 5
```

Solution # 386
```
7 6 4 2 1 3 8 9 5
5 3 8 9 4 7 2 1 6
1 2 9 5 6 8 4 7 3
9 7 3 1 2 6 5 8 4
4 8 1 3 9 5 7 6 2
2 5 6 7 8 4 9 3 1
3 9 5 4 7 1 6 2 8
8 1 2 6 5 9 3 4 7
6 4 7 8 3 2 1 5 9
```

Solution # 387
```
7 5 4 2 3 1 6 9 8
8 2 3 4 9 6 5 7 1
9 1 6 7 5 8 3 4 2
1 7 2 5 8 3 9 6 4
6 3 5 9 2 4 8 1 7
4 8 9 6 1 7 2 5 3
5 6 8 1 7 2 4 3 9
2 9 7 3 4 5 1 8 6
3 4 1 8 6 9 7 2 5
```

Solution # 388
```
5 1 3 7 8 2 9 4 6
4 2 6 1 5 9 7 3 8
8 7 9 4 6 3 1 2 5
9 4 8 5 3 7 6 1 2
3 6 2 8 1 4 5 7 9
7 5 1 2 9 6 3 8 4
1 3 4 6 2 5 8 9 7
2 9 5 3 7 8 4 6 1
6 8 7 9 4 1 2 5 3
```

Solution # 389
```
6 7 5 9 4 8 2 3 1
4 8 2 7 1 3 6 5 9
3 1 9 5 6 2 7 4 8
5 3 1 6 2 4 8 9 7
2 4 6 8 7 9 5 1 3
7 9 8 3 5 1 4 2 6
9 5 7 2 3 6 1 8 4
8 6 4 1 9 5 3 7 2
1 2 3 4 8 7 9 6 5
```

Solution # 390
```
6 8 2 1 9 7 3 4 5
7 4 3 5 2 6 8 9 1
9 1 5 4 8 3 2 7 6
2 7 8 3 5 9 6 1 4
3 5 6 8 4 1 7 2 9
1 9 4 6 7 2 5 3 8
4 2 7 9 6 8 1 5 3
5 6 1 7 3 4 9 8 2
8 3 9 2 1 5 4 6 7
```

Solution # 391
```
7 4 5 2 8 6 3 1 9
2 3 1 4 7 9 5 6 8
6 8 9 5 1 3 4 2 7
3 1 4 7 2 5 8 9 6
8 9 2 1 6 4 7 3 5
5 7 6 3 9 8 2 4 1
9 5 3 6 4 7 1 8 2
1 6 7 8 3 2 9 5 4
4 2 8 9 5 1 6 7 3
```

Solution # 392
```
2 4 5 1 8 3 6 7 9
6 3 1 7 9 4 5 8 2
8 7 9 2 6 5 3 1 4
4 1 7 9 5 8 2 6 3
5 8 3 6 2 1 9 4 7
9 6 2 3 4 7 1 5 8
7 2 8 5 3 6 4 9 1
1 9 6 4 7 2 8 3 5
3 5 4 8 1 9 7 2 6
```

Solution # 393
```
7 8 5 4 9 3 6 2 1
4 6 9 1 2 8 3 7 5
3 2 1 6 7 5 9 4 8
5 3 6 2 4 7 8 1 9
2 9 8 5 6 1 4 3 7
1 7 4 8 3 9 5 6 2
8 4 2 7 5 6 1 9 3
6 5 3 9 1 2 7 8 4
9 1 7 3 8 4 2 5 6
```

Solution # 394
```
3 8 9 1 5 4 6 2 7
6 5 4 8 2 7 3 9 1
2 7 1 6 3 9 5 4 8
9 3 6 4 1 5 8 7 2
4 2 5 7 9 8 1 3 6
8 1 7 2 6 3 9 5 4
5 4 2 3 8 6 7 1 9
7 6 3 9 4 1 2 8 5
1 9 8 5 7 2 4 6 3
```

Solution # 395
```
9 3 6 1 5 8 4 2 7
2 7 8 6 9 4 1 5 3
4 1 5 3 7 2 6 8 9
6 4 7 2 3 1 5 9 8
5 8 3 7 6 9 2 1 4
1 2 9 8 4 5 7 3 6
7 5 1 9 8 6 3 4 2
8 6 2 4 1 3 9 7 5
3 9 4 5 2 7 8 6 1
```

Solution # 396
```
7 4 5 6 3 9 2 1 8
2 9 3 1 8 7 6 5 4
8 6 1 5 4 2 7 9 3
6 5 4 8 7 1 9 3 2
1 3 7 9 2 5 4 8 6
9 2 8 3 6 4 5 7 1
4 7 9 2 1 3 8 6 5
3 8 2 7 5 6 1 4 9
5 1 6 4 9 8 3 2 7
```

Solution # 397
```
8 3 2 6 5 9 4 7 1
9 4 7 8 1 2 3 5 6
5 6 1 4 3 7 9 8 2
6 1 5 3 4 8 7 2 9
3 7 8 9 2 6 5 1 4
2 9 4 5 7 1 6 3 8
1 5 3 2 9 4 8 6 7
7 8 9 1 6 3 2 4 5
4 2 6 7 8 5 1 9 3
```

Solution # 398
```
6 3 2 5 8 1 4 9 7
8 7 4 6 3 9 5 1 2
1 5 9 4 7 2 6 8 3
5 1 7 2 6 3 8 4 9
3 4 6 9 1 8 2 7 5
9 2 8 7 5 4 1 6 3
2 8 1 3 9 6 7 5 4
7 6 3 8 4 5 9 2 1
4 9 5 1 2 7 3 6 8
```

Solution # 399
```
5 1 3 4 9 2 6 8 7
2 7 9 6 1 8 3 5 4
4 8 6 5 7 3 9 2 1
9 3 7 1 8 6 5 4 2
1 4 5 9 2 7 8 6 3
6 2 8 3 4 5 7 1 9
3 9 4 8 6 1 2 7 5
8 5 2 7 3 4 1 9 6
7 6 1 2 5 9 4 3 8
```

Solution # 400
```
8 2 1 4 5 7 6 9 3
5 6 7 2 9 3 8 4 1
9 3 4 1 6 8 5 7 2
4 9 8 3 1 5 2 6 7
2 5 6 9 7 4 3 1 8
1 7 3 8 2 6 9 5 4
3 1 5 6 4 2 7 8 9
7 4 2 5 8 9 1 3 6
6 8 9 7 3 1 4 2 5
```

Solution # 401
```
3 5 6 4 8 1 2 7 9
4 7 9 3 5 2 8 1 6
1 8 2 6 7 9 4 3 5
9 6 1 5 2 7 3 4 8
5 4 7 1 3 8 6 9 2
8 2 3 9 6 4 7 5 1
2 9 8 7 4 5 1 6 3
7 3 5 8 1 6 9 2 4
6 1 4 2 9 3 5 8 7
```

Solution # 402
```
8 1 2 6 7 5 4 3 9
3 9 4 1 2 8 5 6 7
6 7 5 9 3 4 2 8 1
9 4 7 5 1 6 3 2 8
5 8 3 2 9 7 6 1 4
2 6 1 4 8 3 9 7 5
1 3 9 7 5 2 8 4 6
4 5 8 3 6 1 7 9 2
7 2 6 8 4 9 1 5 3
```

Solution # 403
```
7 5 3 4 1 6 8 2 9
9 2 1 3 7 8 4 5 6
4 8 6 2 9 5 7 1 3
3 1 7 9 5 2 6 8 4
5 9 4 8 6 1 3 7 2
2 6 8 7 4 3 5 9 1
6 3 9 1 8 7 2 4 5
1 7 5 6 2 4 9 3 8
8 4 2 5 3 9 1 6 7
```

Solution # 404
```
9 7 1 2 3 5 8 6 4
4 2 3 8 7 6 1 9 5
5 6 8 9 4 1 2 3 7
1 5 2 4 6 9 7 8 3
6 8 4 3 2 7 5 1 9
7 3 9 1 5 8 4 2 6
2 9 6 5 8 4 3 7 1
3 1 5 7 9 2 6 4 8
8 4 7 6 1 3 9 5 2
```

Solution # 405
```
9 2 3 7 1 6 8 5 4
1 5 8 3 2 4 6 9 7
6 7 4 9 5 8 2 3 1
8 9 7 2 3 1 4 6 5
2 6 1 4 9 5 7 8 3
3 4 5 8 6 7 1 2 9
5 8 6 1 7 3 9 4 2
7 3 9 6 4 2 5 1 8
4 1 2 5 8 9 3 7 6
```

Solution # 406
```
5 1 9 3 8 7 2 4 6
2 7 6 4 9 1 3 5 8
8 4 3 2 6 5 7 9 1
6 3 5 7 4 8 1 2 9
1 8 2 5 3 9 6 7 4
4 9 7 1 2 6 5 8 3
9 5 4 6 1 2 8 3 7
3 2 1 8 7 4 9 6 5
7 6 8 9 5 3 4 1 2
```

Solution # 407
```
5 8 1 9 3 6 4 2 7
2 6 3 7 5 4 1 8 9
4 9 7 8 2 1 6 3 5
7 3 4 6 1 8 5 9 2
6 2 5 3 9 7 8 4 1
9 1 8 2 4 5 7 6 3
1 4 2 5 6 9 3 7 8
3 7 6 1 8 2 9 5 4
8 5 9 4 7 3 2 1 6
```

Solution # 408
```
5 2 7 4 1 3 9 8 6
9 1 3 6 2 8 7 4 5
4 6 8 5 7 9 3 2 1
6 7 4 1 5 2 8 3 9
3 8 1 7 9 4 6 5 2
2 9 5 8 3 6 1 7 4
1 5 2 9 8 7 4 6 3
8 4 9 3 6 5 2 1 7
7 3 6 2 4 1 5 9 8
```

Solution # 409
```
6 4 7 5 8 2 9 1 3
5 2 9 1 4 3 6 8 7
3 8 1 7 9 6 5 2 4
1 5 8 3 2 9 7 4 6
9 6 3 4 1 7 2 5 8
2 7 4 6 5 8 3 9 1
8 3 5 9 7 4 1 6 2
4 9 6 2 3 1 8 7 5
7 1 2 8 6 5 4 3 9
```

Solution # 410
```
7 2 9 3 6 8 4 1 5
3 5 1 9 4 2 6 7 8
8 6 4 1 7 5 9 3 2
2 1 8 7 5 9 3 4 6
6 9 5 8 3 4 7 2 1
4 3 7 6 2 1 8 5 9
9 7 2 5 8 3 1 6 4
5 8 3 4 1 6 2 9 7
1 4 6 2 9 7 5 8 3
```

Solution # 411
```
1 9 3 7 6 4 2 8 5
4 5 7 2 9 8 1 3 6
8 2 6 5 1 3 4 7 9
3 6 8 4 5 7 9 1 2
2 7 9 6 3 1 5 4 8
5 1 4 8 2 9 3 6 7
6 4 2 1 7 5 8 9 3
9 8 5 3 4 6 7 2 1
7 3 1 9 8 2 6 5 4
```

Solution # 412
```
4 2 1 6 8 9 5 3 7
5 3 6 4 2 7 8 9 1
7 9 8 5 1 3 4 6 2
1 8 7 3 9 4 2 5 6
2 6 9 8 5 1 7 4 3
3 4 5 2 7 6 1 8 9
8 1 2 9 6 5 3 7 4
6 7 4 1 3 8 9 2 5
9 5 3 7 4 2 6 1 8
```

Solution # 413
```
6 9 8 7 2 5 4 3 1
5 7 1 8 4 3 6 9 2
2 3 4 6 9 1 8 5 7
3 6 7 4 8 2 9 1 5
8 1 5 9 3 6 2 7 4
9 4 2 5 1 7 3 8 6
7 8 9 2 5 4 1 6 3
1 2 6 3 7 9 5 4 8
4 5 3 1 6 8 7 2 9
```

Solution # 414
```
6 2 9 5 1 8 4 7 3
3 5 1 6 4 7 8 9 2
8 4 7 9 2 3 6 5 1
7 9 8 2 3 1 5 6 4
2 3 5 4 6 9 1 8 7
4 1 6 7 8 5 2 3 9
5 8 4 3 9 2 7 1 6
1 6 3 8 7 4 9 2 5
9 7 2 1 5 6 3 4 8
```

Solution # 415
```
7 5 9 2 3 6 1 4 8
1 3 2 5 8 4 9 6 7
4 8 6 9 7 1 3 5 2
6 1 8 7 4 9 2 3 5
2 9 5 6 1 3 7 8 4
3 4 7 8 2 5 6 1 9
8 2 1 3 5 7 4 9 6
5 6 4 1 9 2 8 7 3
9 7 3 4 6 8 5 2 1
```

Solution # 416
```
9 8 4 2 1 3 5 7 6
2 7 1 5 4 6 9 8 3
5 6 3 7 9 8 2 4 1
1 2 8 4 3 5 7 6 9
4 9 5 6 7 2 3 1 8
7 3 6 1 8 9 4 5 2
8 5 9 3 6 4 1 2 7
6 1 2 9 5 7 8 3 4
3 4 7 8 2 1 6 9 5
```

Solution # 417
```
9 5 7 1 8 6 2 4 3
6 2 4 9 3 7 8 5 1
3 8 1 2 4 5 6 9 7
1 9 5 3 6 4 7 8 2
8 6 2 5 7 1 9 3 4
4 7 3 8 2 9 5 1 6
2 4 9 6 5 3 1 7 8
7 1 6 4 9 8 3 2 5
5 3 8 7 1 2 4 6 9
```

Solution # 418
```
2 3 8 5 4 9 1 6 7
1 9 6 7 8 3 4 2 5
7 5 4 2 1 6 9 8 3
8 1 9 6 5 2 3 7 4
3 6 5 1 7 4 2 9 8
4 2 7 9 3 8 5 1 6
6 4 1 8 2 5 7 3 9
9 7 3 4 6 1 8 5 2
5 8 2 3 9 7 6 4 1
```

Solution # 419
```
4 3 7 6 8 2 5 9 1
2 9 1 4 3 5 6 8 7
6 8 5 7 1 9 4 2 3
9 4 2 1 6 7 3 5 8
3 1 6 5 9 8 7 4 2
5 7 8 3 2 4 1 6 9
1 5 9 2 4 3 8 7 6
7 2 3 8 5 6 9 1 4
8 6 4 9 7 1 2 3 5
```

Solution # 420
```
3 4 9 8 6 7 1 5 2
1 7 5 2 9 4 6 3 8
8 2 6 1 5 3 4 9 7
5 8 3 6 1 2 9 7 4
4 6 1 9 7 5 8 2 3
2 9 7 4 3 8 5 6 1
9 1 4 7 2 6 3 8 5
7 3 8 5 4 9 2 1 6
6 5 2 3 8 1 7 4 9
```

Solution # 421
```
8 4 7 9 3 6 1 5 2
9 5 2 8 7 1 6 4 3
6 1 3 2 5 4 8 7 9
3 8 6 4 1 5 9 2 7
7 9 4 3 8 2 5 6 1
1 2 5 6 9 7 4 3 8
5 3 8 7 6 9 2 1 4
2 7 1 5 4 8 3 9 6
4 6 9 1 2 3 7 8 5
```

Solution # 422
```
3 9 2 4 6 5 7 1 8
7 4 1 8 2 3 5 9 6
5 8 6 7 9 1 4 2 3
6 7 4 9 5 8 2 3 1
1 5 3 6 7 2 9 8 4
8 2 9 3 1 4 6 7 5
4 1 5 2 3 7 8 6 9
2 6 8 1 4 9 3 5 7
9 3 7 5 8 6 1 4 2
```

Solution # 423
```
3 7 5 1 2 9 4 8 6
8 1 6 3 5 4 9 2 7
9 4 2 8 6 7 1 5 3
4 2 7 6 8 1 5 3 9
1 3 8 2 9 5 6 7 4
6 5 9 4 7 3 8 1 2
2 8 3 5 4 6 7 9 1
7 6 1 9 3 8 2 4 5
5 9 4 7 1 2 3 6 8
```

Solution # 424
```
1 7 3 2 6 4 9 5 8
6 5 9 3 7 8 1 2 4
8 2 4 9 5 1 3 7 6
2 6 7 5 8 9 4 1 3
3 9 1 7 4 2 6 8 5
5 4 8 1 3 6 7 9 2
9 8 5 6 1 3 2 4 7
4 1 6 8 2 7 5 3 9
7 3 2 4 9 5 8 6 1
```

Solution # 425
```
5 8 2 3 1 9 7 4 6
7 6 1 8 4 2 5 9 3
3 4 9 5 6 7 2 8 1
1 2 8 9 5 6 4 3 7
6 7 3 2 8 4 1 5 9
4 9 5 7 3 1 6 2 8
8 5 6 4 7 3 9 1 2
9 1 4 6 2 8 3 7 5
2 3 7 1 9 5 8 6 4
```

Solution # 426
```
9 3 2 1 6 5 4 7 8
1 7 6 4 8 9 2 5 3
8 5 4 3 2 7 6 1 9
7 8 1 2 9 3 5 4 6
6 2 3 7 5 4 9 8 1
4 9 5 8 1 6 7 3 2
2 4 8 6 7 1 3 9 5
5 1 7 9 3 2 8 6 4
3 6 9 5 4 8 1 2 7
```

Solution # 427
```
4 7 5 8 6 1 2 3 9
9 6 1 2 3 7 4 8 5
8 3 2 4 5 9 7 6 1
1 8 7 5 2 4 3 9 6
3 9 4 6 7 8 1 5 2
2 5 6 1 9 3 8 4 7
5 4 9 3 1 2 6 7 8
6 2 3 7 8 5 9 1 4
7 1 8 9 4 6 5 2 3
```

Solution # 428
```
8 7 5 1 4 2 9 3 6
2 1 6 5 3 9 4 7 8
3 4 9 8 6 7 1 2 5
6 3 4 9 8 5 2 1 7
9 5 1 7 2 4 6 8 3
7 2 8 3 1 6 5 4 9
4 8 7 6 9 1 3 5 2
1 9 3 2 5 8 7 6 4
5 6 2 4 7 3 8 9 1
```

Solution # 429
```
2 8 9 5 4 6 7 1 3
5 1 7 2 3 8 4 6 9
6 3 4 9 1 7 5 2 8
9 2 6 7 8 4 3 5 1
7 5 1 3 6 2 9 8 4
3 4 8 1 5 9 2 7 6
4 7 2 8 9 1 6 3 5
1 9 3 6 2 5 8 4 7
8 6 5 4 7 3 1 9 2
```

Solution # 430
```
5 4 6 2 7 3 9 8 1
7 9 2 8 1 6 3 4 5
1 3 8 5 9 4 2 6 7
3 6 1 7 2 5 8 9 4
2 8 5 9 4 1 6 7 3
9 7 4 6 3 8 1 5 2
8 2 3 4 6 7 5 1 9
6 1 7 3 5 9 4 2 8
4 5 9 1 8 2 7 3 6
```

Solution # 431
```
1 9 2 6 3 7 4 5 8
4 8 7 5 2 9 3 1 6
3 6 5 4 8 1 2 9 7
7 3 1 8 9 2 6 4 5
5 2 9 3 6 4 8 7 1
6 4 8 7 1 5 9 2 3
9 5 6 1 4 3 7 8 2
8 1 4 2 7 6 5 3 9
2 7 3 9 5 8 1 6 4
```

Solution # 432
```
2 9 4 8 5 6 1 3 7
8 7 1 4 3 9 2 5 6
5 6 3 2 7 1 9 8 4
7 2 9 5 4 8 3 6 1
4 3 5 6 1 7 8 9 2
1 8 6 3 9 2 7 4 5
3 4 7 9 2 5 6 1 8
9 1 8 7 6 4 5 2 3
6 5 2 1 8 3 4 7 9
```

Solution # 433
```
6 4 3 2 8 1 7 5 9
2 1 7 4 5 9 3 8 6
5 8 9 3 7 6 4 2 1
4 6 8 5 3 7 1 9 2
3 7 2 9 1 8 5 6 4
1 9 5 6 2 4 8 3 7
9 3 6 1 4 5 2 7 8
8 2 4 7 9 3 6 1 5
7 5 1 8 6 2 9 4 3
```

Solution # 434
```
1 9 7 6 4 8 2 3 5
2 3 6 9 5 1 7 4 8
5 4 8 2 3 7 6 1 9
8 6 3 5 1 2 4 9 7
7 5 1 4 9 6 3 8 2
9 2 4 8 7 3 1 5 6
6 8 5 3 2 4 9 7 1
4 1 9 7 6 5 8 2 3
3 7 2 1 8 9 5 6 4
```

Solution # 435
```
1 3 9 5 6 8 7 4 2
5 4 6 2 7 1 3 8 9
7 8 2 4 3 9 5 6 1
6 9 3 1 8 2 4 5 7
2 5 4 7 9 6 8 1 3
8 7 1 3 4 5 2 9 6
4 1 5 9 2 7 6 3 8
9 2 8 6 5 3 1 7 4
3 6 7 8 1 4 9 2 5
```

Solution # 436
```
8 7 2 4 5 9 6 1 3
5 1 9 6 3 8 7 4 2
3 6 4 2 7 1 5 8 9
2 3 7 1 6 4 8 9 5
9 5 6 3 8 7 4 2 1
1 4 8 9 2 5 3 6 7
6 9 3 5 4 2 1 7 8
7 2 5 8 1 6 9 3 4
4 8 1 7 9 3 2 5 6
```

Solution # 437
```
2 6 1 3 5 4 9 8 7
8 3 7 2 6 9 5 4 1
5 4 9 8 7 1 6 2 3
9 5 4 7 1 8 2 3 6
7 2 6 4 3 5 8 1 9
3 1 8 6 9 2 7 5 4
4 7 5 1 2 6 3 9 8
6 8 2 9 4 3 1 7 5
1 9 3 5 8 7 4 6 2
```

Solution # 438
```
6 3 1 8 2 5 4 9 7
4 8 5 9 1 7 2 3 6
9 7 2 4 6 3 5 1 8
1 5 7 3 8 4 9 6 2
8 9 4 2 5 6 1 7 3
3 2 6 1 7 9 8 4 5
7 6 8 5 4 1 3 2 9
5 1 9 6 3 2 7 8 4
2 4 3 7 9 8 6 5 1
```

Solution # 439
```
2 4 7 9 6 1 5 8 3
6 5 9 2 3 8 4 7 1
3 1 8 4 5 7 2 9 6
4 3 2 8 9 6 1 5 7
1 8 6 5 7 3 9 4 2
7 9 5 1 4 2 3 6 8
8 6 4 3 2 5 7 1 9
5 2 1 7 8 9 6 3 4
9 7 3 6 1 4 8 2 5
```

Solution # 440
```
5 7 9 2 6 4 3 8 1
4 1 2 3 8 5 7 9 6
6 3 8 9 1 7 4 5 2
8 2 5 6 4 9 1 3 7
1 9 4 7 2 3 8 6 5
7 6 3 8 5 1 9 2 4
3 4 6 1 9 2 5 7 8
2 5 7 4 3 8 6 1 9
9 8 1 5 7 6 2 4 3
```

Solution # 441
```
7 1 3 5 2 4 9 6 8
4 2 8 6 9 7 1 3 5
5 6 9 3 1 8 4 2 7
9 8 7 2 4 6 3 5 1
3 4 2 7 5 1 6 8 9
6 5 1 8 3 9 2 7 4
2 9 6 4 8 5 7 1 3
8 7 4 1 6 3 5 9 2
1 3 5 9 7 2 8 4 6
```

Solution # 442
```
8 5 3 9 4 7 1 2 6
4 9 2 8 6 1 5 3 7
1 6 7 2 3 5 4 8 9
3 4 9 6 7 8 2 1 5
2 7 1 3 5 9 8 6 4
6 8 5 4 1 2 9 7 3
9 2 4 7 8 3 6 5 1
5 3 6 1 2 4 7 9 8
7 1 8 5 9 6 3 4 2
```

Solution # 443
```
3 7 8 2 9 4 1 5 6
6 5 9 3 8 1 2 4 7
1 2 4 7 5 6 3 9 8
7 8 3 9 1 2 4 6 5
9 4 1 5 6 8 7 2 3
2 6 5 4 7 3 8 1 9
4 3 6 8 2 9 5 7 1
8 1 7 6 4 5 9 3 2
5 9 2 1 3 7 6 8 4
```

Solution # 444
```
4 2 6 3 8 5 1 9 7
1 5 9 7 4 2 8 3 6
3 7 8 6 1 9 2 5 4
9 3 7 2 6 8 4 1 5
8 1 2 4 5 7 3 6 9
5 6 4 1 9 3 7 2 8
7 4 5 9 2 1 6 8 3
6 8 1 5 3 4 9 7 2
2 9 3 8 7 6 5 4 1
```

Solution # 445
```
1 7 5 2 9 3 6 8 4
3 2 6 4 5 8 1 9 7
8 9 4 1 7 6 2 5 3
6 5 8 3 1 9 7 4 2
9 3 2 6 4 7 8 1 5
7 4 1 8 2 5 3 6 9
2 1 7 9 6 4 5 3 8
5 8 9 7 3 1 4 2 6
4 6 3 5 8 2 9 7 1
```

Solution # 446
```
3 1 5 4 9 8 7 2 6
8 4 7 2 6 1 3 5 9
6 2 9 7 3 5 8 4 1
2 7 8 6 4 9 5 1 3
5 3 4 1 2 7 6 9 8
1 9 6 8 5 3 2 7 4
4 5 2 9 8 6 1 3 7
9 8 1 3 7 2 4 6 5
7 6 3 5 1 4 9 8 2
```

Solution # 447
```
3 7 5 6 2 1 4 9 8
4 1 2 3 8 9 6 7 5
8 9 6 5 4 7 2 3 1
2 5 3 7 6 8 9 1 4
9 6 4 1 5 2 3 8 7
1 8 7 4 9 3 5 6 2
5 3 9 8 1 4 7 2 6
7 4 8 2 3 6 1 5 9
6 2 1 9 7 5 8 4 3
```

Solution # 448
```
9 2 5 8 7 3 6 1 4
1 8 7 6 4 9 2 5 3
3 6 4 2 5 1 9 7 8
5 7 8 3 9 2 4 6 1
6 9 3 4 1 7 8 2 5
2 4 1 5 8 6 3 9 7
8 1 6 9 3 5 7 4 2
7 3 2 1 6 4 5 8 9
4 5 9 7 2 8 1 3 6
```

Solution # 449
```
8 2 6 9 7 1 3 5 4
9 1 7 3 4 5 2 6 8
3 4 5 2 8 6 9 7 1
4 8 9 7 5 2 1 3 6
6 5 2 8 1 3 4 9 7
7 3 1 4 6 9 8 2 5
2 6 8 1 3 7 5 4 9
1 7 3 5 9 4 6 8 2
5 9 4 6 2 8 7 1 3
```

Solution # 450
```
2 9 1 8 3 4 5 6 7
8 4 5 7 1 6 9 2 3
3 6 7 5 2 9 8 4 1
7 5 6 3 9 8 2 1 4
9 3 2 6 4 1 7 5 8
1 8 4 2 5 7 3 9 6
5 7 9 1 6 3 4 8 2
6 2 8 4 7 5 1 3 9
4 1 3 9 8 2 6 7 5
```

Solution # 451
```
5 6 1 4 2 7 3 8 9
8 3 4 9 6 1 7 5 2
2 7 9 3 8 5 4 6 1
1 5 6 2 7 4 8 9 3
4 9 8 1 3 6 2 7 5
3 2 7 5 9 8 1 4 6
6 4 2 8 5 3 9 1 7
9 8 5 7 1 2 6 3 4
7 1 3 6 4 9 5 2 8
```

Solution # 452
```
9 7 4 2 6 3 5 1 8
2 8 1 7 5 4 3 6 9
3 6 5 1 8 9 2 7 4
6 2 3 8 4 5 1 9 7
7 4 9 3 1 2 8 5 6
5 1 8 9 7 6 4 2 3
8 5 7 6 2 1 9 4 3
1 9 6 4 3 8 7 2 5
4 3 2 5 9 7 6 8 1
```

Solution # 453
```
3 6 8 1 2 7 4 9 5
9 7 4 3 6 5 8 1 2
2 5 1 8 4 9 7 6 3
5 3 7 9 8 1 6 2 4
1 9 2 4 7 6 5 3 8
4 8 6 2 5 3 9 7 1
6 2 5 7 1 8 3 4 9
8 1 3 6 9 4 2 5 7
7 4 9 5 3 2 1 8 6
```

Solution # 454
```
4 8 5 3 2 7 9 1 6
7 2 6 5 1 9 4 8 3
1 9 3 8 6 4 2 5 7
2 4 1 9 7 3 8 6 5
9 5 7 6 4 8 3 2 1
3 6 8 2 5 1 7 9 4
5 3 4 1 9 2 6 7 8
6 7 2 4 8 5 1 3 9
8 1 9 7 3 6 5 4 2
```

Solution # 455
```
1 2 5 3 7 9 6 8 4
7 3 9 4 8 6 5 2 1
6 8 4 5 1 2 3 9 7
5 1 2 9 4 3 7 6 8
3 6 8 1 2 7 4 5 9
4 9 7 6 5 8 1 3 2
8 4 6 7 9 5 2 1 3
2 5 1 8 3 4 9 7 6
9 7 3 2 6 1 8 4 5
```

Solution # 456
```
1 8 5 7 3 9 4 2 6
3 7 2 8 6 4 9 5 1
4 9 6 2 5 1 7 3 8
9 6 4 1 2 5 3 8 7
2 1 7 9 8 3 5 6 4
8 5 3 4 7 6 1 9 2
5 2 9 6 1 7 8 4 3
6 3 1 5 4 8 2 7 9
7 4 8 3 9 2 6 1 5
```

Solution # 457
```
9 4 3 5 7 2 8 1 6
5 8 1 6 3 4 2 9 7
2 6 7 9 8 1 4 5 3
3 5 8 4 9 6 7 2 1
1 7 4 2 5 3 9 6 8
6 2 9 8 1 7 3 4 5
8 3 5 1 4 9 6 7 2
4 1 2 7 6 8 5 3 9
7 9 6 3 2 5 1 8 4
```

Solution # 458
```
7 6 9 2 1 4 8 3 5
8 5 2 9 6 3 1 4 7
3 4 1 5 7 8 2 6 9
6 2 5 8 4 1 9 7 3
1 7 4 3 9 2 6 5 8
9 8 3 6 5 7 4 2 1
5 1 8 7 2 6 3 9 4
2 3 7 4 8 9 5 1 6
4 9 6 1 3 5 7 8 2
```

Solution # 459
```
1 4 3 6 5 8 2 9 7
7 2 6 3 1 9 8 5 4
9 8 5 7 4 2 6 3 1
6 1 4 9 8 5 7 2 3
2 5 7 1 6 3 4 8 9
3 9 8 2 7 4 1 6 5
5 6 2 4 9 7 3 1 8
4 3 9 8 2 1 5 7 6
8 7 1 5 3 6 9 4 2
```

Solution # 460
```
8 3 2 4 6 7 9 1 5
5 6 4 3 1 9 7 2 8
9 1 7 2 8 5 4 3 6
2 5 3 9 7 8 1 6 4
7 4 1 5 3 6 2 8 9
6 9 8 1 2 4 5 7 3
3 8 9 7 5 2 6 4 1
4 2 5 6 9 1 8 7 3
1 7 6 8 4 3 5 9 2
```

Solution # 461
```
9 4 3 5 7 8 2 1 6
5 1 2 3 9 6 4 7 8
6 7 8 2 4 1 9 5 3
3 9 7 4 8 2 1 6 5
2 5 4 1 6 3 7 8 9
8 6 1 9 5 7 3 2 4
4 8 9 7 1 5 6 3 2
1 3 6 8 2 4 5 9 7
7 2 5 6 3 9 8 4 1
```

Solution # 462
```
8 1 2 3 4 9 6 5 7
3 5 9 8 7 6 1 4 2
6 4 7 5 1 2 3 8 9
5 3 1 9 2 7 8 6 4
9 8 4 6 3 5 2 7 1
2 7 6 1 8 4 5 9 3
7 2 5 4 6 3 9 1 8
1 6 3 7 9 8 4 2 5
4 9 8 2 5 1 7 3 6
```

Solution # 463
```
5 2 6 9 1 3 7 4 8
4 9 3 8 6 7 5 1 2
1 8 7 2 4 5 6 9 3
2 5 4 6 7 8 1 3 9
6 3 9 5 2 1 8 7 4
8 7 1 4 3 9 2 6 5
3 4 5 1 8 6 9 2 7
7 1 8 3 9 2 4 5 6
9 6 2 7 5 4 3 8 1
```

Solution # 464
```
6 5 9 4 3 1 2 7 8
3 4 7 2 9 8 5 6 1
2 8 1 5 6 7 9 3 4
5 2 4 9 8 3 7 1 6
8 7 6 1 4 2 3 9 5
1 9 3 7 5 6 4 8 2
4 6 5 8 7 9 1 2 3
9 3 2 6 1 4 8 5 7
7 1 8 3 2 5 6 4 9
```

Solution # 465
```
4 1 6 8 3 7 5 9 2
7 3 5 2 1 9 4 6 8
2 8 9 6 5 4 7 3 1
1 9 3 5 4 6 8 2 7
5 2 4 7 9 8 6 1 3
8 6 7 3 2 1 9 5 4
6 4 1 9 7 3 2 8 5
3 5 8 4 6 2 1 7 9
9 7 2 1 8 5 3 4 6
```

Solution # 466
```
7 6 8 9 1 5 2 4 3
1 4 2 3 7 6 5 9 8
3 5 9 4 2 8 6 7 1
9 1 3 2 5 7 4 8 6
6 2 5 8 4 1 9 3 7
8 7 4 6 9 3 1 2 5
5 9 1 7 8 4 3 6 2
2 8 6 1 3 9 7 5 4
4 3 7 5 6 2 8 1 9
```

Solution # 467
```
3 7 2 5 6 4 9 8 1
4 6 1 8 3 9 7 5 2
8 9 5 2 1 7 6 3 4
2 8 6 1 7 3 4 9 5
1 3 9 4 5 6 2 7 8
7 5 4 9 8 2 1 6 3
6 4 7 3 2 8 5 1 9
5 2 8 7 9 1 3 4 6
9 1 3 6 4 5 8 2 7
```

Solution # 468
```
2 1 7 3 8 6 5 4 9
5 9 4 7 2 1 6 3 8
6 8 3 4 5 9 1 2 7
4 6 9 5 3 7 2 8 1
8 3 1 6 9 2 4 7 5
7 5 2 1 4 8 9 6 3
3 4 6 9 7 5 8 1 2
1 2 5 8 6 3 7 9 4
9 7 8 2 1 4 3 5 6
```

Solution # 469
```
5 9 7 8 4 6 3 1 2
3 6 4 2 5 1 9 7 8
2 1 8 7 3 9 6 4 5
9 8 6 5 7 2 1 3 4
7 2 3 6 1 4 5 8 9
1 4 5 9 8 3 2 6 7
6 5 1 4 2 8 7 9 3
8 7 9 3 6 5 4 2 1
4 3 2 1 9 8 7 5 6
```

Solution # 470
```
5 2 6 4 7 9 3 1 8
4 3 7 8 2 1 5 6 9
8 9 1 5 3 6 4 2 7
1 7 4 6 5 2 9 8 3
9 6 8 7 4 3 1 5 2
2 5 3 9 1 8 7 4 6
3 8 9 1 6 4 2 7 5
7 4 2 3 8 5 6 9 1
6 1 5 2 9 7 8 3 4
```

Solution # 471
```
4 6 3 2 5 7 8 9 1
7 1 2 3 9 8 5 6 4
8 5 9 1 6 4 7 2 3
5 8 4 9 7 3 6 1 2
3 2 6 4 1 5 9 7 8
9 7 1 6 8 2 4 3 5
2 4 7 5 3 6 1 8 9
6 9 5 8 2 1 3 4 7
1 3 8 7 4 9 2 5 6
```

Solution # 472
```
6 5 8 9 7 2 4 3 1
2 1 7 3 4 8 6 9 5
3 9 4 5 6 1 2 7 8
5 4 6 7 8 3 9 1 2
1 2 3 6 9 5 7 8 4
8 7 9 2 1 4 5 6 3
7 3 5 8 2 9 1 4 6
9 8 1 4 5 6 3 2 7
4 6 2 1 3 7 8 5 9
```

Solution # 473
```
1 9 3 7 8 5 2 6 4
4 5 7 3 6 2 8 1 9
6 2 8 4 1 9 3 7 5
5 3 9 2 7 8 1 4 6
8 6 4 5 3 1 9 2 7
7 1 2 9 4 6 5 8 3
9 4 6 1 2 3 7 5 8
2 7 5 8 9 4 6 3 1
3 8 1 6 5 7 4 9 2
```

Solution # 474
```
1 2 5 7 8 6 4 9 3
8 3 7 4 9 5 6 2 1
9 6 4 2 1 3 5 8 7
2 5 1 6 7 9 3 4 8
6 9 8 5 3 4 7 1 2
7 4 3 8 2 1 9 6 5
3 7 6 1 4 2 8 5 9
5 8 2 9 6 7 1 3 4
4 1 9 3 5 8 2 7 6
```

Solution # 475
```
4 2 9 8 6 5 1 3 7
5 7 8 4 3 1 6 2 9
1 3 6 7 9 2 4 8 5
3 5 1 9 7 6 2 4 8
9 8 2 3 1 4 7 5 6
7 6 4 2 5 8 3 9 1
2 9 7 1 8 3 5 6 4
8 4 5 6 2 7 9 1 3
6 1 3 5 4 9 8 7 2
```

Solution # 476
```
3 9 1 4 5 8 7 2 6
2 6 4 1 9 7 5 8 3
8 5 7 2 3 6 1 4 9
1 7 6 8 4 5 9 3 2
4 2 5 9 7 3 8 6 1
9 8 3 6 1 2 4 7 5
5 3 8 7 2 1 6 9 4
6 1 9 3 8 4 2 5 7
7 4 2 5 6 9 3 1 8
```

Solution # 477
```
4 5 1 3 9 6 7 2 8
2 3 9 5 7 8 4 1 6
6 8 7 2 4 1 5 9 3
1 7 4 8 6 2 9 3 5
9 6 3 4 1 5 2 8 7
8 2 5 7 3 9 6 4 1
3 4 2 1 5 7 8 6 9
7 1 6 9 8 4 3 5 2
5 9 8 6 2 3 1 7 4
```

Solution # 478
```
2 4 8 1 3 9 7 6 5
9 6 5 4 7 8 3 2 1
1 3 7 5 2 6 8 9 4
6 1 3 2 8 5 4 7 9
7 5 2 9 1 4 6 3 8
8 9 4 7 6 3 1 5 2
3 7 9 8 5 1 2 4 6
4 8 6 3 9 2 5 1 7
5 2 1 6 4 7 9 8 3
```

Solution # 479
```
7 3 6 2 4 1 5 8 9
9 1 2 5 8 3 7 6 4
5 4 8 7 9 6 1 2 3
2 6 7 4 3 5 8 9 1
4 8 1 6 7 9 2 3 5
3 9 5 1 2 8 6 4 7
6 2 4 3 1 7 9 5 8
8 7 3 9 5 2 4 1 6
1 5 9 8 6 4 3 7 2
```

Solution # 480
```
2 8 9 1 6 5 4 7 3
6 7 1 2 4 3 8 9 5
3 4 5 8 9 7 2 1 6
5 2 6 4 7 8 9 3 1
8 1 4 3 5 9 6 2 7
9 3 7 6 1 2 5 4 8
4 5 2 7 3 6 1 8 9
1 9 3 5 8 4 7 6 2
7 6 8 9 2 1 3 5 4
```

Solution # 481
```
5 4 8 1 3 6 9 2 7
2 1 9 8 5 7 6 3 4
7 6 3 9 2 4 1 8 5
9 5 1 7 6 8 3 4 2
3 2 6 5 4 9 8 7 1
8 7 4 3 1 2 5 9 6
1 3 7 2 9 5 4 6 8
4 8 5 6 7 3 2 1 9
6 9 2 4 8 1 7 5 3
```

Solution # 482
```
9 2 8 6 7 5 3 1 4
1 3 6 9 2 4 5 7 8
5 4 7 8 1 3 2 9 6
4 6 3 5 9 1 7 8 2
2 9 5 7 4 8 1 6 3
7 8 1 3 6 2 4 5 9
6 5 4 1 3 9 8 2 7
3 1 9 2 8 7 6 4 5
8 7 2 4 5 6 9 3 1
```

Solution # 483
```
5 8 2 3 1 6 9 4 7
1 4 3 8 7 9 5 6 2
9 6 7 4 5 2 3 8 1
7 5 4 6 9 1 8 2 3
2 3 8 7 4 5 6 1 9
6 1 9 2 3 8 7 5 4
8 9 1 5 2 7 4 3 6
3 7 6 1 8 4 2 9 5
4 2 5 9 6 3 1 7 8
```

Solution # 484
```
4 9 1 6 7 5 3 2 8
2 5 6 3 4 8 9 7 1
7 3 8 1 9 2 4 6 5
6 7 3 4 8 1 2 5 9
8 4 9 2 5 7 6 1 3
5 1 2 9 6 3 7 8 4
3 2 4 5 1 6 8 9 7
1 6 7 8 3 9 5 4 2
9 8 5 7 2 4 1 3 6
```

Solution # 485
```
2 5 7 4 9 1 6 3 8
1 9 8 3 2 6 5 4 7
3 4 6 7 5 8 1 2 9
6 7 1 2 4 3 9 8 5
4 8 3 5 1 9 7 6 2
9 2 5 8 6 7 4 1 3
5 6 9 1 3 2 8 7 4
8 3 4 6 7 5 2 9 1
7 1 2 9 8 4 3 5 6
```

Solution # 486
```
1 6 2 3 8 7 5 4 9
7 4 8 9 5 2 6 3 1
5 9 3 1 4 6 8 7 2
3 5 4 6 1 8 9 2 7
6 8 7 4 2 9 1 5 3
9 2 1 5 7 3 4 6 8
8 1 5 7 3 4 2 9 6
4 7 6 2 9 1 3 8 5
2 3 9 8 6 5 7 1 4
```

Solution # 487
```
6 1 5 3 9 4 7 8 2
8 9 2 7 1 6 5 4 3
7 4 3 2 8 5 1 9 6
1 6 8 9 4 3 2 5 7
9 3 7 5 2 1 8 6 4
2 5 4 8 6 7 9 3 1
5 8 1 4 3 2 6 7 9
3 7 6 1 5 9 4 2 8
4 2 9 6 7 8 3 1 5
```

Solution # 488
```
7 2 8 3 4 1 5 9 6
9 4 1 7 5 6 2 3 8
5 3 6 2 8 9 4 1 7
1 8 3 4 6 2 7 5 9
4 6 7 9 3 5 8 2 1
2 9 5 1 7 8 6 4 3
8 5 4 6 1 3 9 7 2
3 7 9 8 2 4 1 6 5
6 1 2 5 9 7 3 8 4
```

Solution # 489
```
1 7 5 3 4 6 8 2 9
2 8 3 7 5 9 6 1 4
4 9 6 8 2 1 7 5 3
7 5 1 6 9 4 2 3 8
9 3 8 2 7 5 4 6 1
6 2 4 1 3 8 9 7 5
3 1 9 4 6 2 5 8 7
5 6 7 9 8 3 1 4 2
8 4 2 5 1 7 3 9 6
```

Solution # 490
```
1 8 6 4 9 2 5 3 7
9 2 4 7 3 5 6 8 1
5 3 7 8 6 1 4 2 9
7 4 8 1 5 6 3 9 2
2 1 5 3 4 9 8 7 6
3 6 9 2 7 8 1 5 4
4 5 1 9 8 7 2 6 3
8 9 3 6 2 4 7 1 5
6 7 2 5 1 3 9 4 8
```

Solution # 491
```
7 8 4 1 9 6 2 3 5
2 9 1 3 5 8 4 6 7
3 5 6 2 7 4 9 1 8
9 3 2 5 4 7 1 8 6
5 4 8 9 6 1 3 7 2
1 6 7 8 2 3 5 9 4
4 1 3 7 8 5 6 2 9
8 2 5 6 3 9 7 4 1
6 7 9 4 1 2 8 5 3
```

Solution # 492
```
6 3 2 9 7 5 4 8 1
4 8 7 1 3 6 5 9 2
9 1 5 4 2 8 3 6 7
1 6 3 2 5 7 9 4 8
2 5 4 3 8 9 1 7 6
7 9 8 6 1 4 2 3 5
5 2 6 8 9 3 7 1 4
3 4 1 7 6 2 8 5 9
8 7 9 5 4 1 6 2 3
```

Solution # 493
```
9 3 4 2 8 6 5 7 1
8 7 5 3 4 1 6 2 9
1 2 6 5 7 9 4 8 3
6 9 3 4 5 8 7 1 2
5 4 8 7 1 2 9 3 6
7 1 2 9 6 3 8 4 5
2 8 7 6 3 5 1 9 4
3 6 1 8 9 4 2 5 7
4 5 9 1 2 7 3 6 8
```

Solution # 494
```
4 1 2 5 6 7 8 3 9
6 3 5 2 9 8 1 4 7
7 8 9 4 1 3 5 2 6
8 6 1 7 2 4 9 5 3
5 2 4 3 8 9 7 6 1
9 7 3 6 5 1 2 8 4
2 4 7 9 3 5 6 1 8
3 5 8 1 7 6 4 9 2
1 9 6 8 4 2 3 7 5
```

Solution # 495
```
8 5 9 6 2 3 7 4 1
7 4 2 1 8 9 5 3 6
1 3 6 5 4 7 2 8 9
3 1 5 4 7 2 6 9 8
6 2 8 9 3 5 4 1 7
4 9 7 8 1 6 3 5 2
2 8 4 3 6 1 9 7 5
5 7 3 2 9 8 1 6 4
9 6 1 7 5 4 8 2 3
```

Solution # 496
```
6 4 5 9 1 7 8 2 3
2 3 9 4 6 8 5 7 1
8 7 1 2 5 3 9 6 4
7 2 4 3 9 5 6 1 8
9 6 8 7 4 1 3 5 2
5 1 3 8 2 6 4 9 7
4 5 2 1 3 9 7 8 6
1 9 7 6 8 4 2 3 5
3 8 6 5 7 2 1 4 9
```

Solution # 497
```
4 1 2 5 3 8 6 9 7
9 3 6 4 2 7 8 1 5
7 8 5 6 1 9 4 2 3
1 2 4 8 7 6 3 5 9
8 5 9 1 4 3 2 7 6
3 6 7 9 5 2 1 4 8
5 9 8 2 6 4 7 3 1
2 7 1 3 8 5 9 6 4
6 4 3 7 9 1 5 8 2
```

Solution # 498
```
2 1 6 5 4 9 3 8 7
4 5 3 7 2 8 9 1 6
7 9 8 3 6 1 5 4 2
6 4 5 9 1 2 8 7 3
9 8 7 4 5 3 6 2 1
3 2 1 6 8 7 4 9 5
8 3 2 1 9 6 7 5 4
5 7 9 2 3 4 1 6 8
1 6 4 8 7 5 2 3 9
```

Solution # 499
```
7 6 5 9 4 1 8 2 3
4 3 2 7 8 5 9 1 6
1 9 8 2 3 6 5 7 4
2 5 9 4 1 7 6 3 8
3 4 1 8 6 9 2 5 7
8 7 6 3 5 2 4 9 1
9 2 4 6 7 3 1 8 5
5 8 7 1 2 4 3 6 9
6 1 3 5 9 8 7 4 2
```

Solution # 500
```
4 9 1 7 2 5 6 8 3
6 5 3 1 8 4 2 7 9
8 2 7 6 9 3 1 4 5
7 4 9 2 6 8 5 3 1
5 6 2 4 3 1 8 9 7
3 1 8 5 7 9 4 6 2
9 3 5 8 1 2 7 6 4
2 7 4 3 5 6 9 1 8
1 8 6 9 4 7 3 5 2
```

Solution # 501
```
4 3 6 8 9 1 2 5 7
5 2 9 7 6 4 1 8 3
1 7 8 3 5 2 4 9 6
2 1 3 4 8 6 9 7 5
8 4 5 1 7 9 6 3 2
6 9 7 5 2 3 8 1 4
9 8 4 2 3 7 5 6 1
7 5 2 6 1 8 3 4 9
3 6 1 9 4 5 7 2 8
```

Solution # 502
```
2 7 9 3 6 4 8 5 1
4 6 3 8 1 5 2 9 7
1 5 8 2 9 7 3 6 4
9 2 6 4 8 3 7 1 5
7 8 1 9 5 2 6 4 3
3 4 5 6 7 1 9 8 2
5 9 2 1 3 6 4 7 8
6 1 4 7 2 8 5 3 9
8 3 7 5 4 9 1 2 6
```

Solution # 503
```
7 2 8 1 5 4 3 9 6
4 3 9 6 7 2 1 8 5
5 1 6 3 9 8 2 4 7
1 7 4 2 8 3 6 5 9
6 5 2 7 4 9 8 1 3
8 9 3 5 6 1 7 2 4
9 6 1 8 3 5 4 7 2
2 4 7 9 1 6 5 3 8
3 8 5 4 2 7 9 6 1
```

Solution # 504
```
4 8 3 1 6 9 2 7 5
1 2 6 5 7 4 9 8 3
5 7 9 2 8 3 4 1 6
8 6 1 7 5 2 3 9 4
3 5 7 9 4 6 8 2 1
2 9 4 8 3 1 5 6 7
6 1 8 3 9 5 7 4 2
7 4 5 6 2 8 1 3 9
9 3 2 4 1 7 6 5 8
```

Solution # 505
```
6 3 5 8 4 9 2 7 1
2 4 9 7 5 1 6 3 8
7 1 8 2 3 6 5 9 4
8 6 2 5 9 7 1 4 3
4 7 3 6 1 8 9 5 2
9 5 1 4 2 3 7 8 6
3 2 7 9 6 4 8 1 5
5 8 6 1 7 2 4 3 9
1 9 4 3 8 5 7 6 2
```

Solution # 506
```
1 9 2 5 4 8 6 7 3
6 5 8 2 7 3 4 1 9
3 4 7 1 9 6 5 2 8
9 8 5 7 3 4 1 6 2
4 7 6 9 1 2 8 3 5
2 3 1 8 6 5 9 4 7
8 2 4 3 5 1 7 9 6
5 1 9 6 2 7 3 8 4
7 6 3 4 8 9 2 5 1
```

Solution # 507
```
3 5 2 4 9 1 7 6 8
7 6 9 5 2 8 3 1 4
8 4 1 6 7 3 2 9 5
6 9 4 1 5 7 8 2 3
5 3 8 9 4 2 6 7 1
2 1 7 3 8 6 4 5 9
9 8 5 2 6 4 1 3 7
4 2 3 7 1 5 9 8 6
1 7 6 8 3 9 5 4 2
```

Solution # 508
```
7 3 1 5 2 8 4 9 6
9 5 4 6 1 7 8 2 3
2 6 8 4 9 3 7 5 1
5 4 2 9 3 1 6 7 8
8 1 6 7 4 5 2 3 9
3 9 7 2 8 6 1 4 5
6 7 3 1 5 4 9 8 2
4 2 5 8 6 9 3 1 7
1 8 9 3 7 2 5 6 4
```

Solution # 509
```
3 6 5 1 9 4 7 2 8
9 7 2 5 3 8 4 1 6
8 1 4 2 7 6 5 9 3
5 2 7 4 6 1 8 3 9
4 9 8 3 2 7 1 6 5
1 3 6 8 5 9 2 4 7
2 4 3 6 8 5 9 7 1
6 5 9 7 1 2 3 8 4
7 8 1 9 4 3 6 5 2
```

Solution # 510
```
9 2 4 5 3 6 7 1 8
3 6 7 1 2 8 9 5 4
5 8 1 4 9 7 3 6 2
2 1 5 3 6 4 8 7 9
6 7 3 9 8 2 5 4 1
4 9 8 7 5 1 2 3 6
8 4 2 6 7 3 1 9 5
1 3 9 8 4 5 6 2 7
7 5 6 2 1 9 4 8 3
```

Solution # 511
```
2 5 9 7 4 6 3 8 1
7 8 4 1 3 5 2 9 6
3 1 6 2 8 9 4 5 7
1 3 2 4 7 8 5 6 9
4 9 7 6 5 2 1 3 8
5 6 8 3 9 1 7 4 2
6 4 1 9 2 3 8 7 5
8 2 3 5 6 7 9 1 4
9 7 5 8 1 4 6 2 3
```

Solution # 512
```
1 2 8 5 4 7 6 3 9
3 9 5 1 6 2 8 4 7
4 7 6 9 3 8 1 2 5
6 1 3 2 8 9 5 7 4
7 8 4 3 5 1 2 9 6
2 5 9 4 7 6 3 1 8
8 3 2 7 9 5 4 6 1
5 4 7 6 1 3 9 8 2
9 6 1 8 2 4 7 5 3
```

Solution # 513
```
5 1 4 8 2 9 3 6 7
8 3 2 7 4 6 5 1 9
7 9 6 1 5 3 2 8 4
2 5 1 9 6 4 8 7 3
3 6 8 2 7 5 9 4 1
4 7 9 3 8 1 6 5 2
1 8 5 4 3 2 7 9 6
9 2 7 6 1 8 4 3 5
6 4 3 5 9 7 1 2 8
```

Solution # 514
```
2 1 9 7 8 4 3 6 5
4 3 7 6 5 1 9 2 8
6 8 5 3 9 2 7 1 4
7 9 6 8 2 3 5 4 1
3 2 1 4 7 5 8 9 6
8 5 4 9 1 6 2 7 3
5 7 2 1 4 8 6 3 9
1 6 8 2 3 9 4 5 7
9 4 3 5 6 7 1 8 2
```

Solution # 515
```
1 6 2 3 9 8 5 4 7
8 7 9 1 4 5 3 6 2
3 4 5 6 7 2 9 1 8
2 1 4 7 3 6 8 5 9
6 5 7 8 1 9 2 3 4
9 3 8 5 2 4 6 7 1
7 2 1 9 6 3 4 8 5
4 8 3 2 5 1 7 9 6
5 9 6 4 8 7 1 2 3
```

Solution # 516
```
2 1 9 6 4 5 7 3 8
7 8 5 9 3 2 4 1 6
6 3 4 8 1 7 5 9 2
3 6 1 5 8 9 2 4 7
9 7 2 1 6 4 3 8 5
5 4 8 7 2 3 9 6 1
1 9 7 4 2 8 6 5 3
8 5 3 7 9 6 1 2 4
4 2 6 3 5 1 8 7 9
```

Solution # 517
```
6 2 7 8 4 3 9 1 5
8 9 5 1 7 2 3 6 4
3 4 1 5 9 6 8 7 2
1 7 4 6 2 9 5 8 3
2 8 3 4 5 7 1 9 6
9 5 6 3 1 8 2 4 7
4 1 2 9 6 5 7 3 8
5 3 9 7 8 4 6 2 1
7 6 8 2 3 1 4 5 9
```

Solution # 518
```
4 7 8 5 1 2 6 3 9
5 1 3 4 6 9 2 7 8
6 2 9 3 8 7 4 1 5
7 3 4 8 2 1 9 5 6
2 8 1 9 5 6 3 4 7
9 6 5 7 4 3 8 2 1
1 9 2 6 7 4 5 8 3
8 4 6 1 3 5 7 9 2
3 5 7 2 9 8 1 6 4
```

Solution # 519
```
3 6 7 8 2 5 1 4 9
2 4 9 7 3 1 8 5 6
5 8 1 4 6 9 7 2 3
6 5 8 9 1 4 2 3 7
1 2 4 5 7 3 6 9 8
9 7 3 6 8 2 5 1 4
4 9 6 2 5 8 3 7 1
7 3 5 1 4 6 9 8 2
8 1 2 3 9 7 4 6 5
```

Solution # 520
```
1 8 5 6 9 4 7 3 2
6 4 7 3 2 1 8 9 5
2 9 3 7 5 8 6 4 1
3 2 6 9 1 7 5 8 4
7 1 9 4 8 5 3 2 6
4 5 8 2 3 6 9 1 7
9 7 2 1 6 3 4 5 8
5 3 4 8 7 2 1 6 9
8 6 1 5 4 9 2 7 3
```

Solution # 521
```
5 6 1 4 9 8 2 7 3
8 2 4 3 1 7 6 5 9
7 3 9 5 2 6 8 1 4
2 9 5 7 6 4 1 3 8
6 8 3 1 5 9 7 4 2
1 4 7 2 8 3 9 6 5
4 5 2 1 7 9 3 8 6
9 7 8 6 3 2 5 4 1
3 1 6 8 4 5 7 9 2
```

Solution # 522
```
4 5 7 9 3 6 1 8 2
2 9 1 4 5 8 3 7 6
8 3 6 7 2 1 9 5 4
5 6 4 3 7 9 2 1 8
1 7 9 8 6 2 5 4 3
3 8 2 5 1 4 6 9 7
7 4 3 6 9 5 8 2 1
6 2 5 1 8 7 4 3 9
9 1 8 2 4 3 7 6 5
```

Solution # 523
```
6 4 9 1 8 5 2 3 7
2 3 1 4 9 7 8 6 5
5 8 7 3 6 2 1 4 9
1 2 5 6 3 4 7 9 8
3 7 4 5 2 9 6 1 8
9 6 3 8 7 1 5 2 4
4 5 6 2 1 8 9 7 3
8 9 2 7 5 6 4 8 1
7 1 8 9 4 3 6 5 2
```

Solution # 524
```
3 9 8 2 7 1 5 6 4
2 5 1 6 8 4 7 3 9
7 4 6 5 3 9 2 1 8
4 8 2 3 1 6 9 5 7
1 7 3 9 2 5 8 4 6
9 6 5 7 4 8 3 2 1
5 3 4 1 9 7 6 8 2
6 1 7 8 5 2 4 9 3
8 2 9 4 6 3 1 7 5
```

Solution # 525
```
3 5 9 1 4 7 6 8 2
4 2 6 3 8 9 7 5 1
8 7 1 5 2 6 3 4 9
5 4 2 7 9 3 8 1 6
9 8 3 4 6 1 5 2 7
1 6 7 8 5 2 4 9 3
7 3 5 9 1 4 2 6 8
6 9 8 2 7 5 1 3 4
2 1 4 6 3 8 9 7 5
```

Solution # 526
```
1 8 5 3 7 2 6 9 4
3 2 4 9 5 6 7 1 8
6 7 9 4 1 8 3 2 5
2 4 7 6 9 3 8 5 1
5 1 3 2 8 7 4 6 9
8 9 6 1 4 5 2 7 3
4 6 2 5 3 9 1 8 7
9 3 8 7 2 1 5 4 6
7 5 1 8 6 4 9 3 2
```

Solution # 527
```
2 5 1 8 3 6 9 4 7
6 7 9 1 4 2 8 5 3
3 4 8 7 5 9 2 6 1
7 2 3 4 9 8 6 1 5
9 6 5 2 1 3 4 7 8
8 1 4 5 6 7 3 2 9
5 3 7 6 8 4 1 9 2
4 9 2 3 7 1 5 8 6
1 8 6 9 2 5 7 3 4
```

Solution # 528
```
6 7 9 8 4 2 5 3 1
2 5 8 3 9 1 4 7 6
3 1 4 5 6 7 2 8 9
7 9 2 1 3 4 8 6 5
8 3 6 7 5 9 1 2 4
1 4 5 2 8 6 3 9 7
9 8 3 6 1 5 7 4 2
5 6 7 4 2 3 9 1 8
4 2 1 9 7 8 6 5 3
```

Solution # 529
```
5 7 8 2 1 4 6 3 9
6 1 4 3 9 8 7 2 5
3 2 9 5 7 6 1 8 4
8 6 2 9 4 3 5 7 1
1 5 3 6 8 7 9 4 2
4 9 7 1 2 5 3 6 8
9 3 5 4 6 2 8 1 7
7 4 6 8 5 1 2 9 3
2 8 1 7 3 9 4 5 6
```

Solution # 530
```
5 9 7 1 6 2 3 8 4
6 3 8 5 4 9 1 2 7
2 4 1 7 8 3 6 9 5
8 5 6 9 7 1 2 4 3
4 7 9 3 2 6 8 5 1
3 1 2 4 5 8 7 6 9
1 8 3 2 9 5 4 7 6
9 2 4 6 3 7 5 1 8
7 6 5 8 1 4 9 3 2
```

Solution # 531
```
8 5 7 3 6 9 1 4 2
2 3 4 1 8 5 7 6 9
6 1 9 2 7 4 8 5 3
7 2 8 5 1 3 6 9 4
5 4 1 6 9 2 3 7 8
3 9 6 8 4 7 2 1 5
1 6 2 4 5 8 9 3 7
4 7 3 9 2 1 5 8 6
9 8 5 7 3 6 4 2 1
```

Solution # 532
```
2 4 9 6 5 8 3 1 7
7 6 8 3 4 1 9 2 5
5 1 3 9 2 7 8 6 4
6 3 4 2 8 5 1 7 9
1 8 5 7 9 4 6 3 2
9 2 7 1 3 6 4 5 8
8 7 6 4 1 2 5 9 3
3 5 2 8 6 9 7 4 1
4 9 1 5 7 3 2 8 6
```

Solution # 533
```
4 1 7 2 3 8 9 6 5
2 3 5 9 6 1 4 7 8
9 6 8 7 4 5 1 2 3
1 2 9 4 8 7 5 3 6
5 4 6 1 2 3 7 8 9
7 8 3 5 9 6 2 1 4
3 5 4 8 1 2 6 9 7
6 9 2 3 7 4 8 5 1
8 7 1 6 5 9 3 4 2
```

Solution # 534
```
9 1 5 4 2 7 6 3 8
8 3 2 5 1 6 9 7 4
7 6 4 9 8 3 2 1 5
3 2 6 1 4 5 7 8 9
4 8 1 7 9 2 3 5 6
5 9 7 3 6 8 1 4 2
1 4 8 2 7 9 5 6 3
2 7 3 6 5 4 8 9 1
6 5 9 8 3 1 4 2 7
```

Solution # 535
```
2 6 9 3 4 7 8 1 5
8 4 5 9 2 1 6 7 3
3 1 7 8 5 6 4 9 2
9 3 1 4 7 8 5 2 6
5 8 2 1 6 9 7 3 4
4 7 6 5 3 2 1 8 9
1 9 3 6 8 4 2 5 7
6 2 8 7 9 5 3 4 1
7 5 4 2 1 3 9 6 8
```

Solution # 536
```
7 4 3 9 2 1 5 6 8
9 8 6 3 4 5 7 1 2
1 5 2 7 8 6 3 9 4
2 3 9 5 1 8 6 4 7
4 6 5 2 7 3 9 8 1
8 1 7 4 6 9 2 3 5
3 7 4 1 9 2 8 5 6
5 2 8 6 3 4 1 7 9
6 9 1 8 5 7 4 2 3
```

Solution # 537
```
2 6 9 7 3 5 8 4 1
3 4 8 2 6 1 5 9 7
5 1 7 8 4 9 3 6 2
4 5 2 3 8 6 7 1 9
6 9 3 4 1 7 2 5 8
8 7 1 9 5 2 4 3 6
1 8 5 6 7 3 9 2 4
7 2 6 5 9 4 1 8 3
9 3 4 1 2 8 6 7 5
```

Solution # 538
```
1 3 2 5 4 6 9 8 7
9 7 5 8 1 3 2 6 4
8 4 6 2 9 7 1 3 5
2 6 3 7 8 1 5 4 9
5 1 9 6 2 4 8 7 3
4 8 7 3 5 9 6 2 1
6 5 4 1 7 2 3 9 8
3 9 8 4 6 5 7 1 2
7 2 1 9 3 8 4 5 6
```

Solution # 539
```
5 2 6 1 7 4 3 9 8
4 7 8 2 9 3 1 5 6
9 1 3 6 8 5 4 2 7
7 8 1 4 5 6 9 3 2
2 9 4 3 1 8 6 7 5
6 3 5 9 2 7 8 4 1
3 4 2 5 6 1 7 8 9
8 6 9 7 3 2 5 1 4
1 5 7 8 4 9 2 6 3
```

Solution # 540
```
3 9 8 4 5 6 7 1 2
6 5 7 2 1 3 8 4 9
2 1 4 8 9 7 6 3 5
4 2 6 3 7 5 1 9 8
5 7 9 1 6 8 4 2 3
1 8 3 9 4 2 5 7 6
9 6 5 7 3 4 2 8 1
7 3 2 6 8 1 9 5 4
8 4 1 5 2 9 3 6 7
```

Solution # 541
```
6 9 8 7 4 5 3 1 2
2 1 7 3 8 6 9 4 5
4 3 5 2 1 9 8 6 7
7 8 3 9 6 4 5 2 1
5 6 2 8 7 1 4 3 9
1 4 9 5 2 3 7 8 6
9 2 4 6 3 7 1 5 8
3 7 6 1 5 8 2 9 4
8 5 1 4 9 2 6 7 3
```

Solution # 542
```
9 3 5 4 8 1 2 7 6
4 1 2 9 7 6 3 5 8
7 6 8 3 2 5 4 1 9
1 7 9 5 3 8 6 4 2
5 2 6 1 9 4 8 3 7
3 8 4 7 6 2 1 9 5
6 4 1 8 5 7 9 2 3
8 9 7 2 1 3 5 6 4
2 5 3 6 4 9 7 8 1
```

Solution # 543
```
3 6 4 9 8 2 5 1 7
2 1 9 7 5 6 4 3 8
7 5 8 4 1 3 9 2 6
5 9 6 2 3 1 7 8 4
1 8 2 5 7 4 3 6 9
4 3 7 8 6 9 1 5 2
9 4 3 6 2 5 8 7 1
8 2 5 1 9 7 6 4 3
6 7 1 3 4 8 2 9 5
```

Solution # 544
```
5 8 1 3 6 9 7 4 2
3 7 2 4 5 8 6 9 1
4 6 9 1 2 7 3 5 8
2 4 3 5 8 6 9 1 7
9 5 8 7 4 1 2 3 6
6 1 7 9 3 2 4 8 5
1 2 6 8 9 4 5 7 3
7 9 5 6 1 3 8 2 4
8 3 4 2 7 5 1 6 9
```

Solution # 545
```
5 2 7 4 1 3 6 9 8
3 9 4 6 8 2 5 1 7
6 1 8 5 9 7 3 2 4
1 4 2 3 6 5 8 7 9
7 8 3 9 4 1 2 6 5
9 5 6 7 2 8 4 3 1
8 6 1 2 5 9 7 4 3
2 7 5 1 3 4 9 8 6
4 3 9 8 7 6 1 5 2
```

Solution # 546
```
2 8 3 6 9 7 5 4 1
1 6 7 4 5 8 9 3 2
4 9 5 3 2 1 6 8 7
9 7 4 5 3 6 1 2 8
5 3 1 2 8 9 4 7 6
6 2 8 7 1 4 3 5 9
8 5 9 1 7 3 2 6 4
7 4 2 9 6 5 8 1 3
3 1 6 8 4 2 7 9 5
```

Solution # 547
```
8 1 7 6 9 4 3 2 5
5 2 6 8 7 3 9 4 1
4 3 9 5 2 1 8 6 7
2 9 1 4 3 7 6 5 8
6 5 3 2 1 8 7 9 4
7 4 8 9 5 6 2 1 3
1 6 4 7 8 9 5 3 2
3 8 2 1 6 5 4 7 9
9 7 5 3 4 2 1 8 6
```

Solution # 548
```
2 6 1 3 7 9 4 5 8
5 9 3 1 4 8 2 6 7
7 4 8 2 5 6 1 3 9
4 3 5 9 8 1 6 7 2
1 2 9 7 6 3 5 8 4
6 8 7 4 2 5 3 9 1
3 5 2 8 9 4 7 1 6
9 7 6 5 1 2 8 4 3
8 1 4 6 3 7 9 2 5
```

Solution # 549
```
2 1 3 9 4 6 5 7 8
9 4 8 7 5 1 6 2 3
7 5 6 2 8 3 9 1 4
1 7 2 8 6 5 4 3 9
6 3 4 1 9 7 2 8 5
5 8 9 3 2 4 7 6 1
4 2 5 6 1 8 3 9 7
3 9 1 5 7 2 8 4 6
8 6 7 4 3 9 1 5 2
```

Solution # 550
```
6 5 1 2 4 8 9 7 3
8 3 9 7 6 1 2 4 5
4 7 2 9 5 3 6 1 8
7 8 6 3 2 9 4 5 1
2 9 5 1 7 4 8 3 6
1 4 3 6 8 5 7 9 2
5 2 8 4 3 7 1 6 9
3 1 7 8 9 6 5 2 4
9 6 4 5 1 2 3 8 7
```

Solution # 551
```
3 4 5 2 9 6 1 7 8
2 6 8 1 7 3 5 4 9
1 7 9 5 4 8 2 6 3
7 2 6 3 5 1 8 9 4
9 5 4 8 2 7 6 3 1
8 3 1 4 6 9 7 5 2
5 1 7 9 3 2 4 8 6
6 9 2 7 8 4 3 1 5
4 8 3 6 1 5 9 2 7
```

Solution # 552
```
5 4 6 3 2 9 1 8 7
1 2 3 4 7 8 6 5 9
8 9 7 1 6 5 4 2 3
3 6 4 2 8 7 5 9 1
7 1 2 9 5 4 3 6 8
9 5 8 6 1 3 7 4 2
2 7 9 5 3 6 8 1 4
6 8 1 7 4 2 9 3 5
4 3 5 8 9 1 2 7 6
```

Solution # 553
```
5 2 9 4 1 8 3 7 6
1 7 4 5 6 3 2 9 8
8 3 6 9 7 2 4 1 5
2 6 5 3 9 4 1 8 7
9 1 8 6 2 7 5 4 3
7 4 3 8 5 1 6 2 9
4 5 7 2 8 6 9 3 1
6 8 2 1 3 9 7 5 4
3 9 1 7 4 5 8 6 2
```

Solution # 554
```
4 1 2 9 8 5 7 3 6
3 5 9 1 6 7 4 2 8
6 7 8 4 2 3 9 5 1
5 9 6 7 4 2 8 1 3
1 8 4 3 9 6 5 7 2
2 3 7 5 1 8 6 4 9
9 4 3 8 5 1 2 6 7
8 2 1 6 7 4 3 9 5
7 6 5 2 3 9 1 8 4
```

Solution # 555
```
3 9 4 5 6 7 8 2 1
1 7 5 8 3 2 9 6 4
2 8 6 1 9 4 7 5 3
8 1 7 6 2 9 3 4 5
5 6 3 4 8 1 2 7 9
9 4 2 7 5 3 6 1 8
6 2 9 3 1 5 4 8 7
4 5 8 9 7 6 1 3 2
7 3 1 2 4 8 5 9 6
```

Solution # 556
```
9 6 8 2 7 1 5 3 4
5 4 2 3 8 6 7 9 1
1 3 7 5 4 9 6 2 8
6 5 3 1 2 7 8 4 9
7 1 9 8 5 4 2 6 3
8 2 4 9 6 3 1 7 5
3 9 6 7 1 8 4 5 2
2 7 1 4 9 5 3 8 6
4 8 5 6 3 2 9 1 7
```

Solution # 557
```
9 2 5 4 3 7 1 6 8
3 4 1 2 8 6 9 5 7
7 6 8 9 5 1 3 4 2
2 8 4 6 7 3 5 9 1
6 1 7 5 4 9 8 2 3
5 9 3 8 1 2 6 7 4
1 5 6 3 2 4 7 8 9
4 7 9 1 6 8 2 3 5
8 3 2 7 9 5 4 1 6
```

Solution # 558
```
1 5 8 7 9 2 6 3 4
7 6 9 3 4 5 2 1 8
3 2 4 1 8 6 9 5 7
5 9 6 4 2 1 7 8 3
8 4 1 9 7 3 5 6 2
2 3 7 5 6 8 4 9 1
4 1 3 6 5 7 8 2 9
6 7 2 8 3 9 1 4 5
9 8 5 2 1 4 3 7 6
```

Solution # 559
```
5 2 9 1 7 3 4 8 6
4 8 1 2 6 9 7 5 3
6 7 3 5 4 8 1 2 9
2 1 5 4 9 7 3 6 8
8 4 7 3 2 6 5 9 1
9 3 6 8 1 5 2 7 4
7 5 8 6 3 1 9 4 2
1 9 4 7 8 2 6 3 5
3 6 2 9 5 4 8 1 7
```

Solution # 560
```
1 2 3 7 8 6 5 4 9
4 8 7 9 5 2 1 3 6
9 5 6 4 3 1 2 7 8
6 9 2 3 1 7 4 8 5
8 3 4 2 6 5 9 1 7
5 7 1 8 9 4 6 2 3
2 1 8 5 7 9 3 6 4
7 4 5 6 2 3 8 9 1
3 6 9 1 4 8 7 5 2
```

Solution # 561
```
6 7 1 5 2 9 4 3 8
5 9 2 3 4 8 6 7 1
4 3 8 1 6 7 5 9 2
9 6 7 4 8 5 2 1 3
1 4 3 6 7 2 8 5 9
8 2 5 9 3 1 7 4 6
7 8 4 2 1 3 9 6 5
2 1 9 7 5 6 3 8 4
3 5 6 8 9 4 1 2 7
```

Solution # 562
```
8 4 5 7 6 2 3 9 1
6 2 1 3 9 5 4 7 8
3 9 7 4 8 1 6 5 2
9 7 4 2 3 8 5 1 6
2 5 8 1 4 6 9 3 7
1 3 6 5 7 9 2 8 4
7 6 2 9 1 3 8 4 5
5 1 9 8 2 4 7 6 3
4 8 3 6 5 7 1 2 9
```

Solution # 563
```
6 4 5 2 8 9 1 3 7
1 2 3 6 5 7 4 8 9
7 9 8 4 3 1 6 2 5
3 8 9 1 2 4 7 5 6
5 1 2 8 7 6 3 9 4
4 6 7 3 9 5 2 1 8
2 5 1 7 4 8 9 6 3
8 7 6 9 1 3 5 4 2
9 3 4 5 6 2 8 7 1
```

Solution # 564
```
7 1 2 3 4 5 9 6 8
6 4 3 2 8 9 1 5 7
8 5 9 1 7 6 4 2 3
2 3 4 9 6 1 7 8 5
1 8 7 4 5 3 2 9 6
9 6 5 8 2 7 3 1 4
3 2 8 6 1 4 5 7 9
5 9 6 7 3 2 8 4 1
4 7 1 5 9 8 6 3 2
```

Solution # 565
```
4 1 2 6 3 9 8 7 5
8 3 6 4 5 7 2 1 9
9 5 7 2 1 8 3 4 6
5 8 1 3 7 4 9 6 2
2 9 3 5 6 1 7 8 4
6 7 4 8 9 2 5 3 1
3 2 8 9 4 6 1 5 7
7 4 5 1 2 3 6 9 8
1 6 9 7 8 5 4 2 3
```

Solution # 566
```
1 6 7 4 3 2 5 8 9
4 8 3 5 9 1 7 6 2
9 2 5 8 6 7 3 4 1
3 7 6 2 5 9 4 1 8
2 9 8 1 4 3 6 5 7
5 1 4 7 8 6 2 9 3
7 5 2 9 1 4 8 3 6
6 4 9 3 7 8 1 2 5
8 3 1 6 2 5 9 7 4
```

Solution # 567
```
3 8 1 4 5 7 9 2 6
4 6 7 2 3 9 8 5 1
9 2 5 6 1 8 4 7 3
8 9 3 7 6 5 1 4 2
2 5 6 9 4 1 3 8 7
7 1 4 8 2 3 6 9 5
5 3 8 1 7 4 2 6 9
6 7 9 3 8 2 5 1 4
1 4 2 5 9 6 7 3 8
```

Solution # 568
```
6 9 1 5 7 4 2 3 8
3 5 7 1 2 8 6 4 9
2 8 4 3 9 6 5 7 1
8 1 5 9 4 2 3 6 7
9 2 3 6 5 7 8 1 4
7 4 6 8 1 3 9 2 5
5 7 9 2 6 1 4 8 3
4 3 2 7 8 9 1 5 6
1 6 8 4 3 5 7 9 2
```

Solution # 569
```
5 2 6 4 9 3 1 7 8
4 7 8 5 1 6 2 3 9
3 9 1 8 2 7 4 5 6
9 8 3 2 5 4 6 1 7
7 1 4 6 3 9 8 2 5
2 6 5 1 7 8 9 4 3
8 4 2 3 6 5 7 9 1
6 3 9 7 4 1 5 8 2
1 5 7 9 8 2 3 6 4
```

Solution # 570
```
6 1 9 8 4 7 2 5 3
8 3 4 2 5 9 7 6 1
7 5 2 3 1 6 8 9 4
3 9 7 6 2 4 5 1 8
4 2 1 9 8 5 3 7 6
5 8 6 7 3 1 9 4 2
9 6 3 4 7 2 1 8 5
1 4 8 5 9 3 6 2 7
2 7 5 1 6 8 4 3 9
```

Solution # 571
```
6 7 9 4 1 2 5 3 8
5 4 2 8 6 3 1 9 7
1 8 3 5 7 9 2 6 4
3 2 4 1 9 6 8 7 5
8 5 6 3 4 7 9 1 2
7 9 1 2 8 5 3 4 6
4 3 7 9 5 8 6 2 1
2 6 5 7 3 1 4 8 9
9 1 8 6 2 4 7 5 3
```

Solution # 572
```
8 4 5 6 9 3 2 1 7
2 1 9 4 5 7 6 8 3
3 7 6 2 8 1 4 9 5
5 9 7 8 4 6 3 2 1
4 2 3 7 1 5 9 6 8
1 6 8 3 2 9 5 7 4
7 3 4 1 6 2 8 5 9
9 8 2 5 7 4 1 3 6
6 5 1 9 3 8 7 4 2
```

Solution # 573
```
2 8 1 4 6 7 9 3 5
5 7 6 3 9 2 1 8 4
9 3 4 5 8 1 7 2 6
3 5 2 1 4 6 8 7 9
7 4 9 8 2 3 5 6 1
6 1 8 7 5 9 3 4 2
4 9 5 6 7 8 2 1 3
1 2 7 9 3 4 6 5 8
8 6 3 2 1 5 4 9 7
```

Solution # 574
```
2 8 9 5 3 6 7 1 4
4 5 6 7 9 1 8 2 3
7 1 3 4 8 2 6 9 5
5 7 8 6 4 9 2 3 1
1 9 2 3 5 7 4 8 6
3 6 4 1 2 8 9 5 7
9 4 7 2 1 5 3 6 8
6 2 5 8 7 3 1 4 9
8 3 1 9 6 4 5 7 2
```

Solution # 575
```
9 2 8 4 3 5 1 6 7
1 5 6 2 7 8 3 4 9
7 3 4 6 1 9 2 8 5
6 8 1 3 9 2 5 7 4
2 9 5 7 4 6 8 3 1
4 7 3 8 5 1 9 2 6
3 1 2 5 6 4 7 9 8
5 6 7 9 8 3 4 1 2
8 4 9 1 2 7 6 5 3
```

Solution # 576
```
3 7 1 6 4 8 2 5 9
4 9 8 5 2 7 6 1 3
6 5 2 3 9 1 4 8 7
1 6 3 9 7 4 5 2 8
8 4 9 2 1 5 3 7 6
7 2 5 8 3 6 9 4 1
5 1 4 7 6 9 8 3 2
9 3 7 4 8 2 1 6 5
2 8 6 1 5 3 7 9 4
```

Solution # 577
```
5 6 4 2 1 8 3 7 9
8 9 2 5 3 7 4 6 1
1 7 3 6 4 9 5 8 2
6 4 8 3 5 2 1 9 7
9 1 5 8 7 6 2 4 3
2 3 7 1 9 4 8 5 6
7 5 9 4 2 3 6 1 8
4 2 6 7 8 1 9 3 5
3 8 1 9 6 5 7 2 4
```

Solution # 578
```
4 5 8 6 1 9 7 2 3
1 6 2 3 7 5 9 8 4
7 9 3 2 4 8 5 6 1
6 8 9 7 5 4 3 1 2
3 2 4 1 9 6 8 5 7
5 7 1 8 2 3 4 9 6
8 4 7 5 6 2 1 3 9
2 1 5 9 3 7 6 4 8
9 3 6 4 8 1 2 7 5
```

Solution # 579
```
3 1 5 2 7 4 6 9 8
2 6 7 9 3 8 5 4 1
9 4 8 5 6 1 7 3 2
6 9 4 3 2 7 8 1 5
7 2 1 8 9 5 4 6 3
8 5 3 1 4 6 9 2 7
4 3 2 7 8 9 1 5 6
5 7 6 4 1 3 2 8 9
1 8 9 6 5 2 3 7 4
```

Solution # 580
```
6 1 4 3 9 8 5 7 2
7 9 5 1 2 6 4 8 3
3 8 2 5 7 4 1 9 6
8 6 7 2 3 5 9 1 4
5 2 9 4 1 7 3 6 8
4 3 1 6 8 9 2 5 7
9 4 3 8 6 1 7 2 5
1 5 6 7 4 2 8 3 9
2 7 8 9 5 3 6 4 1
```

Solution # 581
```
2 8 3 1 7 5 4 9 6
7 4 5 9 6 3 1 8 2
6 9 1 8 2 4 3 5 7
5 7 2 6 3 1 8 4 9
4 3 8 2 5 9 6 7 1
9 1 6 7 4 8 5 2 3
8 2 4 3 9 6 7 1 5
1 6 7 5 8 2 9 3 4
3 5 9 4 1 7 2 6 8
```

Solution # 582
```
9 6 4 7 2 8 3 5 1
3 8 7 9 1 5 2 4 6
5 2 1 3 4 6 8 7 9
8 4 5 2 6 7 1 9 3
2 1 9 4 8 3 7 6 5
6 7 3 1 5 9 4 8 2
1 9 2 6 7 4 5 3 8
4 5 6 8 3 2 9 1 7
7 3 8 5 9 1 6 2 4
```

Solution # 583
```
7 9 2 3 8 5 4 6 1
8 6 3 4 1 9 2 5 7
4 5 1 7 2 6 8 3 9
5 8 9 2 6 7 1 4 3
1 3 4 5 9 8 7 2 6
6 2 7 1 4 3 9 8 5
2 4 5 6 7 1 3 9 8
3 7 8 9 5 4 6 1 2
9 1 6 8 3 2 5 7 4
```

Solution # 584
```
7 9 2 6 5 1 4 3 8
5 3 4 7 8 9 6 1 2
8 6 1 3 4 2 7 5 9
6 5 3 9 7 4 2 8 1
1 4 9 8 2 5 3 7 6
2 7 8 1 6 3 9 4 5
4 1 7 5 9 6 8 2 3
9 8 5 2 3 7 1 6 4
3 2 6 4 1 8 5 9 7
```

Solution # 585
```
6 1 2 5 7 4 8 3 9
9 5 3 6 2 8 4 7 1
7 4 8 9 3 1 2 6 5
8 7 6 1 9 2 3 5 4
3 2 4 7 6 5 1 9 8
1 9 5 8 4 3 6 2 7
5 3 7 4 8 6 9 1 2
2 8 1 3 5 9 7 4 6
4 6 9 2 1 7 5 8 3
```

Solution # 586
```
2 3 9 4 7 1 5 8 6
5 8 4 6 9 3 2 1 7
1 6 7 5 2 8 4 9 3
6 2 8 9 1 5 3 7 4
9 4 5 3 6 7 1 2 8
3 7 1 2 8 4 6 5 9
4 1 2 7 3 9 8 6 5
7 5 6 8 4 2 9 3 1
8 9 3 1 5 6 7 4 2
```

Solution # 587
```
1 8 7 6 3 4 9 5 2
9 6 2 7 8 5 3 4 1
5 3 4 2 9 1 7 8 6
7 5 1 9 6 2 4 3 8
8 4 9 5 1 3 6 2 7
3 2 6 4 7 8 1 9 5
4 9 8 1 2 6 5 7 3
6 7 3 8 5 9 2 1 4
2 1 5 3 4 7 8 6 9
```

Solution # 588
```
5 9 6 2 7 1 3 4 8
2 3 8 4 6 9 7 5 1
1 4 7 8 3 5 2 9 6
7 2 3 5 1 6 9 8 4
8 1 9 3 2 4 5 6 7
4 6 5 9 8 7 1 2 3
6 5 1 7 4 2 8 3 9
3 7 2 6 9 8 4 1 5
9 8 4 1 5 3 6 7 2
```

Solution # 589
```
5 2 3 9 7 1 8 4 6
9 1 7 6 8 4 2 5 3
6 4 8 5 3 2 1 7 9
4 9 5 7 2 3 6 1 8
8 3 2 1 6 5 7 9 4
7 6 1 4 9 8 5 3 2
2 7 4 3 5 6 9 8 1
3 5 6 8 1 9 4 2 7
1 8 9 2 4 7 3 6 5
```

Solution # 590
```
2 4 6 1 5 8 9 7 3
3 5 9 6 4 7 2 8 1
1 8 7 9 2 3 4 5 6
9 6 4 5 7 1 8 3 2
5 3 8 2 6 9 7 1 4
7 1 2 8 3 4 5 6 9
6 2 1 4 8 5 3 9 7
8 9 3 7 1 2 6 4 5
4 7 5 3 9 6 1 2 8
```

Solution # 591
```
9 3 2 8 4 5 6 7 1
7 4 1 9 2 6 8 5 3
6 5 8 1 7 3 2 9 4
2 1 5 6 9 4 7 3 8
4 8 9 3 5 7 1 6 2
3 7 6 2 8 1 5 4 9
5 2 7 4 3 8 9 1 6
1 9 3 5 6 2 4 8 7
8 6 4 7 1 9 3 2 5
```

Solution # 592
```
9 8 4 5 1 7 6 2 3
2 7 3 6 8 9 4 1 5
5 6 1 3 4 2 8 7 9
8 1 5 4 7 3 2 9 6
4 2 6 1 9 5 3 8 7
3 9 7 8 2 6 5 4 1
6 4 9 2 5 1 7 3 8
7 3 2 9 6 8 1 5 4
1 5 8 7 3 4 9 6 2
```

Solution # 593
```
2 6 1 9 4 8 7 5 3
9 8 5 6 3 7 4 1 2
3 4 7 5 2 1 9 8 6
5 9 2 7 6 4 8 3 1
8 7 4 3 1 2 5 6 9
1 3 6 8 9 5 2 4 7
6 5 3 4 7 9 1 2 8
7 2 8 1 5 6 3 9 4
4 1 9 2 8 3 6 7 5
```

Solution # 594
```
7 1 4 2 5 9 3 6 8
2 9 3 1 8 6 5 4 7
8 5 6 3 4 7 9 1 2
1 3 5 6 9 8 2 7 4
6 7 9 5 2 4 1 8 3
4 8 2 7 3 1 6 9 5
5 6 7 8 1 3 4 2 9
3 4 1 9 7 2 8 5 6
9 2 8 4 6 5 7 3 1
```

Solution # 595
```
1 5 7 6 9 4 3 2 8
9 2 3 1 8 7 5 6 4
8 6 4 2 3 5 9 7 1
7 9 2 5 4 6 1 8 3
5 3 6 7 1 8 4 9 2
4 1 8 9 2 3 6 5 7
6 4 1 8 5 2 7 3 9
3 8 5 4 7 9 2 1 6
2 7 9 3 6 1 8 4 5
```

Solution # 596
```
6 9 3 1 5 2 4 7 8
4 5 2 8 9 7 1 6 3
8 7 1 6 3 4 2 9 5
9 1 7 5 2 3 6 8 4
5 4 6 7 8 9 3 1 2
3 2 8 4 1 6 7 5 9
7 6 9 3 4 5 8 2 1
2 8 4 9 7 1 5 3 6
1 3 5 2 6 8 9 4 7
```

Solution # 597
```
4 6 9 3 5 7 2 8 1
5 8 7 2 4 1 3 9 6
1 3 2 9 6 8 4 5 7
2 5 3 8 7 6 9 1 4
6 1 8 4 3 9 5 7 2
9 7 4 5 1 2 8 6 3
7 4 5 1 8 3 6 2 9
3 9 1 6 2 5 7 4 8
8 2 6 7 9 4 1 3 5
```

Solution # 598
```
4 1 8 3 9 6 7 2 5
7 5 6 2 4 8 3 1 9
2 9 3 7 5 1 6 8 4
9 3 4 6 2 5 8 7 1
6 2 7 1 8 4 9 5 3
5 8 1 9 7 3 4 6 2
8 6 5 4 1 9 2 3 7
3 4 2 5 6 7 1 9 8
1 7 9 8 3 2 5 4 6
```

Solution # 599
```
4 3 8 5 6 7 1 2 9
7 9 2 8 1 4 5 3 6
6 1 5 2 3 9 4 7 8
5 7 3 9 2 8 6 1 4
8 2 6 4 5 1 7 9 3
9 4 1 3 7 6 2 8 5
1 8 9 7 4 5 3 6 2
3 5 7 6 8 2 9 4 1
2 6 4 1 9 3 8 5 7
```

Solution # 600
```
4 9 7 1 5 6 3 2 8
5 8 6 3 7 2 9 1 4
1 3 2 4 9 8 7 5 6
7 4 5 8 6 9 2 3 1
2 1 3 5 4 7 6 8 9
9 6 8 2 1 3 4 7 5
3 5 1 6 2 4 8 9 7
8 7 4 9 3 5 1 6 2
6 2 9 7 8 1 5 4 3
```

Solution # 601
```
3 4 1 5 7 6 8 9 2
7 8 2 9 4 3 5 6 1
5 6 9 2 8 1 4 7 3
9 2 4 8 6 5 3 1 7
8 1 5 7 3 2 9 4 6
6 3 7 4 1 9 2 5 8
2 7 8 1 9 4 6 3 5
4 5 6 3 2 7 1 8 9
1 9 3 6 5 8 7 2 4
```

Solution # 602
```
2 3 8 4 5 7 6 1 9
9 1 4 3 2 6 8 5 7
6 5 7 9 1 8 4 2 3
7 6 1 2 8 3 5 9 4
5 9 2 1 7 4 3 6 8
8 4 3 6 9 5 1 7 2
1 8 5 7 4 2 9 3 6
4 7 6 5 3 9 2 8 1
3 2 9 8 6 1 7 4 5
```

Solution # 603
```
8 1 9 3 6 4 2 5 7
3 7 2 5 8 9 4 6 1
6 4 5 7 2 1 3 8 9
2 6 7 9 3 8 1 4 5
4 9 8 1 5 2 7 3 6
1 5 3 4 7 6 9 2 8
5 8 4 2 9 7 6 1 3
7 2 6 8 1 3 5 9 4
9 3 1 6 4 5 8 7 2
```

Solution # 604
```
1 3 9 2 4 6 7 8 5
5 4 8 1 7 9 2 6 3
6 2 7 5 3 8 9 1 4
4 6 5 9 1 2 3 7 8
3 8 1 4 5 7 6 9 2
9 7 2 8 6 3 5 4 1
7 1 4 6 2 5 8 3 9
2 9 3 7 8 4 1 5 6
8 5 6 3 9 1 4 2 7
```

Solution # 605
```
4 9 8 3 6 5 2 1 7
5 2 6 8 7 1 3 9 4
7 1 3 9 2 4 8 6 5
8 7 1 5 4 9 6 3 2
2 4 9 6 1 3 5 7 8
6 3 5 2 8 7 9 4 1
3 8 7 1 9 2 4 5 6
9 6 4 7 5 8 1 2 3
1 5 2 4 3 6 7 8 9
```

Solution # 606
```
1 7 2 5 9 3 4 6 8
6 9 3 4 1 8 2 5 7
5 8 4 6 7 2 3 1 9
9 5 8 2 4 1 6 7 3
2 3 1 8 6 7 5 9 4
7 4 6 3 5 9 8 2 1
4 2 9 7 3 5 1 8 6
3 1 5 9 8 6 7 4 2
8 6 7 1 2 4 9 3 5
```

Solution # 607
```
9 5 2 1 7 3 8 6 4
6 4 7 9 8 5 1 3 2
1 8 3 2 6 4 9 7 5
5 6 8 7 3 1 2 4 9
3 1 9 6 4 2 5 8 7
7 2 4 8 5 9 6 1 3
4 3 1 5 9 6 7 2 8
8 9 6 4 2 7 3 5 1
2 7 5 3 1 8 4 9 6
```

Solution # 608
```
7 4 3 8 1 6 5 2 9
5 2 6 3 9 7 4 1 8
9 8 1 4 2 5 3 6 7
2 9 7 6 5 8 1 3 4
1 5 4 9 3 2 8 7 6
3 6 8 1 7 4 2 9 5
8 3 9 5 6 1 7 4 2
4 1 2 7 8 9 6 5 3
6 7 5 2 4 3 9 8 1
```

Solution # 609
```
7 5 4 1 6 9 8 2 3
9 8 1 3 5 2 6 7 4
2 6 3 7 8 4 1 9 5
4 9 8 5 1 3 2 6 7
5 1 6 2 9 7 4 3 8
3 7 2 6 4 8 9 5 1
1 4 5 9 7 6 3 8 2
8 2 9 4 3 5 7 1 6
6 3 7 8 2 1 5 4 9
```

Solution # 610
```
7 5 4 9 1 2 3 6 8
1 2 9 6 8 3 7 5 4
3 6 8 4 7 5 9 2 1
5 4 2 3 9 7 1 8 6
8 7 6 5 2 1 4 3 9
6 9 5 7 4 8 2 1 3
9 1 3 2 6 4 8 7 5
4 3 7 1 5 9 6 2 ...
2 8 1 ...
```

Solution # 611
```
5 2 1 4 3 6 9 7 8
4 6 9 7 2 8 1 3 5
8 3 7 9 1 5 2 4 6
9 1 4 2 6 7 5 8 3
3 7 8 1 5 9 6 2 4
6 5 2 3 8 4 7 1 9
1 8 6 5 7 3 4 9 2
2 9 5 8 4 1 3 6 7
7 4 3 6 9 2 8 5 1
```

Solution # 612
```
2 8 1 6 5 9 3 7 4
6 5 3 7 4 2 8 9 1
4 9 7 8 3 1 6 2 5
5 7 6 4 1 8 9 3 2
9 4 2 3 7 6 1 5 8
3 1 8 9 2 5 4 6 7
8 3 5 1 9 7 2 4 6
1 2 9 5 6 4 7 8 3
7 6 4 2 8 3 5 1 9
```

Solution # 613
```
6 4 1 3 8 2 7 9 5
8 9 2 5 6 7 4 1 3
5 7 3 9 1 4 6 8 2
7 6 5 4 9 8 3 2 1
4 2 9 1 5 3 8 7 6
3 1 8 7 2 6 9 5 4
9 5 7 6 3 1 2 4 8
2 3 4 8 7 5 1 6 9
1 8 6 2 4 9 5 3 7
```

Solution # 614
```
5 6 8 9 7 1 2 4 3
3 7 9 4 6 2 5 8 1
1 4 2 3 8 5 6 9 7
2 5 3 1 9 6 8 7 4
8 1 4 2 5 7 9 3 6
6 9 7 8 3 4 1 2 5
7 3 1 6 2 9 4 5 8
4 2 5 7 1 8 3 6 9
9 8 6 5 4 3 7 1 2
```

Solution # 615
```
2 6 3 4 8 7 9 5 1
9 8 5 3 6 1 7 2 4
1 4 7 2 9 5 3 6 8
4 9 2 6 1 8 5 7 3
6 5 1 7 3 2 4 8 9
7 3 8 5 4 9 2 1 6
3 2 4 1 7 6 8 9 5
5 1 9 8 2 3 6 4 7
8 7 6 9 5 4 1 3 2
```

Solution # 616
```
6 8 1 3 2 9 4 5 7
4 9 2 7 5 1 8 3 6
3 7 5 8 4 6 2 9 1
2 5 3 1 8 7 6 4 9
1 6 8 2 9 4 5 7 3
7 4 9 5 6 3 1 8 2
9 2 4 6 3 5 7 1 8
8 3 7 4 1 2 9 6 5
5 1 6 9 7 8 3 2 4
```

Solution # 617
```
3 9 8 4 1 5 6 2 7
1 6 7 8 3 2 4 5 9
4 5 2 6 9 7 8 1 3
7 1 6 2 4 8 3 9 5
2 8 3 9 5 6 7 4 1
9 4 5 1 7 3 2 6 8
5 7 1 3 2 4 9 8 6
8 2 9 7 6 1 5 3 4
6 3 4 5 8 9 1 7 2
```

Solution # 618
```
2 1 3 4 9 8 5 6 7
5 6 9 7 3 2 8 1 4
8 7 4 6 5 1 3 9 2
9 4 6 3 8 5 2 7 1
7 3 2 9 1 6 4 5 8
1 5 8 2 4 7 9 3 6
6 9 1 8 2 3 7 4 5
4 8 7 5 6 9 1 2 3
3 2 5 1 7 4 6 8 9
```

Solution # 619
```
5 1 4 9 2 6 8 7 3
7 2 8 5 1 3 6 9 4
6 9 3 7 8 4 2 5 1
8 6 5 4 7 9 3 1 2
9 7 2 6 3 1 4 8 5
4 3 1 8 5 2 7 6 9
2 5 6 1 4 7 9 3 8
3 8 7 2 9 5 1 4 6
1 4 9 3 6 8 5 2 7
```

Solution # 620
```
6 2 7 4 8 1 3 5 9
1 9 3 5 2 7 4 8 6
5 4 8 6 9 3 7 2 1
7 8 6 9 4 5 1 3 2
9 5 4 1 3 2 8 6 7
3 1 2 8 7 6 5 9 4
8 3 9 7 6 4 2 1 5
4 6 1 2 5 8 9 7 3
2 7 5 3 1 9 6 4 8
```

Solution # 621
```
1 8 7 3 6 4 2 9 5
5 3 9 1 7 2 6 8 4
6 2 4 8 5 9 1 3 7
8 7 2 5 9 3 4 6 1
9 1 3 2 4 6 7 5 8
4 6 5 7 1 8 9 2 3
2 4 8 6 3 7 5 1 9
3 9 1 4 2 5 8 7 6
7 5 6 9 8 1 3 4 2
```

Solution # 622
```
4 2 7 3 1 8 5 9 6
1 3 9 4 5 6 7 8 2
6 5 8 7 9 2 3 4 1
7 4 3 2 6 5 9 1 8
9 1 5 8 4 3 6 2 7
8 6 2 9 7 1 4 3 5
5 9 6 1 2 4 8 7 3
3 7 1 5 8 9 2 6 4
2 8 4 6 3 7 1 5 9
```

Solution # 623
```
4 2 6 7 3 1 9 8 5
5 7 3 2 8 9 1 4 6
9 1 8 5 6 4 2 3 7
3 6 7 4 5 2 8 9 1
8 9 5 3 1 7 4 6 2
2 4 1 8 9 6 5 7 3
7 3 2 9 4 5 6 1 8
1 5 9 6 7 8 3 2 4
6 8 4 1 2 3 7 5 9
```

Solution # 624
```
6 7 8 1 3 9 5 2 4
5 3 4 8 7 2 9 1 6
1 2 9 4 5 6 8 7 3
9 1 3 6 2 4 7 5 8
8 6 2 7 1 5 4 3 9
4 5 7 3 9 8 2 6 1
7 9 6 2 8 1 3 4 5
3 4 5 9 6 7 1 8 2
2 8 1 5 4 3 6 9 7
```

Solution # 625
```
2 3 5 7 9 1 4 6 8
1 6 9 2 8 4 7 3 5
8 7 4 5 3 6 1 2 9
5 2 6 3 4 7 8 9 1
7 9 1 8 6 5 3 4 2
3 4 8 9 1 2 5 7 6
6 5 7 4 2 8 9 1 3
9 8 2 1 7 3 6 5 4
4 1 3 6 5 9 2 8 7
```

Solution # 626
```
4 8 5 7 3 2 6 1 9
3 1 7 4 6 9 5 2 8
2 6 9 5 8 1 4 7 3
1 7 8 9 4 3 2 6 5
6 5 2 8 1 7 3 9 4
9 4 3 6 2 5 7 8 1
7 9 6 1 5 4 8 3 2
5 3 1 2 7 8 9 4 6
8 2 4 3 9 6 1 5 7
```

Solution # 627
```
9 4 8 1 6 5 2 3 7
6 2 7 4 3 9 8 1 5
3 1 5 7 2 8 6 4 9
2 5 1 8 9 4 7 6 3
8 6 3 2 7 1 5 9 4
7 9 4 3 5 6 1 2 8
5 8 9 6 4 2 3 7 1
4 7 6 5 1 3 9 8 2
1 3 2 9 8 7 4 5 6
```

Solution # 628
```
5 2 7 8 6 4 1 9 3
9 6 3 2 1 7 5 4 8
1 8 4 5 3 9 6 2 7
3 1 9 6 7 2 8 5 4
2 4 6 3 5 8 7 1 9
7 5 8 4 9 1 3 6 2
4 3 2 1 8 5 9 7 6
6 7 5 9 4 3 2 8 1
8 9 1 7 2 6 4 3 5
```

Solution # 629
```
4 7 9 5 8 2 6 1 3
3 2 5 9 6 1 7 4 8
6 8 1 3 7 4 9 2 5
9 5 7 4 1 3 2 8 6
2 1 6 7 9 8 5 3 4
8 4 3 2 5 6 1 7 9
5 6 4 1 3 7 8 9 2
1 3 8 6 2 9 4 5 7
7 9 2 8 4 5 3 6 1
```

Solution # 630
```
5 6 7 8 3 9 2 4 1
1 3 8 6 2 4 5 7 9
2 9 4 5 1 7 3 8 6
3 5 9 7 4 2 6 1 8
8 2 6 9 5 1 7 3 4
4 7 1 3 6 8 9 2 5
7 4 5 1 9 3 8 6 2
9 8 2 4 7 6 1 5 3
6 1 3 2 8 5 4 9 7
```

Solution # 631
```
1 4 9 2 6 3 7 5 8
7 3 5 4 1 8 6 2 9
8 6 2 5 9 7 3 4 1
4 5 7 6 3 9 1 8 2
2 9 8 7 4 1 5 3 6
3 1 6 8 2 5 9 7 4
9 8 1 3 7 4 2 6 5
6 7 4 1 5 2 8 9 3
5 2 3 9 8 6 4 1 7
```

Solution # 632
```
1 3 9 6 2 5 8 4 7
8 2 4 3 9 7 5 1 6
6 5 7 4 8 1 2 3 9
5 9 2 1 4 3 6 7 8
7 4 6 2 5 8 1 9 3
3 8 1 7 6 9 4 2 5
2 6 3 5 7 4 9 8 1
4 1 8 9 3 6 7 5 2
9 7 5 8 1 2 3 6 4
```

Solution # 633
```
9 5 6 3 4 7 1 8 2
2 1 7 9 8 6 4 5 3
4 8 3 1 5 2 9 7 6
3 4 9 7 1 8 6 2 5
8 6 1 2 9 5 3 4 7
5 7 2 6 3 4 8 1 9
7 2 4 8 6 9 5 3 1
6 3 8 5 2 1 7 9 4
1 9 5 4 7 3 2 6 8
```

Solution # 634
```
6 7 9 8 5 3 2 4 1
3 2 5 1 9 4 8 7 6
8 4 1 6 2 7 9 5 3
5 9 2 4 7 6 1 3 8
7 8 3 5 1 2 4 6 9
1 6 4 9 3 8 5 2 7
2 1 6 7 4 9 3 8 5
4 5 7 3 8 1 6 9 2
9 3 8 2 6 5 7 1 4
```

Solution # 635
```
9 4 7 1 8 2 5 3 6
8 5 2 7 3 6 4 1 9
1 3 6 4 9 5 8 7 2
3 9 4 5 7 1 2 6 8
2 8 1 3 6 4 9 5 7
6 7 5 9 2 8 1 4 3
4 6 3 2 5 9 7 8 1
7 1 9 8 4 3 6 2 5
5 2 8 6 1 7 3 9 4
```

Solution # 636
```
8 2 5 7 1 9 3 6 4
7 6 9 3 4 2 1 8 5
1 3 4 8 6 5 9 2 7
3 4 2 5 8 1 6 7 9
9 5 8 6 3 7 4 1 2
6 7 1 2 9 4 5 3 8
2 1 3 4 5 8 7 9 6
5 8 6 9 7 3 2 4 1
4 9 7 1 2 6 8 5 3
```

Solution # 637
```
3 9 4 2 6 7 5 1 8
1 6 5 4 8 3 9 2 7
8 2 7 1 9 5 4 6 3
9 3 6 8 2 4 7 5 1
7 5 8 9 3 1 2 4 6
2 4 1 5 7 6 8 3 9
5 8 3 7 1 2 6 9 4
6 7 2 3 4 9 1 8 5
4 1 9 6 5 8 3 7 2
```

Solution # 638
```
9 8 5 2 1 3 6 7 4
2 4 3 8 6 7 1 5 9
1 6 7 9 5 4 3 2 8
4 5 2 3 8 6 9 1 7
6 9 8 7 2 1 5 4 3
3 7 1 5 4 9 2 8 6
7 2 6 4 9 5 8 3 1
8 3 9 1 7 2 4 6 5
5 1 4 6 3 8 7 9 2
```

Solution # 639
```
5 3 7 2 9 8 4 1 6
6 4 8 7 3 1 2 5 9
9 2 1 5 4 6 3 7 8
1 9 2 8 7 3 6 4 5
4 7 6 9 2 5 1 8 3
3 8 5 1 6 4 9 2 7
8 1 9 3 5 2 7 6 4
7 5 4 6 1 9 8 3 2
2 6 3 4 8 7 5 9 1
```

Solution # 640
```
5 4 6 7 3 1 9 2 8
2 7 3 5 8 9 4 6 1
9 8 1 4 6 2 5 7 3
1 9 7 2 5 8 6 3 4
3 5 4 6 9 7 8 1 2
6 2 8 1 4 3 7 9 5
7 3 5 8 1 6 2 4 9
4 1 2 9 7 5 3 8 6
8 6 9 3 2 4 1 5 7
```

Solution # 641
```
7 6 8 3 4 1 9 2 5
2 5 4 8 6 9 7 1 3
3 1 9 5 7 2 4 8 6
9 8 3 6 1 7 5 4 2
1 7 2 9 5 4 3 6 8
6 4 5 2 3 8 1 9 7
4 9 6 7 2 5 8 3 1
8 3 7 1 9 6 2 5 4
5 2 1 4 8 3 6 7 9
```

Solution # 642
```
1 2 5 8 9 4 6 3 7
4 7 6 2 3 5 8 9 1
3 8 9 6 7 1 5 4 2
6 9 8 1 2 3 4 7 5
7 1 4 5 8 9 3 2 6
5 3 2 4 6 7 9 1 8
2 5 3 9 1 8 7 6 4
9 4 1 7 5 6 2 8 3
8 6 7 3 4 2 1 5 9
```

Solution # 643
```
6 4 5 1 3 7 9 8 2
2 7 9 5 8 4 1 6 3
3 1 8 6 2 9 5 4 7
9 8 6 7 5 2 4 3 1
7 3 2 4 6 1 8 9 5
4 5 1 3 9 8 7 2 6
5 6 4 8 7 3 2 1 9
8 2 7 9 1 6 3 5 4
1 9 3 2 4 5 6 7 8
```

Solution # 644
```
2 7 4 1 5 3 6 9 8
3 6 9 4 8 2 7 5 1
1 5 8 9 6 7 4 3 2
5 4 7 8 3 9 1 2 6
6 8 3 2 4 1 5 7 9
9 2 1 6 7 5 8 4 3
7 1 2 5 9 6 3 8 4
4 9 5 3 1 8 2 6 7
8 3 6 7 2 4 9 1 5
```

Solution # 645
```
8 6 5 4 7 1 2 3 9
4 3 7 9 2 5 6 8 1
2 9 1 3 6 8 4 7 5
5 7 8 6 4 2 1 9 3
3 2 4 1 5 9 7 6 8
9 1 6 7 8 3 5 2 4
6 8 2 5 9 4 3 1 7
7 5 3 8 1 6 9 4 2
1 4 9 2 3 7 8 5 6
```

Solution # 646
```
8 2 3 1 6 4 5 7 9
6 7 5 9 2 3 4 1 8
9 4 1 7 5 8 6 2 3
2 6 9 3 8 1 7 4 5
7 3 4 6 9 5 2 8 1
1 5 8 4 7 2 9 3 6
4 8 6 5 1 7 3 9 2
3 9 2 8 4 6 1 5 7
5 1 7 2 3 9 8 6 4
```

Solution # 647
```
9 4 6 3 5 7 2 1 8
5 1 8 4 2 6 3 7 9
7 2 3 9 1 8 5 4 6
3 6 1 2 4 5 8 9 7
4 5 9 7 8 1 6 3 2
8 7 2 6 3 9 4 5 1
1 8 7 5 6 4 9 2 3
2 9 4 8 7 3 1 6 5
6 3 5 1 9 2 7 8 4
```

Solution # 648
```
7 6 5 9 8 2 3 4 1
8 2 3 1 4 5 9 6 7
9 4 1 7 6 3 2 8 5
2 3 6 4 1 8 5 7 9
4 1 9 5 7 6 8 3 2
5 7 8 2 3 9 6 1 4
1 8 2 6 9 7 4 5 3
3 5 4 8 2 1 7 9 6
6 9 7 3 5 4 1 2 8
```

Solution # 649
```
6 4 5 8 9 2 1 7 3
9 8 1 7 6 3 4 5 2
3 2 7 1 5 4 9 8 6
2 7 3 5 4 9 6 1 8
8 1 4 6 2 7 3 9 5
5 9 6 3 1 8 2 4 7
4 5 8 2 3 1 7 6 9
1 6 2 9 7 5 8 3 4
7 3 9 4 8 6 5 2 1
```

Solution # 650
```
7 9 1 4 2 8 5 6 3
6 4 3 7 5 9 8 2 1
8 5 2 1 3 6 4 7 9
9 1 7 5 8 4 2 3 6
2 8 5 9 6 3 1 4 7
3 6 4 2 7 1 9 5 8
5 7 6 8 1 2 3 9 4
1 3 9 6 4 5 7 8 2
4 2 8 3 9 7 6 1 5
```

Solution # 651
```
5 4 6 8 9 7 3 1 2
1 9 7 2 3 5 6 4 8
3 8 2 4 6 1 9 5 7
9 5 8 7 1 3 4 2 6
2 7 1 6 5 4 8 3 9
6 3 4 9 2 8 5 7 1
4 2 9 3 7 6 1 8 5
7 1 3 5 8 9 2 6 4
8 6 5 1 4 2 7 9 3
```

Solution # 652
```
8 9 1 7 2 6 4 3 5
5 6 7 8 3 4 2 9 1
2 3 4 5 9 1 7 8 6
1 8 5 6 4 7 9 2 3
3 4 2 1 8 9 6 5 7
6 7 9 3 5 2 1 4 8
7 2 3 9 6 8 5 1 4
4 1 8 2 7 5 3 6 9
9 5 6 4 1 3 8 7 2
```

Solution # 653
```
3 9 8 1 4 6 2 5 7
7 5 6 3 8 2 9 4 1
2 4 1 9 7 5 8 6 3
9 3 5 7 2 4 6 1 8
6 1 2 5 9 8 7 3 4
8 7 4 6 1 3 5 2 9
4 8 9 2 6 1 3 7 5
1 6 3 8 5 7 4 9 2
5 2 7 4 3 9 1 8 6
```

Solution # 654
```
1 3 8 4 7 5 2 9 6
7 9 4 1 6 2 3 8 5
5 6 2 9 3 8 4 7 1
9 8 6 5 2 1 7 3 4
4 2 7 6 8 3 5 1 9
3 5 1 7 4 9 6 2 8
2 1 9 3 5 6 8 4 7
8 4 5 2 1 7 9 6 3
6 7 3 8 9 4 1 5 2
```

Solution # 655
```
3 9 2 5 1 7 4 6 8
4 6 5 8 3 2 7 9 1
8 1 7 6 4 9 2 5 3
6 7 9 2 8 1 3 4 5
5 8 3 7 9 4 6 1 2
2 4 1 3 5 6 9 8 7
9 3 4 1 2 8 5 7 6
7 5 8 9 6 3 1 2 4
1 2 6 4 7 5 8 3 9
```

Solution # 656
```
4 3 1 7 9 5 2 8 6
8 5 9 6 2 1 4 7 3
2 7 6 3 4 8 5 9 1
7 1 2 5 8 4 6 3 9
9 8 4 1 3 6 7 2 5
3 6 5 2 7 9 1 4 8
5 9 3 4 6 7 8 1 2
6 2 7 8 1 3 9 5 4
1 4 8 9 5 2 3 6 7
```

Solution # 657
```
8 5 6 2 9 3 1 7 4
2 4 1 6 7 8 3 9 5
7 3 9 5 4 1 8 6 2
6 7 5 3 8 9 2 4 1
1 2 4 7 5 6 9 3 8
9 8 3 1 2 4 6 5 7
4 6 7 8 3 2 5 1 9
5 1 8 9 6 7 4 2 3
3 9 2 4 1 5 7 8 6
```

Solution # 658
```
1 2 4 6 9 3 7 5 8
5 3 9 4 8 7 6 1 2
7 6 8 5 2 1 9 4 3
9 4 5 1 7 8 3 2 6
6 1 7 2 3 4 8 9 5
2 8 3 9 6 5 1 7 4
4 5 6 3 1 9 2 8 7
8 9 2 7 4 6 5 3 1
3 7 1 8 5 2 4 6 9
```

Solution # 659
```
6 5 4 1 3 8 2 7 9
2 1 3 6 9 7 4 5 8
7 9 8 4 2 5 6 1 3
4 2 7 3 6 1 8 9 5
8 3 5 2 7 9 1 6 4
9 6 1 8 5 4 3 2 7
3 7 2 5 8 6 9 4 1
5 4 6 9 1 3 7 8 2
1 8 9 7 4 2 5 3 6
```

Solution # 660
```
5 8 1 9 7 6 3 4 2
9 3 7 2 5 4 8 6 1
4 2 6 8 1 3 9 5 7
8 4 9 5 3 2 1 7 6
2 7 5 6 9 1 4 8 3
6 1 3 4 8 7 5 2 9
7 5 8 1 6 9 2 3 4
3 9 4 7 2 5 6 1 8
1 6 2 3 4 8 7 9 5
```

Solution # 661
```
2 1 7 5 4 6 9 3 8
8 6 5 3 9 1 2 4 7
4 3 9 7 2 8 1 6 5
5 9 6 4 7 2 3 8 1
3 4 8 6 1 9 7 5 2
7 2 1 8 5 3 6 9 4
1 5 3 2 6 4 8 7 9
6 7 2 9 8 5 4 1 3
9 8 4 1 3 7 5 2 6
```

Solution # 662
```
1 2 3 4 6 5 8 9 7
5 6 8 7 9 2 3 4 1
9 7 4 1 3 8 5 2 6
4 8 9 5 1 7 2 6 3
6 1 5 9 2 3 7 8 4
7 3 2 6 8 4 1 5 9
3 5 7 2 4 9 6 1 8
8 9 6 3 5 1 4 7 2
2 4 1 8 7 6 9 3 5
```

Solution # 663
```
1 8 9 6 2 3 5 4 7
4 2 7 8 9 5 6 1 3
3 6 5 7 4 1 2 8 9
6 5 8 2 1 9 7 3 4
2 7 4 3 6 8 9 5 1
9 1 3 5 7 4 8 2 6
5 4 6 1 8 7 3 9 2
7 3 1 9 5 2 4 6 8
8 9 2 4 3 6 1 7 5
```

Solution # 664
```
4 2 9 8 5 1 7 6 3
8 6 5 3 9 7 2 4 1
7 1 3 6 4 2 9 8 5
2 3 7 4 8 9 5 1 6
6 9 4 5 1 3 8 2 7
1 5 8 7 2 6 3 9 4
9 7 6 2 3 4 1 5 8
3 8 2 1 6 5 4 7 9
5 4 1 9 7 8 6 3 2
```

Solution # 665
```
5 6 1 3 2 8 4 9 7
4 7 8 5 9 6 1 3 2
3 2 9 4 1 7 6 5 8
1 9 3 2 6 5 7 8 4
6 4 5 8 7 9 3 2 1
2 8 7 1 4 3 9 6 5
8 3 4 6 5 1 2 7 9
7 1 6 9 8 2 5 4 3
9 5 2 7 3 4 8 1 6
```

Solution # 666
```
3 7 6 5 4 9 8 1 2
8 4 5 2 1 3 7 6 9
9 1 2 8 6 7 3 5 4
5 9 7 4 3 1 6 2 8
6 2 8 9 7 5 4 3 1
4 3 1 6 8 2 5 9 7
1 6 9 7 5 4 2 8 3
7 5 3 1 2 8 9 4 6
2 8 4 3 9 6 1 7 5
```

Solution # 667
```
3 2 1 6 5 7 4 9 8
8 6 4 2 1 9 3 5 7
5 9 7 3 8 4 1 2 6
7 8 2 5 4 1 6 3 9
6 4 3 8 9 2 5 7 1
1 5 9 7 6 3 8 4 2
4 1 5 9 2 8 7 6 3
2 3 8 4 7 6 9 1 5
9 7 6 1 3 5 2 8 4
```

Solution # 668
```
4 1 7 3 2 9 5 8 6
6 2 3 7 5 8 9 4 1
5 8 9 4 6 1 7 2 3
1 6 5 9 4 2 3 7 8
9 4 8 1 3 7 2 6 5
3 7 2 5 8 6 4 1 9
7 9 4 8 1 5 6 3 2
2 5 1 6 7 3 8 9 4
8 3 6 2 9 4 1 5 7
```

Solution # 669
```
7 9 3 5 6 1 4 8 2
5 8 2 9 3 4 6 1 7
4 6 1 2 7 8 9 5 3
9 2 6 3 8 5 1 7 4
1 5 4 6 2 7 8 3 9
8 3 7 4 1 9 5 2 6
3 1 8 7 4 6 2 9 5
2 4 9 1 5 3 7 6 8
6 7 5 8 9 2 3 4 1
```

Solution # 670
```
6 3 7 9 5 8 2 4 1
9 2 5 7 4 1 3 6 8
4 8 1 2 6 3 7 9 5
2 1 6 3 8 7 9 5 4
3 9 4 5 2 6 1 8 7
7 5 8 1 9 4 6 2 3
1 4 2 6 7 5 8 3 9
5 6 3 8 1 9 4 7 2
8 7 9 4 3 2 5 1 6
```

Solution # 671
```
2 6 3 4 9 1 8 5 7
4 7 9 5 8 3 2 6 1
1 8 5 7 2 6 3 4 9
3 9 2 8 4 5 1 7 6
5 4 7 6 1 2 9 3 8
6 1 8 3 7 9 4 2 5
9 3 4 1 5 7 6 8 2
7 2 6 9 3 8 5 1 4
8 5 1 2 6 4 7 9 3
```

Solution # 672
```
5 8 3 4 1 6 2 7 9
7 1 4 5 9 2 3 6 8
2 6 9 3 7 8 5 4 1
8 3 1 7 6 4 9 5 2
9 2 6 1 8 5 4 3 7
4 7 5 2 3 9 8 1 6
3 9 7 8 4 1 6 2 5
6 4 2 9 5 7 1 8 3
1 5 8 6 2 3 7 9 4
```

Solution # 673
```
9 5 3 4 7 2 8 6 1
2 7 6 8 1 3 9 5 4
1 8 4 5 6 9 2 3 7
5 2 1 3 4 6 7 8 9
6 9 8 2 5 7 1 4 3
4 3 7 9 8 1 5 2 6
7 1 5 6 2 4 3 9 8
8 4 9 7 3 5 6 1 2
3 6 2 1 9 8 4 7 5
```

Solution # 674
```
4 8 3 6 9 2 1 7 5
1 2 6 5 8 7 3 4 9
7 9 5 3 4 1 2 8 6
2 3 8 9 1 6 7 5 4
9 5 4 7 3 8 6 2 1
6 7 1 4 2 5 9 3 8
5 1 2 8 6 3 4 9 7
3 4 7 1 5 9 8 6 2
8 6 9 2 7 4 5 1 3
```

Solution # 675
```
5 2 1 3 7 6 4 9 8
6 4 7 9 1 8 5 2 3
8 3 9 2 5 4 6 7 1
2 9 5 4 8 7 1 3 6
4 1 3 6 2 9 8 5 7
7 8 6 1 3 5 2 4 9
1 7 8 5 9 2 3 6 4
3 5 4 7 6 1 9 8 2
9 6 2 8 4 3 7 1 5
```

Solution # 676
```
4 6 7 9 3 5 2 1 8
9 5 2 8 1 7 6 4 3
1 8 3 4 2 6 9 7 5
5 7 4 3 9 1 8 2 6
8 3 9 6 4 2 1 5 7
2 1 6 5 7 8 3 9 4
7 2 5 1 6 3 4 8 9
3 4 8 2 5 9 7 6 1
6 9 1 7 8 4 5 3 2
```

Solution # 677
```
8 9 2 7 3 1 4 6 5
3 5 4 9 2 6 1 8 7
6 7 1 5 4 8 3 2 9
9 2 6 8 1 3 5 7 4
7 4 3 2 5 9 8 1 6
1 8 5 4 6 7 2 9 3
5 6 8 3 9 2 7 4 1
2 3 9 1 7 4 6 5 8
4 1 7 6 8 5 9 3 2
```

Solution # 678
```
7 4 2 6 9 3 8 5 1
9 5 8 2 1 7 6 4 3
1 3 6 4 5 8 2 7 9
3 6 9 8 4 5 7 1 2
5 7 4 9 2 1 3 6 8
8 2 1 3 7 6 4 9 5
2 1 5 7 3 4 9 8 6
6 9 7 1 8 2 5 3 4
4 8 3 5 6 9 1 2 7
```

Solution # 679
```
7 6 4 5 1 3 9 8 2
3 2 5 7 8 9 6 4 1
9 1 8 2 4 6 7 3 5
6 3 2 8 9 4 1 5 7
1 4 9 6 7 5 3 2 8
8 5 7 3 2 1 4 6 9
2 9 1 4 6 8 5 7 3
5 7 6 9 3 2 8 1 4
4 8 3 1 5 7 2 9 6
```

Solution # 680
```
2 1 6 9 4 7 8 3 5
7 8 9 5 6 3 1 4 2
4 5 3 1 2 8 9 6 7
6 7 5 8 1 9 4 2 3
8 2 1 6 3 4 5 7 9
3 9 4 7 5 2 6 8 1
5 3 2 4 9 6 7 1 8
1 4 8 2 7 5 3 9 6
9 6 7 3 8 1 2 5 4
```

Solution # 681
```
7 6 8 5 3 1 4 2 9
5 9 3 2 4 6 8 7 1
1 2 4 7 9 8 3 6 5
2 8 6 4 1 3 9 5 7
4 3 1 9 7 5 2 8 6
9 5 7 6 8 2 1 4 3
6 1 5 8 2 9 7 3 4
8 7 9 3 6 4 5 1 2
3 4 2 1 5 7 6 9 8
```

Solution # 682
```
4 6 7 2 3 5 8 1 9
1 8 3 4 9 7 6 2 5
5 9 2 8 6 1 3 7 4
8 3 9 1 5 2 7 4 6
2 7 4 9 8 6 1 5 3
6 5 1 3 7 4 2 9 8
3 2 8 7 4 9 5 6 1
7 4 6 5 1 8 9 3 2
9 1 5 6 2 3 4 8 7
```

Solution # 683
```
8 5 9 7 6 4 1 2 3
7 6 1 5 3 2 9 8 4
3 2 4 8 9 1 5 7 6
4 7 8 9 1 3 6 5 2
1 3 6 2 5 7 4 9 8
2 9 5 4 8 6 3 7 1
9 4 2 1 7 8 3 6 5
6 8 7 3 4 5 2 1 9
5 1 3 6 2 9 8 4 7
```

Solution # 684
```
6 9 2 8 3 1 7 5 4
4 1 7 9 5 6 3 8 2
5 8 3 4 7 2 9 6 1
9 2 5 7 8 3 4 1 6
1 3 8 5 6 4 2 7 9
7 4 6 2 1 9 8 3 5
3 5 4 1 9 8 6 2 7
8 7 9 6 2 5 1 4 3
2 6 1 3 4 7 5 9 8
```

Solution # 685
```
2 5 7 4 6 8 9 1 3
4 3 1 7 9 2 8 5 6
6 9 8 3 1 5 4 2 7
1 2 3 5 8 9 6 7 4
8 6 5 1 7 4 3 9 2
9 7 4 2 3 6 5 8 1
3 8 2 6 5 1 7 4 9
5 1 6 9 4 7 2 3 8
7 4 9 8 2 3 1 6 5
```

Solution # 686
```
4 9 8 6 5 2 1 7 3
3 2 5 7 9 1 6 4 8
1 6 7 4 8 3 9 2 5
9 4 2 3 1 7 8 5 6
8 7 1 5 6 9 2 3 4
6 5 3 2 4 8 7 1 9
2 1 6 9 3 5 4 8 7
7 3 4 8 2 6 5 9 1
5 8 9 1 7 4 3 6 2
```

Solution # 687
```
8 3 6 7 5 1 2 4 9
1 5 4 2 6 9 7 8 3
2 9 7 8 3 4 6 5 1
4 8 2 5 9 7 3 1 6
3 6 5 4 1 2 8 9 7
7 1 9 6 8 3 4 2 5
5 4 8 1 7 6 9 3 2
6 2 3 9 4 5 1 7 8
9 7 1 3 2 8 5 6 4
```

Solution # 688
```
8 1 6 3 4 2 5 9 7
3 4 7 5 1 9 2 6 8
9 2 5 6 7 8 3 4 1
2 9 1 8 5 7 4 3 6
5 6 4 9 3 1 7 8 2
7 8 3 2 6 4 9 1 5
6 7 2 1 9 3 8 5 4
1 3 8 4 2 5 6 7 9
4 5 9 7 8 6 1 2 3
```

Solution # 689
```
3 6 5 4 7 9 1 2 8
4 9 7 2 1 8 3 6 5
2 1 8 6 5 3 4 7 9
1 2 9 5 4 6 7 8 3
7 4 3 8 2 1 5 9 6
5 8 6 3 9 7 2 1 4
6 5 2 7 8 4 9 3 1
8 7 1 9 3 5 6 4 2
9 3 4 1 6 2 8 5 7
```

Solution # 690
```
2 8 3 6 1 9 7 5 4
9 5 4 2 7 3 1 8 6
6 1 7 4 8 5 2 9 3
8 4 5 1 2 6 3 7 9
7 2 9 3 4 8 5 6 1
3 6 1 5 9 7 8 4 2
5 9 2 8 3 4 6 1 7
4 3 6 7 5 1 9 2 8
1 7 8 9 6 2 4 3 5
```

Solution # 691
```
3 6 4 7 5 8 9 2 1
2 8 1 9 4 6 5 7 3
5 9 7 3 2 1 8 4 6
8 1 2 5 3 4 6 9 7
6 4 9 1 8 7 3 5 2
7 3 5 2 6 9 1 8 4
9 2 8 4 1 3 7 6 5
1 5 6 8 7 2 4 3 9
4 7 3 6 9 5 2 1 8
```

Solution # 692
```
6 4 2 8 3 9 5 7 1
9 7 3 6 1 5 2 4 8
5 8 1 2 7 4 9 6 3
8 2 4 5 6 1 7 3 9
3 9 7 4 8 2 6 1 5
1 5 6 3 9 7 4 8 2
2 3 8 9 4 6 1 5 7
4 1 9 7 5 8 3 2 6
7 6 5 1 2 3 8 9 4
```

Solution # 693
```
3 1 2 7 8 5 9 4 6
8 9 6 2 1 4 5 3 7
4 7 5 3 9 6 2 8 1
9 4 7 8 5 2 6 1 3
5 8 3 4 6 1 7 2 9
6 2 1 9 3 7 4 5 8
7 6 8 5 4 3 1 9 2
2 3 4 1 7 9 8 6 5
1 5 9 6 2 8 3 7 4
```

Solution # 694
```
2 6 1 4 5 9 3 8 7
7 8 9 2 1 3 4 6 5
4 3 5 7 6 8 1 9 2
3 1 6 8 2 7 5 4 9
9 2 8 1 4 5 7 3 6
5 7 4 3 9 6 2 1 8
1 5 3 6 8 2 9 7 4
8 4 2 9 7 1 6 5 3
6 9 7 5 3 4 8 2 1
```

Solution # 695
```
8 7 9 6 5 2 4 3 1
5 2 4 3 1 7 6 8 9
1 6 3 9 4 8 2 7 5
3 4 6 1 2 5 8 9 7
2 9 5 8 7 3 1 6 4
7 1 8 4 6 9 5 2 3
4 8 7 5 3 6 9 1 2
6 5 2 7 9 1 3 4 8
9 3 1 2 8 4 7 5 6
```

Solution # 696
```
7 8 4 3 2 6 9 1 5
3 5 9 8 1 7 2 4 6
1 2 6 4 5 9 8 3 7
4 9 7 1 6 8 3 5 2
2 3 8 7 9 5 1 6 4
6 1 5 2 3 4 7 9 8
9 7 3 5 4 2 6 8 1
8 4 1 6 7 3 5 2 9
5 6 2 9 8 1 4 7 3
```

Solution # 697
```
5 2 8 9 6 7 1 3 4
3 9 6 2 4 1 8 7 5
4 7 1 8 3 5 2 9 6
9 4 7 6 5 8 3 2 1
8 1 2 4 7 3 6 5 9
6 3 5 1 2 9 4 8 7
2 6 3 7 9 4 5 1 8
1 5 9 3 8 6 7 4 2
7 8 4 5 1 2 9 6 3
```

Solution # 698
```
8 6 4 1 2 9 5 3 7
2 7 3 8 5 4 6 9 1
1 9 5 7 6 3 2 8 4
9 5 8 6 4 7 1 2 3
4 2 1 5 3 8 7 6 9
7 3 6 2 9 1 4 5 8
5 4 9 3 1 6 8 7 2
3 8 2 4 7 5 9 1 6
6 1 7 9 8 2 3 4 5
```

Solution # 699
```
6 9 2 1 7 4 3 5 8
1 4 8 5 6 3 7 2 9
3 5 7 9 8 2 1 6 4
9 1 6 8 3 7 2 4 5
5 7 4 6 2 1 9 8 3
8 2 3 4 5 9 6 7 1
2 3 1 7 4 5 8 9 6
7 8 5 3 9 6 4 1 2
4 6 9 2 1 8 5 3 7
```

Solution # 700
```
1 5 2 6 4 7 8 3 9
8 6 4 3 9 2 5 7 1
9 7 3 1 5 8 4 6 2
2 4 5 8 1 3 7 9 6
7 8 6 5 2 9 3 1 4
3 9 1 7 6 4 2 8 5
4 2 8 9 3 1 6 5 7
5 1 7 4 8 6 9 2 3
6 3 9 2 7 5 1 4 8
```

Solution # 701
```
3 5 4 9 7 8 2 6 1
2 7 8 6 3 1 9 5 4
9 1 6 4 5 2 8 3 7
7 8 2 3 6 4 5 1 9
5 9 3 1 2 7 4 8 6
4 6 1 8 9 5 3 7 2
8 4 7 2 1 3 6 9 5
6 2 5 7 8 9 1 4 3
1 3 9 5 4 6 7 2 8
```

Solution # 702
```
4 3 6 1 2 7 5 9 8
9 2 8 5 3 6 1 7 4
5 1 7 8 4 9 2 3 6
6 9 4 2 5 8 3 1 7
3 8 5 9 7 1 6 4 2
2 7 1 3 6 4 9 8 5
7 4 3 6 9 5 8 2 1
1 6 9 7 8 2 4 5 3
8 5 2 4 1 3 7 6 9
```

Solution # 703
```
3 4 1 7 6 2 5 8 9
6 2 5 9 3 8 7 1 4
9 8 7 5 1 4 2 6 3
5 6 9 4 2 3 8 7 1
2 7 8 1 9 6 3 4 5
4 1 3 8 7 5 9 2 6
8 9 6 2 5 1 4 3 7
7 3 4 6 8 9 1 5 2
1 5 2 3 4 7 6 9 8
```

Solution # 704
```
4 5 6 7 8 3 2 9 1
2 1 7 5 4 9 6 3 8
3 8 9 2 1 6 4 7 5
6 4 3 8 9 7 1 5 2
1 2 5 6 3 4 7 8 9
9 7 8 1 5 2 3 4 6
7 9 2 4 6 8 5 1 3
5 3 4 9 2 1 8 6 7
8 6 1 3 7 5 9 2 4
```

Solution # 705
```
2 6 7 1 9 8 4 5 3
8 9 4 5 3 6 2 1 7
3 1 5 4 7 2 9 6 8
4 5 2 3 6 9 8 7 1
7 8 6 2 1 5 3 4 9
9 3 1 7 8 4 5 2 6
1 4 8 6 2 3 7 9 5
6 2 9 8 5 7 1 3 4
5 7 3 9 4 1 6 8 2
```

Solution # 706
```
3 4 7 2 6 8 5 1 9
9 8 2 3 1 5 7 6 4
6 1 5 7 4 9 2 8 3
2 6 4 8 9 1 3 5 7
5 7 9 6 2 3 8 4 1
1 3 8 5 7 4 9 2 6
7 2 1 9 8 6 4 3 5
4 9 3 1 5 2 6 7 8
8 5 6 4 3 7 1 9 2
```

Solution # 707
```
6 3 2 9 1 4 5 8 7
5 8 4 7 3 2 9 6 1
7 1 9 5 6 8 4 3 2
2 9 5 6 8 3 7 1 4
3 6 7 4 5 1 8 2 9
1 4 8 2 9 7 3 5 6
8 2 3 1 4 9 6 7 5
9 7 6 8 2 5 1 4 3
4 5 1 3 7 6 2 9 8
```

Solution # 708
```
1 2 9 8 7 3 6 5 4
6 3 8 2 4 5 7 9 1
5 7 4 6 9 1 2 3 8
4 9 5 1 3 6 8 2 7
7 8 6 9 5 2 4 1 3
2 1 3 4 8 7 5 6 9
9 4 2 5 1 8 3 7 6
3 5 1 7 6 4 9 8 2
8 6 7 3 2 9 1 4 5
```

Solution # 709
```
7 6 2 5 1 9 4 3 8
5 9 3 8 7 4 1 6 2
1 4 8 2 3 6 5 7 9
6 3 1 4 8 5 9 2 7
9 2 4 3 6 7 8 5 1
8 7 5 9 2 1 3 4 6
2 8 9 6 5 3 7 1 4
3 1 6 7 4 8 2 9 5
4 5 7 1 9 2 6 8 3
```

Solution # 710
```
4 1 2 3 8 5 7 9 6
3 7 8 6 2 9 1 5 4
5 6 9 4 7 1 3 2 8
6 8 5 2 9 3 4 7 1
9 3 1 7 6 4 5 8 2
7 2 4 1 5 8 6 3 9
2 4 6 8 3 7 9 1 5
1 9 7 5 4 2 8 6 3
8 5 3 9 1 6 2 4 7
```

Solution # 711
```
9 2 4 8 1 6 3 5 7
8 5 3 9 7 2 4 1 6
1 7 6 4 5 3 9 2 8
6 4 1 3 2 5 8 7 9
5 3 9 7 4 8 1 6 2
7 8 2 6 9 1 5 3 4
2 9 7 5 3 4 6 8 1
4 6 5 1 8 7 2 9 3
3 1 8 2 6 9 7 4 5
```

Solution # 712
```
8 9 6 3 1 7 2 5 4
5 2 7 9 4 6 8 3 1
3 4 1 2 5 8 7 9 6
6 8 2 5 7 9 1 4 3
4 3 5 8 2 1 9 6 7
1 7 9 4 6 3 5 8 2
7 5 4 6 9 2 3 1 8
2 6 8 1 3 5 4 7 9
9 1 3 7 8 4 6 2 5
```

Solution # 713
```
9 1 8 2 6 4 5 7 3
2 4 3 8 5 7 9 1 6
7 5 6 1 9 3 2 4 8
5 3 9 4 7 2 6 8 1
6 7 1 5 3 8 4 2 9
8 2 4 9 1 6 3 5 7
3 8 7 6 2 5 1 9 4
1 6 5 7 4 9 8 3 2
4 9 2 3 8 1 7 6 5
```

Solution # 714
```
2 5 9 1 7 3 4 6 8
6 1 7 4 5 8 2 3 9
4 8 3 6 2 9 1 7 5
5 9 4 8 3 7 6 1 2
7 6 2 5 4 1 8 9 3
8 3 1 9 6 2 7 5 4
1 4 8 3 9 6 5 2 7
3 2 5 7 1 4 9 8 6
9 7 6 2 8 5 3 4 1
```

Solution # 715
```
3 1 5 6 7 8 4 9 2
7 4 2 3 5 9 1 8 6
9 6 8 2 1 4 5 7 3
1 5 6 9 4 7 2 3 8
8 9 7 5 2 3 6 4 1
4 2 3 8 6 1 9 5 7
6 3 9 1 8 5 7 2 4
5 7 1 4 3 2 8 6 9
2 8 4 7 9 6 3 1 5
```

Solution # 716
```
4 9 7 5 3 6 2 1 8
6 3 8 2 1 7 5 9 4
2 1 5 4 9 8 3 6 7
8 5 2 6 4 3 9 7 1
7 4 1 9 2 5 8 3 6
9 6 3 8 7 1 4 2 5
1 2 4 7 5 9 6 8 3
5 7 6 3 8 2 1 4 9
3 8 9 1 6 4 7 5 2
```

Solution # 717
```
2 9 3 5 4 8 7 6 1
6 1 5 3 9 7 8 4 2
7 4 8 2 6 1 3 9 5
5 8 9 7 1 4 6 2 3
3 7 4 6 5 2 9 1 8
1 2 6 8 3 9 5 7 4
9 3 2 1 7 5 4 8 6
4 5 1 9 8 6 2 3 7
8 6 7 4 2 3 1 5 9
```

Solution # 718
```
7 6 1 8 4 2 5 3 9
5 3 9 1 6 7 8 2 4
4 2 8 9 3 5 6 7 1
8 9 2 3 5 6 1 4 7
3 7 4 2 8 1 9 5 6
1 5 6 7 9 4 3 8 2
9 1 5 4 2 8 7 6 3
6 4 3 5 7 9 2 1 8
2 8 7 6 1 3 4 9 5
```

Solution # 719
```
9 3 2 7 4 1 5 6 8
8 4 6 3 2 5 9 1 7
7 1 5 9 6 8 2 4 3
5 9 8 2 3 4 6 7 1
3 2 7 5 1 6 4 8 9
1 6 4 8 9 7 3 2 5
6 5 1 4 8 3 7 9 2
2 8 3 6 7 9 1 5 4
4 7 9 1 5 2 8 3 6
```

Solution # 720
```
5 7 2 3 9 1 8 4 6
4 3 9 6 8 5 2 1 7
8 1 6 2 4 7 5 3 9
6 5 1 8 7 2 4 9 3
9 2 3 5 6 4 7 8 1
7 8 4 1 3 9 6 5 2
3 4 7 9 2 8 1 6 5
2 6 5 4 1 3 9 7 8
1 9 8 7 5 6 3 2 4
```

Solution # 721
```
3 9 2 8 5 4 1 6 7
6 8 1 9 7 3 4 2 5
5 4 7 2 1 6 3 8 9
7 5 8 1 4 9 6 3 2
4 1 6 7 3 2 9 5 8
2 3 9 5 6 8 7 1 4
9 7 3 6 2 5 8 4 1
1 2 4 3 8 7 5 9 6
8 6 5 4 9 1 2 7 3
```

Solution # 722
```
6 5 4 1 7 2 9 8 3
2 9 8 4 5 3 6 1 7
1 7 3 6 9 8 2 4 5
5 4 7 2 8 6 1 3 9
8 3 2 5 1 9 4 7 6
9 6 1 3 4 7 8 5 2
3 8 5 9 6 4 7 2 1
4 1 6 7 2 5 3 9 8
7 2 9 8 3 1 5 6 4
```

Solution # 723
```
7 5 4 8 6 2 1 9 3
9 2 6 1 7 3 4 5 8
1 8 3 4 5 9 6 2 7
2 3 1 7 9 6 5 8 4
6 4 5 3 2 8 7 1 9
8 7 9 5 1 4 3 6 2
4 9 2 6 3 1 8 7 5
3 1 7 2 8 5 9 4 6
5 6 8 9 4 7 2 3 1
```

Solution # 724
```
2 4 6 9 5 8 7 1 3
3 1 8 2 4 7 9 6 5
9 7 5 1 3 6 4 8 2
8 9 7 6 2 4 3 5 1
4 5 1 8 9 3 6 2 7
6 2 3 7 1 5 8 4 9
7 6 2 5 8 9 1 3 4
5 8 4 3 7 1 2 9 6
1 3 9 4 6 2 5 7 8
```

Solution # 725
```
5 2 3 8 7 4 6 1 9
4 7 6 5 9 1 3 8 2
8 9 1 6 3 2 5 7 4
3 6 8 4 1 5 2 9 7
7 5 9 3 2 8 4 6 1
1 4 2 9 6 7 8 5 3
2 8 7 1 5 3 9 4 6
9 1 4 2 8 6 7 3 5
6 3 5 7 4 9 1 2 8
```

Solution # 726
```
2 9 5 6 7 3 1 8 4
8 3 6 4 2 1 5 7 9
7 4 1 9 8 5 2 3 6
5 2 9 3 1 8 6 4 7
3 1 7 5 6 4 9 2 8
4 6 8 2 9 7 3 5 1
6 8 2 7 3 9 4 1 5
9 7 4 1 5 2 8 6 3
1 5 3 8 4 6 7 9 2
```

Solution # 727
```
2 3 8 1 5 6 9 4 7
7 6 5 4 9 3 2 8 1
1 9 4 7 8 2 6 3 5
6 1 3 9 2 5 8 7 4
4 8 9 3 6 7 5 1 2
5 2 7 8 4 1 3 6 9
9 4 1 5 3 8 7 2 6
8 5 2 6 7 4 1 9 3
3 7 6 2 1 9 4 5 8
```

Solution # 728
```
7 1 6 2 3 4 8 9 5
3 8 2 7 9 5 6 1 4
5 4 9 8 1 6 7 2 3
1 2 8 9 4 7 3 5 6
4 5 3 6 2 8 1 7 9
9 6 7 3 5 1 4 8 2
8 9 1 5 6 3 2 4 7
6 7 5 4 8 2 9 3 1
2 3 4 1 7 9 5 6 8
```

Solution # 729
```
6 1 2 4 7 9 3 8 5
4 5 7 3 8 2 1 9 6
8 9 3 6 5 1 2 4 7
5 8 9 1 4 6 7 2 3
7 4 6 5 2 3 8 1 9
2 3 1 8 9 7 6 5 4
1 2 4 7 6 5 9 3 8
3 7 8 9 1 4 5 6 2
9 6 5 2 3 8 4 7 1
```

Solution # 730
```
8 6 1 3 4 2 7 5 9
2 7 4 9 5 1 8 3 6
5 9 3 8 6 7 4 1 2
9 3 5 2 8 4 1 6 7
1 4 7 6 9 3 5 2 8
6 2 8 1 7 5 9 4 3
3 8 2 4 1 9 6 7 5
4 5 9 7 3 6 2 8 1
7 1 6 5 2 8 3 9 4
```

Solution # 731
```
6 2 4 7 1 3 9 5 8
3 7 8 5 4 9 6 2 1
5 9 1 8 2 6 3 4 7
7 5 3 4 9 2 8 1 6
2 8 6 1 3 7 5 9 4
4 1 9 6 8 5 7 3 2
9 6 2 3 7 4 1 8 5
1 3 7 2 5 8 4 6 9
8 4 5 9 6 1 2 7 3
```

Solution # 732
```
4 5 8 3 1 6 2 7 9
2 6 7 4 8 9 5 3 1
1 9 3 2 5 7 4 8 6
6 4 2 7 9 1 3 5 8
5 3 9 8 4 2 6 1 7
8 7 1 5 6 3 9 2 4
9 8 6 1 3 5 7 4 2
7 1 5 9 2 4 8 6 3
3 2 4 6 7 8 1 9 5
```

Solution # 733
```
8 1 6 9 5 7 3 2 4
7 5 3 2 6 4 1 9 8
9 2 4 8 1 3 7 5 6
3 4 1 6 9 5 2 8 7
6 9 7 4 2 8 5 3 1
2 8 5 7 3 1 6 4 9
4 7 2 5 8 6 9 1 3
1 6 9 3 4 2 8 7 5
5 3 8 1 7 9 4 6 2
```

Solution # 734
```
5 2 3 4 8 7 6 9 1
6 7 1 5 9 2 8 4 3
4 8 9 6 3 1 7 2 5
9 5 6 1 2 8 4 3 7
8 3 7 9 4 6 5 1 2
1 4 2 7 5 3 9 6 8
7 9 4 3 1 5 2 8 6
3 6 8 2 7 9 1 5 4
2 1 5 8 6 4 3 7 9
```

Solution # 735
```
1 3 7 4 2 5 8 6 9
9 4 8 1 6 7 3 5 2
2 5 6 9 3 8 4 1 7
4 1 3 5 8 2 7 9 6
6 8 9 7 4 1 5 2 3
7 2 5 3 9 6 1 4 8
8 7 4 2 5 9 6 3 1
3 9 1 6 7 4 2 8 5
5 6 2 8 1 3 9 7 4
```

Solution # 736
```
9 6 8 2 7 4 1 3 5
2 7 5 1 3 6 8 4 9
4 1 3 8 5 9 2 6 7
5 8 1 7 9 3 4 2 6
6 3 2 4 8 5 9 7 1
7 4 9 6 2 1 3 5 8
3 2 7 5 1 8 6 9 4
1 9 6 3 4 5 7 8 2
8 5 4 9 6 2 7 1 3
```

Solution # 737
```
8 1 6 2 3 5 9 4 7
7 5 3 9 8 4 6 2 1
4 2 9 6 1 7 5 3 8
5 6 1 7 9 3 4 8 2
9 7 2 4 6 8 3 1 5
3 4 8 1 5 2 7 9 6
6 3 4 5 2 1 8 7 9
1 9 7 8 4 6 2 5 3
2 8 5 3 7 9 1 6 4
```

Solution # 738
```
7 2 1 9 3 4 8 5 6
3 4 5 8 7 6 9 2 1
9 8 6 5 2 1 7 4 3
2 5 7 6 4 8 3 1 9
8 1 4 3 9 7 2 6 5
6 9 3 2 1 5 4 8 7
5 6 9 4 8 3 1 7 2
1 3 8 7 5 2 6 9 4
4 7 2 1 6 9 5 3 8
```

Solution # 739
```
1 6 4 3 9 2 8 7 5
7 5 8 4 6 1 9 3 2
2 9 3 7 5 8 4 1 6
9 7 6 8 1 5 2 4 3
3 1 2 6 4 9 7 5 8
8 4 5 2 3 7 6 9 1
4 8 9 1 2 3 5 6 7
6 2 1 5 7 4 3 8 9
5 3 7 9 8 6 1 2 4
```

Solution # 740
```
6 7 5 4 9 1 2 8 3
3 1 9 5 2 8 7 6 4
2 4 8 7 6 3 9 5 1
5 3 4 2 8 7 1 9 6
1 9 2 6 4 5 3 7 8
8 6 7 3 1 9 4 2 5
9 2 3 8 5 4 6 1 7
4 8 6 1 7 2 5 3 9
7 5 1 9 3 6 8 4 2
```

Solution # 741
```
7 5 6 4 1 8 2 3 9
3 2 8 9 7 6 1 4 5
9 4 1 2 5 3 7 6 8
6 9 4 7 8 1 3 5 2
1 3 2 5 9 4 6 8 7
8 7 5 3 6 2 4 9 1
2 8 9 6 4 7 5 1 3
5 6 3 1 2 9 8 7 4
4 1 7 8 3 5 9 2 6
```

Solution # 742
```
7 9 4 5 6 3 8 1 2
1 3 6 4 2 8 9 5 7
2 5 8 7 9 1 3 6 4
5 8 1 2 3 6 4 7 9
3 6 9 8 7 4 5 2 1
4 2 7 1 5 9 6 8 3
9 7 2 3 8 5 1 4 6
6 4 5 9 1 2 7 3 8
8 1 3 6 4 7 2 9 5
```

Solution # 743
```
5 6 4 2 1 8 7 9 3
9 2 7 4 6 3 8 1 5
8 3 1 9 5 7 4 6 2
7 8 3 1 4 6 5 2 9
4 1 9 5 3 2 6 8 7
2 5 6 7 8 9 1 3 4
3 4 5 8 2 1 9 7 6
6 9 8 3 7 5 2 4 1
1 7 2 6 9 4 3 5 8
```

Solution # 744
```
1 9 8 2 7 3 5 6 4
2 6 7 5 1 4 8 3 9
3 4 5 8 6 9 1 7 2
9 2 4 3 8 5 7 1 6
8 3 6 7 2 1 4 9 5
5 7 1 9 4 6 3 2 8
7 1 2 4 9 8 6 5 3
6 8 3 1 5 2 9 4 7
4 5 9 6 3 7 2 8 1
```

Solution # 745
```
5 2 6 9 3 8 7 4 1
4 7 8 6 2 1 5 3 9
9 1 3 5 4 7 2 6 8
8 9 1 3 7 4 6 5 2
6 3 5 8 9 2 4 1 7
7 4 2 1 6 5 9 8 3
1 6 4 7 8 9 3 2 5
3 5 7 2 1 6 8 9 4
2 8 9 4 5 3 1 7 6
```

Solution # 746
```
8 6 3 5 9 7 2 4 1
5 2 1 3 8 4 6 7 9
4 9 7 1 6 2 5 3 8
9 8 5 4 7 6 3 1 2
1 7 2 8 3 9 4 6 5
3 4 6 2 1 5 8 9 7
7 3 9 6 2 8 1 5 4
6 5 8 7 4 1 9 2 3
2 1 4 9 5 3 7 8 6
```

Solution # 747
```
3 4 6 2 1 7 9 5 8
9 2 5 6 4 8 3 7 1
8 1 7 5 3 9 4 6 2
2 3 9 7 6 1 5 8 4
5 6 1 8 2 4 7 3 9
7 8 4 9 5 3 2 1 6
6 5 3 1 9 2 8 4 7
1 9 8 4 7 5 6 2 3
4 7 2 3 8 6 1 9 5
```

Solution # 748
```
4 8 6 5 2 1 3 7 9
5 9 7 3 6 4 1 2 8
2 1 3 7 9 8 4 5 6
8 2 4 9 1 7 5 6 3
6 5 1 4 8 3 7 9 2
3 7 9 6 5 2 8 1 4
1 6 2 8 4 5 9 3 7
7 4 5 2 3 9 6 8 1
9 3 8 1 7 6 2 4 5
```

Solution # 749
```
5 9 8 4 6 1 2 7 3
3 1 4 5 7 2 6 9 8
2 6 7 3 8 9 1 5 4
1 7 2 8 9 4 3 6 5
8 5 6 7 2 3 4 1 9
9 4 3 6 1 5 8 2 7
7 8 9 2 4 6 5 3 1
4 2 5 1 3 7 9 8 6
6 3 1 9 5 8 7 4 2
```

Solution # 750
```
9 4 3 5 8 1 2 7 6
6 8 1 7 9 2 4 3 5
7 5 2 4 3 6 8 1 9
3 7 8 9 2 4 6 5 1
1 9 6 8 5 3 7 4 2
4 2 5 6 1 7 9 8 3
8 3 4 2 6 5 1 9 7
2 1 7 3 4 9 5 6 8
5 6 9 1 7 8 3 2 4
```

Solution # 751
```
9 3 7 4 1 2 8 5 6
1 8 4 3 5 6 9 7 2
5 6 2 9 8 7 4 3 1
3 1 5 6 2 8 7 9 4
2 4 8 7 9 5 1 6 3
7 9 6 1 4 3 5 2 8
4 5 1 2 3 9 6 8 7
8 7 3 5 6 1 2 4 9
6 2 9 8 7 4 3 1 5
```

Solution # 752
```
2 4 3 5 9 8 6 7 1
7 6 5 1 2 3 4 8 9
8 9 1 4 6 7 3 2 5
5 7 4 9 3 1 2 6 8
9 3 6 8 5 2 7 1 4
1 2 8 6 7 4 5 9 3
3 8 7 2 4 9 1 5 6
4 5 9 7 1 6 8 3 2
6 1 2 3 8 5 9 4 7
```

Solution # 753
```
3 4 7 6 1 5 9 2 8
8 1 2 9 7 4 6 5 3
6 5 9 3 8 2 1 4 7
1 7 3 2 5 6 8 9 4
5 9 6 7 4 8 2 3 1
4 2 8 1 3 9 7 6 5
2 8 5 4 9 7 3 1 6
9 3 4 8 6 1 5 7 2
7 6 1 5 2 3 4 8 9
```

Solution # 754
```
9 4 8 2 7 6 5 3 1
6 3 7 9 5 1 4 8 2
1 5 2 4 3 8 6 7 9
7 6 9 3 2 5 8 1 4
5 2 4 8 1 9 3 6 7
8 1 3 6 4 7 2 9 5
2 9 6 7 8 4 1 5 3
4 7 5 1 6 3 9 2 8
3 8 1 5 9 2 7 4 6
```

Solution # 755
```
9 6 3 7 4 1 5 2 8
2 8 4 5 6 3 9 1 7
5 1 7 2 8 9 4 3 6
3 4 6 9 5 2 7 8 1
7 2 1 8 3 4 6 9 5
8 9 5 1 7 6 3 4 2
6 3 8 4 1 5 2 7 9
1 5 2 3 9 7 8 6 4
4 7 9 6 2 8 1 5 3
```

Solution # 756
```
9 5 1 2 8 4 7 3 6
4 3 2 1 6 7 5 8 9
8 7 6 3 5 9 1 2 4
2 4 8 5 9 1 3 6 7
5 1 7 4 3 6 2 9 8
3 6 9 8 7 2 4 1 5
1 2 5 9 4 8 6 7 3
6 9 4 7 1 3 8 5 2
7 8 3 6 2 5 9 4 1
```

Solution # 757
```
6 1 7 2 4 8 3 9 5
4 3 8 1 9 5 7 2 6
5 9 2 7 3 6 1 8 4
9 7 1 4 6 2 8 5 3
8 5 6 3 1 7 2 4 9
2 4 3 8 5 9 6 1 7
3 2 5 6 1 4 9 7 8
1 6 4 9 8 7 5 3 2
7 8 9 5 2 3 4 6 1
```

Solution # 758
```
1 4 9 5 6 3 8 2 7
3 5 2 4 8 7 6 1 9
6 7 8 2 9 1 4 5 3
7 9 4 1 2 5 3 6 8
2 6 1 3 4 8 7 9 5
8 3 5 6 7 9 1 4 2
9 1 7 8 5 6 2 3 4
5 2 6 7 3 4 9 8 1
4 8 3 9 1 2 5 7 6
```

Solution # 759
```
4 2 3 8 6 7 1 9 5
8 9 5 1 2 3 4 7 6
7 6 1 9 5 4 8 2 3
6 3 9 4 7 8 2 5 1
2 7 4 3 1 5 9 6 8
1 5 8 2 9 6 3 4 7
9 1 7 5 8 2 6 3 4
5 4 2 6 3 1 7 8 9
3 8 6 7 4 9 5 1 2
```

Solution # 760
```
7 4 3 1 8 9 5 6 2
8 9 5 6 2 4 1 7 3
1 2 6 3 5 7 9 4 8
6 8 4 9 7 3 2 5 1
3 5 1 4 6 2 7 8 9
2 7 9 8 1 5 4 3 6
9 6 8 7 4 1 3 2 5
5 1 7 2 3 6 8 9 4
4 3 2 5 9 8 6 1 7
```

Solution # 761
```
3 9 2 7 8 1 4 5 6
1 7 6 5 4 3 2 8 9
8 4 5 2 6 9 1 3 7
6 3 7 9 2 4 8 1 5
2 5 1 8 7 6 9 4 3
4 8 9 1 3 5 7 6 2
7 6 4 3 9 8 5 2 1
5 2 3 4 1 7 6 9 8
9 1 8 6 5 2 3 7 4
```

Solution # 762
```
1 9 3 7 2 6 4 5 8
5 7 6 8 4 3 1 2 9
8 2 4 1 5 9 6 3 7
6 4 9 2 3 8 5 7 1
2 1 5 4 6 7 9 8 3
7 3 8 9 1 5 2 4 6
3 8 1 5 9 4 7 6 2
4 6 2 3 7 1 8 9 5
9 5 7 6 8 2 3 1 4
```

Solution # 763
```
9 3 2 7 8 4 5 6 1
6 4 1 3 9 5 8 2 7
5 7 8 1 6 2 4 3 9
3 6 9 5 4 1 2 7 8
7 2 5 9 3 8 6 1 4
1 8 4 6 2 7 3 9 5
2 1 7 4 5 3 9 8 6
4 9 3 8 1 6 7 5 2
8 5 6 2 7 9 1 4 3
```

Solution # 764
```
4 2 6 1 3 7 5 8 9
9 1 8 6 4 5 7 3 2
5 7 3 2 9 8 6 4 1
7 9 4 8 1 3 2 6 5
2 8 5 9 7 6 4 1 3
6 3 1 5 2 4 8 9 7
1 6 9 4 5 2 3 7 8
3 4 2 7 8 1 9 5 6
8 5 7 3 6 9 1 2 4
```

Solution # 765
```
2 4 8 9 5 3 6 1 7
3 7 1 4 6 2 5 9 8
9 5 6 7 8 1 3 4 2
7 8 3 2 1 5 9 6 4
5 2 4 3 9 6 8 7 1
6 1 9 8 7 4 2 5 3
4 3 5 1 2 9 7 8 6
1 9 7 6 3 8 4 2 5
8 6 2 5 4 7 1 3 9
```

Solution # 766
```
3 8 1 7 5 6 4 9 2
6 5 9 8 2 4 1 7 3
2 4 7 3 9 1 8 6 5
8 6 4 9 3 2 5 1 7
1 3 5 6 8 7 9 2 4
7 9 2 4 1 5 3 8 6
4 1 6 5 7 8 2 3 9
9 7 8 2 4 3 6 5 1
5 2 3 1 6 9 7 4 8
```

Solution # 767
```
2 1 6 5 8 3 7 4 9
3 8 9 4 7 1 2 5 6
5 4 7 2 6 9 3 1 8
1 9 2 3 5 4 8 6 7
4 6 5 8 9 7 1 3 2
7 3 8 1 2 6 4 9 5
8 2 4 6 3 5 9 7 1
9 5 1 7 4 2 6 8 3
6 7 3 9 1 8 5 2 4
```

Solution # 768
```
2 5 7 9 4 8 1 6 3
8 1 4 6 7 3 2 5 9
3 9 6 5 1 2 4 7 8
6 2 1 7 5 9 3 8 4
5 4 9 8 3 1 7 2 6
7 8 3 2 6 4 5 9 1
1 6 2 3 9 5 8 4 7
4 7 5 1 8 6 9 3 2
9 3 8 4 2 7 6 1 5
```

Solution # 769
```
7 8 5 9 3 2 6 4 1
3 9 1 6 8 4 5 7 2
2 6 4 1 7 5 3 9 8
8 4 9 7 6 1 2 3 5
1 2 7 3 5 8 4 6 9
5 3 6 2 4 9 8 1 7
9 7 8 4 2 6 1 5 3
6 1 2 5 9 3 7 8 4
4 5 3 8 1 7 9 2 6
```

Solution # 770
```
1 4 7 2 9 3 5 8 6
8 3 9 4 5 6 1 2 7
5 6 2 7 8 1 9 4 3
4 7 8 5 1 9 3 6 2
9 5 3 6 2 7 4 1 8
2 1 6 3 4 8 7 9 5
3 2 1 9 6 5 8 7 4
6 9 5 8 7 4 2 3 1
7 8 4 1 3 2 6 5 9
```

Solution # 771
```
5 6 1 8 7 2 3 4 9
3 4 2 9 1 6 5 7 8
9 7 8 4 3 5 6 2 1
8 1 4 3 2 7 9 6 5
2 3 5 6 8 9 7 1 4
7 9 6 5 4 1 8 3 2
4 5 7 1 9 3 2 8 6
1 2 9 7 6 8 4 5 3
6 8 3 2 5 4 1 9 7
```

Solution # 772
```
7 6 3 1 2 5 9 8 4
5 8 2 9 7 4 6 1 3
4 1 9 8 3 6 5 2 7
9 7 6 4 5 1 8 3 2
1 2 5 3 6 8 4 7 9
3 4 8 7 9 2 1 5 6
2 9 4 5 8 7 3 6 1
6 5 1 2 4 3 7 9 8
8 3 7 6 1 9 2 4 5
```

Solution # 773
```
6 7 5 4 2 9 8 1 3
8 4 9 3 1 7 2 6 5
3 2 1 6 5 8 4 9 7
1 9 4 2 3 5 6 7 8
2 6 7 8 4 1 5 3 9
5 8 3 7 9 6 1 2 4
4 1 8 9 6 3 7 5 2
7 3 6 5 8 2 9 4 1
9 5 2 1 7 4 3 8 6
```

Solution # 774
```
3 1 7 4 9 2 5 6 8
4 2 9 6 5 8 1 3 7
8 6 5 7 1 3 9 2 4
9 3 2 8 6 1 4 7 5
5 7 1 3 4 9 6 8 2
6 8 4 2 7 5 3 9 1
7 5 8 1 3 6 2 4 9
1 4 6 9 2 7 8 5 3
2 9 3 5 8 4 7 1 6
```

Solution # 775
```
8 5 6 9 3 2 1 7 4
3 1 9 4 7 8 5 2 6
4 7 2 5 6 1 8 9 3
6 9 5 7 8 3 4 1 2
2 4 8 1 9 6 3 5 7
7 3 1 2 4 5 6 8 9
5 2 4 3 1 7 9 6 8
1 6 3 8 2 9 7 4 5
9 8 7 6 5 4 2 3 1
```

Solution # 776
```
6 8 2 1 4 9 7 5 3
5 3 1 2 8 7 6 9 4
7 9 4 6 5 3 2 1 8
9 5 7 8 6 4 1 3 2
3 2 6 5 7 1 4 8 9
1 4 8 3 9 2 5 6 7
2 6 9 7 3 5 8 4 1
4 7 5 9 1 8 3 2 6
8 1 3 4 2 6 9 7 5
```

Solution # 777
```
3 9 4 1 6 2 8 7 5
7 2 6 5 4 8 3 9 1
1 5 8 7 3 9 2 6 4
5 7 3 2 8 6 4 1 9
8 4 2 9 1 5 6 3 7
6 1 9 3 7 4 5 2 8
2 3 7 4 5 1 9 8 6
9 8 5 6 2 7 1 4 3
4 6 1 8 9 3 7 5 2
```

Solution # 778
```
6 7 3 5 8 1 4 9 2
8 2 1 9 4 3 5 6 7
5 4 9 7 2 6 1 3 8
4 5 7 6 1 9 8 2 3
3 9 2 4 7 8 6 5 1
1 6 8 3 5 2 9 7 4
2 3 4 8 6 5 7 1 9
7 1 6 2 9 4 3 8 5
9 8 5 1 3 7 2 4 6
```

Solution # 779
```
2 3 9 1 7 6 4 5 8
6 7 4 5 2 8 1 9 3
8 1 5 3 9 4 2 6 7
5 6 3 9 1 2 7 8 4
7 9 8 4 5 3 6 1 2
1 4 2 6 8 7 5 3 9
3 2 6 8 4 1 9 7 5
9 8 7 2 6 5 3 4 1
4 5 1 7 3 9 8 2 6
```

Solution # 780
```
9 2 5 3 4 6 8 1 7
7 4 6 8 2 1 5 9 3
8 1 3 5 9 7 4 6 2
3 5 2 4 1 8 9 7 6
6 8 7 2 5 9 3 4 1
4 9 1 6 7 3 2 5 8
1 3 8 9 6 5 7 2 4
2 6 9 7 3 4 1 8 5
5 7 4 1 8 2 6 3 9
```

Solution # 781
```
4 7 1 2 6 9 5 8 3
6 9 3 1 5 8 4 7 2
2 8 5 7 3 4 9 6 1
5 6 9 3 7 2 8 1 4
7 4 2 8 9 1 3 5 6
1 3 8 5 4 6 2 9 7
8 1 7 9 2 3 6 4 5
9 2 4 6 1 5 7 3 8
3 5 6 4 8 7 1 2 9
```

Solution # 782
```
2 3 6 7 1 9 8 4 5
5 9 7 8 2 4 3 6 1
1 8 4 3 5 6 2 9 7
7 5 1 4 9 8 6 3 2
4 2 8 1 6 3 7 5 9
9 6 3 5 7 2 4 1 8
8 4 9 2 3 5 1 7 6
3 1 5 6 8 7 9 2 4
6 7 2 9 4 1 5 8 3
```

Solution # 783
```
3 6 2 8 7 5 4 9 1
1 8 4 3 2 9 6 5 7
9 7 5 1 4 6 2 3 8
4 2 9 5 6 8 7 1 3
6 5 3 9 1 7 8 4 2
7 1 8 4 3 2 9 6 5
2 3 1 7 9 4 5 8 6
5 9 7 6 8 3 1 2 4
8 4 6 2 5 1 3 7 9
```

Solution # 784
```
2 9 6 8 1 7 4 5 3
3 7 4 2 5 6 9 8 1
1 8 5 9 4 3 2 6 7
5 2 8 1 7 4 3 9 6
6 3 1 5 9 2 8 7 4
7 4 9 3 6 8 5 1 2
8 1 7 4 2 5 6 3 9
4 6 3 7 8 9 1 2 5
9 5 2 6 3 1 7 4 8
```

Solution # 785
```
4 1 7 9 8 2 6 5 3
9 6 5 3 4 1 8 2 7
3 8 2 6 7 5 9 4 1
6 3 1 7 5 9 2 8 4
7 9 8 4 2 3 5 1 6
2 5 4 8 1 6 7 3 9
8 4 3 2 9 7 1 6 5
5 2 9 1 6 4 3 7 8
1 7 6 5 3 8 4 9 2
```

Solution # 786
```
6 1 7 9 8 4 3 2 5
9 8 2 7 5 3 4 6 1
4 5 3 6 1 2 7 8 9
5 6 1 4 3 7 8 9 2
3 2 4 5 9 8 6 1 7
7 9 8 2 6 1 5 4 3
8 4 5 3 2 9 1 7 6
2 7 6 1 4 5 9 3 8
1 3 9 8 7 6 2 5 4
```

Solution # 787
```
8 5 3 2 9 7 4 6 1
2 4 9 5 1 6 3 8 7
1 7 6 8 3 4 2 9 5
6 3 2 7 5 8 9 1 4
7 1 8 9 4 2 6 5 3
5 9 4 1 6 3 7 2 8
4 8 7 6 2 5 1 3 9
9 2 5 3 7 1 8 4 6
3 6 1 4 8 9 5 7 2
```

Solution # 788
```
2 6 7 8 5 3 1 4 9
1 5 8 7 4 9 6 2 3
3 9 4 2 1 6 7 5 8
4 1 2 9 6 5 3 8 7
5 3 9 4 8 7 2 1 6
7 8 6 1 3 2 4 9 5
6 2 3 5 9 4 8 7 1
9 7 1 3 2 8 5 6 4
8 4 5 6 7 1 9 3 2
```

Solution # 789
```
4 8 5 2 9 6 1 3 7
2 6 7 3 4 1 9 5 8
1 9 3 7 5 8 4 6 2
6 1 4 5 3 7 8 2 9
5 7 2 4 8 9 6 1 3
8 3 9 1 6 2 5 7 4
9 5 6 8 7 3 2 4 1
7 2 8 6 1 4 3 9 5
3 4 1 9 2 5 7 8 6
```

Solution # 790
```
3 4 6 2 8 5 7 9 1
8 9 5 7 6 1 3 2 4
1 7 2 3 9 4 8 5 6
6 5 1 9 3 8 4 7 2
7 8 3 4 2 6 5 1 9
4 2 9 1 5 7 6 8 3
9 6 7 8 1 3 2 4 5
2 3 8 5 4 9 1 6 7
5 1 4 6 7 2 9 3 8
```

Solution # 791
```
4 6 2 1 9 7 3 8 5
5 9 3 4 2 8 1 7 6
8 1 7 3 5 6 4 2 9
7 3 6 8 4 5 2 9 1
1 4 8 2 7 9 5 6 3
2 5 9 6 1 3 8 4 7
9 7 1 5 8 4 6 3 2
6 2 4 7 3 1 9 5 8
3 8 5 9 6 2 7 1 4
```

Solution # 792
```
3 9 8 2 4 5 7 1 6
5 2 7 1 3 6 8 9 4
4 1 6 8 9 7 3 5 2
7 8 3 4 6 1 9 2 5
6 5 9 3 7 2 1 4 8
2 4 1 5 8 9 6 7 3
1 6 5 7 2 8 4 3 9
8 7 4 9 5 3 2 6 1
9 3 2 6 1 4 5 8 7
```

Solution # 793
```
9 8 5 3 2 7 1 4 6
3 6 2 8 4 1 5 9 7
4 7 1 6 5 9 3 2 8
1 3 8 5 7 2 4 6 9
6 4 7 1 9 3 8 5 2
5 2 9 4 6 8 7 1 3
2 1 4 7 8 6 9 3 5
7 9 3 2 1 5 6 8 4
8 5 6 9 3 4 2 7 1
```

Solution # 794
```
2 1 5 8 3 6 9 7 4
4 6 9 1 5 7 8 2 3
7 3 8 4 2 9 5 6 1
8 9 4 6 1 2 3 5 7
3 2 1 7 8 5 4 9 6
6 5 7 9 4 3 1 8 2
1 7 6 5 9 4 2 3 8
9 4 2 3 6 8 7 1 5
5 8 3 2 7 1 6 4 9
```

Solution # 795
```
5 4 1 2 8 6 7 9 3
6 9 8 3 7 5 2 4 1
2 3 7 4 9 1 6 8 5
9 5 2 8 1 7 4 3 6
4 7 6 9 2 3 1 5 8
8 1 3 5 6 4 9 7 2
3 2 5 6 4 9 8 1 7
7 6 9 1 5 8 3 2 4
1 8 4 7 3 2 5 6 9
```

Solution # 796
```
7 5 6 1 8 3 4 9 2
2 4 1 9 7 6 3 5 8
9 8 3 2 5 4 7 1 6
1 9 8 7 4 5 6 2 3
6 2 4 8 3 1 5 7 9
5 3 7 6 2 9 8 4 1
8 6 5 4 9 2 1 3 7
4 1 9 3 6 7 2 8 5
3 7 2 5 1 8 9 6 4
```

Solution # 797
```
2 6 4 9 1 3 7 5 8
3 8 1 6 5 7 4 2 9
7 9 5 4 2 8 3 1 6
5 2 7 3 9 1 6 8 4
6 4 9 7 8 5 2 3 1
1 3 8 2 4 6 5 9 7
9 1 2 5 7 4 8 6 3
8 7 3 1 6 2 9 4 5
4 5 6 8 3 9 1 7 2
```

Solution # 798
```
2 3 9 1 5 7 4 6 8
1 7 8 3 4 6 2 5 9
5 6 4 8 9 2 3 7 1
7 9 5 2 6 4 8 1 3
8 4 3 7 1 9 6 2 5
6 2 1 5 8 3 9 4 7
4 5 2 9 3 1 7 8 6
3 1 6 4 7 8 5 9 2
9 8 7 6 2 5 1 3 4
```

Solution # 799
```
9 1 7 8 5 6 4 3 2
3 8 2 1 4 7 6 5 9
6 5 4 9 3 2 1 8 7
5 2 9 4 6 3 8 7 1
8 6 3 7 1 5 2 9 4
7 4 1 2 9 8 3 6 5
4 9 8 3 7 1 5 2 6
2 7 5 6 8 4 9 1 3
1 3 6 5 2 9 7 4 8
```

Solution # 800
```
6 3 2 8 1 7 4 5 9
5 7 1 2 4 9 3 8 6
4 8 9 3 5 6 2 7 1
1 4 7 9 6 5 8 3 2
2 6 3 4 8 1 5 9 7
9 5 8 7 2 3 1 6 4
3 1 6 5 7 4 9 2 8
8 9 4 6 3 2 7 1 5
7 2 5 1 9 8 6 4 3
```

Solution # 801
```
6 5 7 9 8 4 2 3 1
8 3 9 7 2 1 4 6 5
2 4 1 3 6 5 9 8 7
4 9 6 2 5 7 3 1 8
7 8 5 4 1 3 6 2 9
1 2 3 8 9 6 5 7 4
5 7 8 6 4 2 1 9 3
9 1 2 5 3 8 7 4 6
3 6 4 1 7 9 8 5 2
```

Solution # 802
```
2 7 9 4 6 3 5 8 1
4 1 5 2 8 7 3 9 6
6 3 8 5 1 9 4 7 2
8 2 1 3 4 5 7 6 9
9 5 3 7 2 6 1 4 8
7 4 6 1 9 8 2 3 5
1 9 7 8 3 2 6 5 4
5 6 2 9 7 4 8 1 3
3 8 4 6 5 1 9 2 7
```

Solution # 803
```
7 5 3 8 6 1 4 9 2
6 2 9 7 4 5 1 3 8
8 4 1 2 3 9 6 5 7
3 8 4 9 1 2 7 6 5
5 1 6 4 7 3 8 2 9
2 9 7 5 8 6 3 1 4
4 3 2 6 9 8 5 7 1
1 7 5 3 2 4 9 8 6
9 6 8 1 5 7 2 4 3
```

Solution # 804
```
4 2 1 9 5 8 3 7 6
5 9 6 1 3 7 4 8 2
3 7 8 6 4 2 1 5 9
2 3 7 5 1 6 8 9 4
6 4 9 8 7 3 5 2 1
1 8 5 4 2 9 6 3 7
9 5 3 2 6 1 7 4 8
7 1 2 3 8 4 9 6 5
8 6 4 7 9 5 2 1 3
```

Solution # 805
```
1 6 7 2 9 8 5 4 3
4 2 3 1 7 5 8 9 6
5 8 9 6 4 3 7 2 1
9 7 1 3 8 4 2 6 5
2 4 8 5 1 6 9 3 7
6 3 5 7 2 9 4 1 8
7 1 4 8 6 2 3 5 9
3 9 6 4 5 7 1 8 2
8 5 2 9 3 1 6 7 4
```

```
Solution # 806        Solution # 807        Solution # 808        Solution # 809        Solution # 810
2 1 4 3 7 9 5 6 8     5 3 8 6 7 9 2 4 1     5 8 2 3 6 9 7 4 1     5 3 8 9 7 1 2 4 6     1 6 2 3 5 9 4 7 8
9 6 5 1 2 8 3 7 4     7 9 1 5 2 4 3 6 8     4 9 6 8 1 7 3 5 2     6 9 1 3 4 2 5 8 7     5 4 9 8 7 2 3 6 1
3 8 7 6 4 5 2 1 9     2 6 4 3 1 8 5 9 7     3 1 7 2 4 5 6 8 9     2 4 7 5 8 6 9 1 3     3 8 7 4 6 1 9 2 5
6 4 3 2 8 1 7 9 5     3 7 5 4 8 6 9 1 2     1 2 3 6 5 8 4 9 7     9 5 3 1 2 4 6 7 8     6 2 8 5 9 7 1 4 3
8 5 9 7 6 4 1 3 2     1 4 9 7 5 2 6 8 3     6 5 9 1 7 4 2 3 8     8 1 2 6 3 7 4 9 5     9 1 5 2 4 3 7 8 6
1 7 2 5 9 3 4 8 6     8 2 6 9 3 1 4 7 5     8 7 4 9 3 2 1 6 5     7 6 4 8 5 9 3 2 1     4 7 3 1 8 6 5 9 2
7 9 1 8 5 2 6 4 3     4 8 2 1 6 5 7 3 9     9 6 5 7 2 3 8 1 4     4 8 5 2 1 3 7 6 9     7 3 1 9 2 8 6 5 4
4 2 6 9 3 7 8 5 1     9 1 3 2 4 7 8 5 6     7 3 8 4 9 1 5 2 6     1 7 9 4 6 5 8 3 2     8 9 4 6 1 5 2 3 7
5 3 8 4 1 6 9 2 7     6 5 7 8 9 3 1 2 4     2 4 1 5 8 6 9 7 3     3 2 6 7 9 8 1 5 4     2 5 6 7 3 4 8 1 9

Solution # 811        Solution # 812        Solution # 813        Solution # 814        Solution # 815
3 2 6 5 1 7 8 9 4     2 8 3 4 7 6 1 5 9     6 1 5 4 9 2 3 7 8     2 7 1 6 4 5 9 3 8     4 1 5 2 8 3 7 6 9
4 8 1 3 2 9 6 7 5     7 9 4 5 1 8 2 6 3     4 9 8 3 7 5 2 6 1     6 8 3 7 9 1 5 2 4     9 8 2 4 6 7 3 1 5
9 7 5 6 8 4 1 2 3     5 1 6 2 3 9 4 8 7     7 3 2 6 1 8 5 4 9     9 4 5 2 3 8 7 1 6     7 6 3 5 1 9 4 8 2
5 1 9 7 4 8 2 3 6     3 2 7 6 5 4 8 9 1     3 8 1 2 6 7 4 9 5     5 6 2 3 8 4 1 9 7     2 7 6 3 4 8 9 5 1
8 3 2 1 5 6 7 4 9     4 6 8 7 9 1 5 3 2     5 6 9 8 3 4 7 1 2     1 9 7 5 2 6 4 8 3     8 5 4 9 2 1 6 3 7
7 6 4 2 9 3 5 1 8     9 5 1 8 2 3 6 7 4     2 4 7 9 5 1 8 3 6     4 3 8 9 1 7 6 5 2     3 9 1 6 7 5 2 4 8
2 4 8 9 6 1 3 5 7     8 4 9 1 6 7 3 2 5     9 7 3 5 2 6 1 8 4     7 1 6 8 5 3 2 4 9     1 3 7 8 9 4 5 2 6
1 9 7 8 3 5 4 6 2     1 7 5 3 8 2 9 4 6     8 2 6 1 4 3 9 5 7     8 2 4 1 6 9 3 7 5     6 4 8 7 5 2 1 9 3
6 5 3 4 7 2 9 8 1     6 3 2 9 4 5 7 1 8     1 5 4 7 8 9 6 2 3     3 5 9 4 7 2 8 6 1     5 2 9 1 3 6 8 7 4

Solution # 816        Solution # 817        Solution # 818        Solution # 819        Solution # 820
8 3 9 5 2 4 6 7 1     6 3 9 8 2 7 4 1 5     7 5 3 2 8 9 4 6 1     2 9 1 7 4 5 6 3 8     2 8 9 6 7 3 5 1 4
4 1 6 9 7 8 2 5 3     8 4 2 5 3 1 7 9 6     1 9 8 7 4 6 5 3 2     3 5 6 9 8 1 7 2 4     7 6 5 4 2 1 8 3 9
5 7 2 1 6 3 4 8 9     5 7 1 4 9 6 2 3 8     4 6 2 3 5 1 8 9 7     8 4 7 6 2 3 1 5 9     4 3 1 9 8 5 2 7 6
7 2 1 3 8 9 5 6 4     9 6 3 1 7 5 8 2 4     9 8 7 1 6 4 2 5 3     1 8 2 4 5 6 3 9 7     3 2 6 1 5 8 4 9 7
6 4 5 2 1 7 9 3 8     1 8 5 6 4 2 9 7 3     3 2 6 9 7 5 1 4 8     5 6 9 3 7 2 8 4 1     8 1 4 7 9 2 6 5 3
3 9 8 4 5 6 7 1 2     4 2 7 9 8 3 6 5 1     5 1 4 8 2 3 9 7 6     7 3 4 8 1 9 2 6 5     9 5 7 3 4 6 1 2 8
1 5 4 6 3 2 8 9 7     3 5 8 7 6 9 1 4 2     6 3 5 4 1 8 7 2 9     4 1 5 2 3 8 9 7 6     5 4 3 8 1 7 9 6 2
2 8 3 7 9 5 1 4 6     2 9 4 3 1 8 5 6 7     8 7 9 5 3 2 6 1 4     6 2 8 5 9 7 4 1 3     6 9 2 5 3 4 7 8 1
9 6 7 8 4 1 3 2 5     7 1 6 2 5 4 3 8 9     2 4 1 6 9 7 3 8 5     9 7 3 1 6 4 5 8 2     1 7 8 2 6 9 3 4 5

Solution # 821        Solution # 822        Solution # 823        Solution # 824        Solution # 825
7 4 9 6 3 1 5 8 2     8 4 3 2 6 5 9 7 1     2 7 9 8 5 1 3 6 4     8 9 3 7 6 5 4 1 2     1 2 8 9 5 3 4 6 7
8 3 1 5 4 2 6 9 7     2 9 1 3 8 7 4 5 6     6 5 8 2 3 4 1 7 9     1 7 4 3 2 9 8 5 6     9 3 6 8 4 7 2 1 5
2 6 5 7 9 8 4 3 1     5 7 6 9 4 1 8 2 3     1 4 3 9 6 7 5 8 2     6 2 5 1 4 8 7 9 3     5 4 7 2 6 1 8 3 9
3 9 6 2 5 7 1 4 8     9 3 4 5 2 8 6 1 7     4 8 2 5 7 3 6 9 1     7 3 2 4 5 1 6 8 9     6 1 9 3 8 4 7 5 2
5 7 4 1 8 9 2 6 3     6 8 7 1 3 4 2 9 5     3 1 6 4 9 2 7 5 8     4 5 6 9 8 3 1 2 7     4 7 2 6 9 5 3 8 1
1 2 8 4 6 3 7 5 9     1 2 5 7 9 6 3 4 8     5 9 7 6 1 8 2 4 3     9 1 8 2 7 6 3 4 5     8 5 3 7 1 2 9 4 6
6 8 2 9 7 5 3 1 4     7 6 9 4 5 3 1 8 2     9 2 4 1 8 6 7 3 5     5 6 7 8 9 4 2 3 1     7 8 5 1 3 9 6 2 4
4 1 3 8 2 6 9 7 5     4 5 8 6 1 2 7 3 9     8 3 1 7 4 5 9 2 6     3 8 9 6 1 2 5 7 4     2 6 4 5 7 8 1 9 3
9 5 7 3 1 4 8 2 6     3 1 2 8 7 9 5 6 4     7 6 5 3 2 9 4 1 8     2 4 1 5 3 7 9 6 8     3 9 1 4 2 6 5 7 8

Solution # 826        Solution # 827        Solution # 828        Solution # 829        Solution # 830
3 5 9 7 8 6 2 1 4     4 2 8 6 9 5 3 7 1     4 7 5 2 9 1 8 3 6     6 9 8 7 1 5 3 4 2     5 7 6 3 9 4 8 2 1
4 2 6 3 5 1 7 9 8     5 3 1 4 7 2 9 6 8     3 9 1 4 8 6 2 5 7     3 2 1 4 9 8 7 6 5     8 4 1 7 2 6 5 9 3
1 7 8 9 4 2 6 5 3     7 6 9 3 1 8 4 5 2     2 8 6 5 7 3 1 9 4     5 7 4 6 3 2 8 9 1     9 2 3 5 8 1 6 4 7
5 4 7 2 1 8 3 6 9     6 7 2 9 5 4 8 1 3     9 4 2 8 3 5 7 6 1     7 3 5 9 8 1 4 2 6     7 6 9 8 1 5 4 3 2
9 6 2 5 7 3 4 8 1     8 9 5 2 3 1 6 4 7     5 1 8 6 2 7 3 4 9     1 6 2 5 4 7 9 8 3     1 8 2 6 4 3 9 7 5
8 1 3 4 6 9 5 7 2     1 4 3 8 6 7 2 9 5     6 3 7 9 1 4 5 2 8     8 4 9 2 6 3 5 1 7     3 5 4 9 7 2 1 6 8
7 9 5 1 2 4 8 3 6     9 5 6 7 2 3 1 8 4     1 2 3 7 4 9 6 8 5     9 5 7 1 2 4 6 3 8     4 9 7 2 5 8 3 1 6
2 8 1 6 3 7 9 4 5     3 8 7 1 4 9 5 2 6     7 5 9 3 6 8 4 1 2     4 1 3 8 5 6 2 7 9     6 1 8 4 3 7 2 5 9
6 3 4 8 9 5 1 2 7     2 1 4 5 8 6 7 3 9     8 6 4 1 5 2 9 7 3     2 8 6 3 7 9 1 5 4     2 3 5 1 6 9 7 8 4

Solution # 831        Solution # 832        Solution # 833        Solution # 834        Solution # 835
6 2 9 4 8 7 5 3 1     9 7 2 5 6 3 1 8 4     2 3 9 1 7 4 8 5 6     7 4 2 5 3 6 8 9 1     1 3 6 5 7 4 8 9 2
8 7 5 2 1 3 9 4 6     3 8 4 2 1 7 6 5 9     4 6 7 5 8 9 2 3 1     9 6 3 4 8 1 2 5 7     4 8 7 3 2 9 5 6 1
3 4 1 9 6 5 8 7 2     1 6 5 9 4 8 2 7 3     5 8 1 3 2 6 4 7 9     5 1 8 9 2 7 4 6 3     2 9 5 6 8 1 7 4 3
9 8 2 3 7 6 1 5 4     5 2 6 3 7 1 4 9 8     7 9 3 6 1 2 5 8 4     4 8 6 3 1 2 5 7 9     7 5 8 9 4 3 1 2 6
4 5 3 1 9 2 6 8 7     4 3 9 8 2 5 7 1 6     1 2 5 8 4 3 6 9 7     3 5 9 7 6 4 1 8 2     9 1 3 2 6 8 4 5 7
7 1 6 5 4 8 2 9 3     8 1 7 4 9 6 3 2 5     8 4 6 9 5 7 3 1 2     2 7 1 8 9 5 3 4 6     6 2 4 1 5 7 9 3 8
1 3 8 6 5 4 7 2 9     2 4 3 7 5 9 8 6 1     9 1 2 4 3 8 7 6 5     6 9 5 2 4 3 7 1 8     3 7 9 4 1 2 6 8 5
2 6 7 8 3 9 4 1 5     7 9 1 6 8 4 5 3 2     6 7 8 2 9 5 1 4 3     1 2 7 6 5 8 9 3 4     5 4 1 8 3 6 2 7 9
5 9 4 7 2 1 3 6 8     6 5 8 1 3 2 9 4 7     3 5 4 7 6 1 9 2 8     8 3 4 1 7 9 6 2 5     8 6 2 7 9 5 3 1 4

Solution # 836        Solution # 837        Solution # 838        Solution # 839        Solution # 840
7 9 8 2 3 1 5 6 4     8 3 2 4 5 6 7 9 1     4 5 6 2 1 8 9 7 3     5 1 8 4 9 3 2 7 6     9 8 1 2 3 4 7 6 5
4 6 2 8 5 7 3 1 9     9 1 6 2 8 7 4 3 5     7 9 1 4 5 3 2 8 6     2 4 7 8 6 5 3 1 9     2 7 4 5 6 1 9 3 8
5 3 1 6 4 9 2 7 8     7 4 5 1 3 9 2 6 8     3 2 8 6 9 7 5 1 4     6 3 9 2 7 1 5 8 4     6 3 5 7 8 9 4 1 2
6 4 3 5 9 2 7 8 1     3 2 4 8 6 5 9 1 7     8 3 4 7 2 9 1 6 5     8 6 1 7 4 2 9 3 5     4 9 8 3 1 7 5 2 6
1 2 5 4 7 8 6 9 3     5 6 7 9 2 1 8 4 3     2 7 5 3 6 1 4 9 8     4 2 3 5 8 9 7 6 1     3 5 7 4 2 6 1 8 9
8 7 9 3 1 6 4 5 2     1 8 9 3 7 4 5 2 6     6 1 9 8 4 5 3 2 7     9 7 5 3 1 6 4 2 8     1 2 6 9 5 8 3 4 7
9 5 4 7 8 3 1 2 6     4 5 8 6 1 2 3 7 9     9 8 3 5 7 2 6 4 1     7 8 6 9 3 4 1 5 2     8 6 3 1 7 5 2 9 4
3 1 6 9 2 5 8 4 7     2 7 1 5 9 3 6 8 4     1 4 7 9 3 6 8 5 2     3 5 4 1 2 8 6 9 7     5 1 9 8 4 2 6 7 3
2 8 7 1 6 4 9 3 5     6 9 3 7 4 8 1 5 2     5 6 2 1 8 4 7 3 9     1 9 2 6 5 7 8 4 3     7 4 2 6 9 3 8 5 1
```

Solution # 841
```
8 6 4 2 9 5 7 3 1
9 7 2 8 1 3 5 4 6
1 3 5 7 6 4 2 8 9
3 1 8 9 7 2 6 5 4
5 2 9 4 3 6 1 7 8
6 4 7 5 8 1 9 2 3
4 9 6 3 2 7 8 1 5
2 5 1 6 4 8 3 9 7
7 8 3 1 5 9 4 6 2
```

Solution # 842
```
7 8 4 3 5 6 2 9 1
2 1 5 4 7 9 3 6 8
6 3 9 1 8 2 5 7 4
3 5 6 7 1 4 8 2 9
4 9 1 6 2 8 7 3 5
8 2 7 9 3 5 1 4 6
9 7 3 5 6 1 4 8 2
1 6 2 8 4 3 9 5 7
5 4 8 2 9 7 6 1 3
```

Solution # 843
```
5 7 3 8 4 6 1 9 2
6 2 4 9 3 1 5 8 7
1 9 8 2 5 7 3 6 4
4 3 9 1 2 5 8 7 6
8 1 7 3 6 9 4 2 5
2 5 6 4 7 8 9 3 1
3 4 5 6 8 2 7 1 9
7 6 1 5 9 3 2 4 8
9 8 2 7 1 4 6 5 3
```

Solution # 844
```
3 1 4 2 8 6 9 7 5
7 8 9 1 4 5 3 2 6
5 2 6 7 9 3 4 8 1
6 9 5 4 3 7 8 1 2
2 4 3 8 6 1 5 9 7
1 7 8 9 5 2 6 4 3
4 3 2 5 7 8 1 6 9
9 6 1 3 2 4 7 5 8
8 5 7 6 1 9 2 3 4
```

Solution # 845
```
4 2 5 7 9 3 1 6 8
1 9 6 4 8 5 2 7 3
3 8 7 2 6 1 5 9 4
6 5 2 9 1 4 3 8 7
7 4 1 3 2 8 6 5 9
9 3 8 6 5 7 4 2 1
2 6 4 8 3 9 7 1 5
8 1 3 5 7 2 9 4 6
5 7 9 1 4 6 8 3 2
```

Solution # 846
```
3 2 4 8 6 7 9 5 1
5 9 8 2 4 1 6 7 3
7 6 1 3 9 5 4 8 2
8 5 6 1 7 4 3 2 9
1 4 3 9 2 8 7 6 5
9 7 2 5 3 6 1 4 8
2 1 7 4 8 9 5 3 6
4 8 5 6 1 3 2 9 7
6 3 9 7 5 2 8 1 4
```

Solution # 847
```
7 1 8 4 9 3 2 5 6
4 2 6 5 7 8 1 3 9
9 3 5 6 1 2 8 4 7
6 4 2 1 5 7 3 9 8
5 8 7 2 3 9 4 6 1
1 9 3 8 4 6 5 7 2
8 5 4 9 6 1 7 2 3
2 7 9 3 8 4 6 1 5
3 6 1 7 2 5 9 8 4
```

Solution # 848
```
1 5 2 9 7 3 8 4 6
7 9 4 8 2 6 5 3 1
6 8 3 5 1 4 7 2 9
8 3 7 4 6 1 2 9 5
2 6 9 7 8 5 3 1 4
5 4 1 3 9 2 6 8 7
3 1 5 2 4 7 9 6 8
9 7 6 1 3 8 4 5 2
4 2 8 6 5 9 1 7 3
```

Solution # 849
```
1 6 2 8 9 3 4 5 7
7 9 3 5 2 4 6 8 1
4 5 8 1 6 7 2 9 3
3 2 5 7 4 1 9 6 8
9 1 6 2 8 5 7 3 4
8 7 4 9 3 6 1 2 5
5 8 7 6 1 2 3 4 9
2 4 9 3 7 8 5 1 6
6 3 1 4 5 9 8 7 2
```

Solution # 850
```
6 7 3 5 1 9 2 4 8
9 5 2 7 8 4 6 3 1
8 4 1 3 6 2 7 5 9
4 2 5 1 7 6 9 8 3
7 3 9 4 5 8 1 2 6
1 8 6 2 9 3 4 7 5
3 1 8 6 4 7 5 9 2
5 9 4 8 2 1 3 6 7
2 6 7 9 3 5 8 1 4
```

Solution # 851
```
3 4 1 2 6 8 7 9 5
2 8 6 7 9 5 4 3 1
9 5 7 3 1 4 2 6 8
7 9 2 5 8 6 3 1 4
8 1 4 9 3 2 6 5 7
5 6 3 1 4 7 8 2 9
1 2 8 4 5 3 9 7 6
6 3 5 8 7 9 1 4 2
4 7 9 6 2 1 5 8 3
```

Solution # 852
```
9 7 1 4 5 2 6 8 3
3 4 6 9 8 7 2 1 5
2 8 5 1 6 3 7 4 9
4 6 2 3 7 9 8 5 1
8 1 7 6 4 5 9 3 2
5 9 3 2 1 8 4 6 7
7 2 8 5 3 4 1 9 6
1 5 9 8 2 6 3 7 4
6 3 4 7 9 1 5 2 8
```

Solution # 853
```
6 1 2 7 5 8 3 4 9
8 7 9 3 2 4 5 6 1
5 4 3 6 9 1 8 2 7
9 6 8 2 4 5 7 1 3
2 3 1 9 8 7 6 5 4
7 5 4 1 6 3 9 8 2
4 9 5 8 7 2 1 3 6
3 2 7 5 1 6 4 9 8
1 8 6 4 3 9 2 7 5
```

Solution # 854
```
9 3 1 2 6 7 4 5 8
4 6 7 5 1 8 3 2 9
5 2 8 3 9 4 6 7 1
7 1 4 9 8 5 2 6 3
6 8 5 4 3 2 9 1 7
2 9 3 6 7 1 8 4 5
1 4 6 8 5 9 7 3 2
8 5 2 7 4 3 1 9 6
3 7 9 1 2 6 5 8 4
```

Solution # 855
```
3 7 5 4 6 8 9 2 1
9 4 2 5 7 1 6 3 8
8 1 6 9 3 2 7 5 4
6 3 7 2 5 4 8 1 9
4 2 8 6 1 9 5 7 3
5 9 1 3 8 7 4 6 2
1 6 3 8 9 5 2 4 7
7 8 4 1 2 6 3 9 5
2 5 9 7 4 3 1 8 6
```

Solution # 856
```
6 1 2 3 8 9 4 5 7
7 9 3 5 6 4 2 1 8
8 4 5 2 7 1 3 9 6
2 7 4 9 1 5 8 6 3
3 6 9 8 2 7 5 4 1
5 8 1 6 4 3 7 2 9
1 3 6 4 5 8 9 7 2
9 5 7 1 3 2 6 8 4
4 2 8 7 9 6 1 3 5
```

Solution # 857
```
2 3 1 6 5 7 4 9 8
8 6 4 1 9 3 2 5 7
9 7 5 4 8 2 6 3 1
6 4 8 3 7 9 5 1 2
3 5 2 8 6 1 9 7 4
7 1 9 2 4 5 8 6 3
5 8 3 7 2 6 1 4 9
4 9 7 5 1 8 3 2 6
1 2 6 9 3 4 7 8 5
```

Solution # 858
```
9 2 4 8 7 5 6 1 3
6 7 3 2 1 9 4 8 5
5 1 8 6 4 3 9 2 7
2 6 9 3 8 4 7 5 1
4 8 1 5 2 7 3 6 9
7 3 5 9 6 1 2 4 8
1 9 7 4 5 2 8 3 6
3 4 6 1 9 8 5 7 2
8 5 2 7 3 6 1 9 4
```

Solution # 859
```
7 5 9 2 8 4 3 1 6
3 2 6 7 9 1 4 5 8
8 4 1 6 3 5 2 7 9
2 3 4 9 5 6 1 8 7
5 6 8 4 1 7 9 2 3
9 1 7 3 2 8 6 4 5
1 9 3 8 7 2 5 6 4
6 7 2 5 4 3 8 9 1
4 8 5 1 6 9 7 3 2
```

Solution # 860
```
7 8 6 9 2 5 1 3 4
4 2 9 8 1 3 5 6 7
3 1 5 6 4 7 2 8 9
8 9 2 1 3 4 7 5 6
1 5 3 7 9 6 4 2 8
6 4 7 2 5 8 9 1 3
2 7 1 3 6 9 8 4 5
5 6 8 4 7 2 3 9 1
9 3 4 5 8 1 6 7 2
```

Solution # 861
```
2 6 9 1 8 4 7 3 5
7 3 1 5 6 9 2 8 4
8 5 4 3 2 7 1 6 9
3 9 2 8 7 1 5 4 6
1 8 6 2 4 5 9 7 3
4 7 5 6 9 3 8 1 2
5 2 3 7 1 6 4 9 8
9 1 8 4 3 2 6 5 7
6 4 7 9 5 8 3 2 1
```

Solution # 862
```
2 7 4 6 8 5 1 3 9
9 6 1 2 3 4 8 5 7
3 8 5 7 1 9 4 6 2
4 1 7 5 6 3 9 2 8
8 3 2 4 9 7 5 1 6
5 9 6 8 2 1 7 4 3
6 2 9 1 4 8 3 7 5
7 4 8 3 5 6 2 9 1
1 5 3 9 7 2 6 8 4
```

Solution # 863
```
5 4 8 6 3 7 9 2 1
1 3 7 9 2 8 5 4 6
6 9 2 4 5 1 3 7 8
2 1 5 7 4 6 8 3 9
4 6 9 2 8 3 7 1 5
7 8 3 5 1 9 4 6 2
9 2 4 3 6 5 1 8 7
8 5 6 1 7 4 2 9 3
3 7 1 8 9 2 6 5 4
```

Solution # 864
```
1 2 6 5 3 7 4 8 9
5 4 3 9 8 2 6 7 1
9 7 8 1 4 6 5 3 2
2 9 4 6 7 8 3 1 5
7 6 1 3 2 5 9 4 8
8 3 5 4 1 9 2 6 7
3 1 2 7 5 4 8 9 6
6 8 7 2 9 3 1 5 4
4 5 9 8 6 1 7 2 3
```

Solution # 865
```
2 8 3 4 9 6 5 1 7
5 1 4 2 3 7 8 6 9
6 9 7 8 5 1 4 2 3
4 2 9 7 1 5 3 8 6
7 3 1 6 8 4 2 9 5
8 5 6 9 2 3 1 7 4
3 6 5 1 7 8 9 4 2
1 4 2 3 6 9 7 5 8
9 7 8 5 4 2 6 3 1
```

Solution # 866
```
2 6 9 4 8 1 3 7 5
5 7 1 6 9 3 4 2 8
8 3 4 7 2 5 1 6 9
3 4 5 8 1 2 7 9 6
9 2 6 5 4 7 8 1 3
7 1 8 3 6 9 2 5 4
1 8 2 9 5 4 6 3 7
6 5 3 1 7 8 9 4 2
4 9 7 2 3 6 5 8 1
```

Solution # 867
```
5 3 2 4 9 7 8 6 1
7 8 9 6 3 1 2 5 4
1 4 6 2 8 5 7 3 9
4 9 8 7 5 3 6 1 2
2 1 3 8 6 4 5 9 7
6 7 5 1 2 9 3 4 8
8 2 4 5 1 6 9 7 3
9 5 1 3 7 8 4 2 6
3 6 7 9 4 2 1 8 5
```

Solution # 868
```
6 5 9 1 8 7 4 2 3
1 4 3 2 5 6 8 7 9
7 8 2 4 9 3 1 6 5
4 9 8 7 6 1 3 5 2
3 2 1 9 4 5 6 8 7
5 7 6 3 2 8 9 4 1
8 1 4 5 3 2 7 9 6
2 6 7 8 1 9 5 3 4
9 3 5 6 7 4 2 1 8
```

Solution # 869
```
8 3 2 9 5 4 1 6 7
5 7 9 1 8 6 4 3 2
6 1 4 7 2 3 5 8 9
4 2 8 5 3 7 6 9 1
1 5 7 8 6 9 2 4 3
3 9 6 2 4 1 7 5 8
2 4 5 3 1 8 9 7 6
7 8 1 6 9 5 3 2 4
9 6 3 4 7 2 8 1 5
```

Solution # 870
```
3 9 8 5 6 1 2 7 4
4 2 5 8 3 7 1 6 9
1 7 6 4 2 9 5 8 3
7 5 4 1 8 2 9 3 6
2 8 9 6 5 3 4 1 7
6 1 3 9 7 4 8 2 5
8 6 1 3 4 5 7 9 2
9 4 7 2 1 6 3 5 8
5 3 2 7 9 8 6 4 1
```

Solution # 871
```
2 7 4 3 6 9 1 8 5
1 9 6 5 8 2 4 7 3
3 5 8 1 7 4 6 9 2
5 4 3 8 2 6 9 1 7
9 6 1 4 3 7 2 5 8
7 8 2 9 5 1 3 4 6
8 1 5 2 9 3 7 6 4
6 3 9 7 4 8 5 2 1
4 2 7 6 1 5 8 3 9
```

Solution # 872
```
8 3 4 7 6 9 2 1 5
5 9 2 8 1 3 4 6 7
1 7 6 5 2 4 3 8 9
2 4 8 1 9 5 7 3 6
7 5 3 2 8 6 9 4 1
6 1 9 4 3 7 5 2 8
3 2 1 9 7 8 6 5 4
4 6 7 3 5 1 8 9 2
9 8 5 6 4 2 1 7 3
```

Solution # 873
```
6 9 1 5 2 4 3 8 7
2 5 4 3 8 7 1 9 6
3 7 8 9 1 6 4 2 5
8 4 3 2 6 5 9 7 1
9 2 6 4 7 1 5 3 8
7 1 5 8 9 3 6 4 2
4 3 7 1 5 8 2 6 9
5 8 2 6 3 9 7 1 4
1 6 9 7 4 2 8 5 3
```

Solution # 874
```
3 2 9 6 5 8 4 7 1
6 1 7 3 2 4 8 9 5
8 5 4 9 7 1 2 3 6
1 8 6 4 3 5 7 2 9
4 7 3 2 1 9 6 5 8
2 9 5 7 8 6 1 4 3
9 4 1 5 6 2 3 8 7
5 3 8 1 4 7 9 6 2
7 6 2 8 9 3 5 1 4
```

Solution # 875
```
9 4 1 6 7 3 8 2 5
8 6 2 4 9 5 1 3 7
7 5 3 2 8 1 4 6 9
4 2 8 7 3 6 9 5 1
3 1 9 5 2 4 7 8 6
5 7 6 8 1 9 3 4 2
1 8 7 3 6 2 5 9 4
6 9 4 1 5 8 2 7 3
2 3 5 9 4 7 6 1 8
```

Solution # 876
```
3 8 7 1 6 5 9 4 2
2 1 6 9 4 3 5 7 8
5 4 9 8 7 2 1 6 3
8 9 5 4 2 7 3 1 6
1 7 3 5 9 6 2 8 4
4 6 2 3 1 8 7 9 5
6 3 1 7 5 4 8 2 9
7 5 4 2 8 9 6 3 1
9 2 8 6 3 1 4 5 7
```

Solution # 877
```
4 7 6 1 9 8 5 3 2
5 1 8 7 3 2 6 4 9
3 2 9 4 6 5 8 7 1
2 5 4 9 8 7 3 1 6
8 3 1 5 4 6 9 2 7
6 9 7 2 1 3 4 8 5
1 6 5 3 7 4 2 9 8
9 4 2 8 5 1 7 6 3
7 8 3 6 2 9 1 5 4
```

Solution # 878
```
6 5 3 1 9 7 4 8 2
7 8 4 5 2 3 6 9 1
2 9 1 6 4 8 7 3 5
4 3 7 9 8 1 5 2 6
1 2 8 7 5 6 9 4 3
5 6 9 2 3 4 1 7 8
3 1 6 8 7 9 2 5 4
9 4 2 3 1 5 8 6 7
8 7 5 4 6 2 3 1 9
```

Solution # 879
```
4 5 7 8 2 3 6 9 1
6 9 3 1 5 7 8 2 4
1 8 2 9 6 4 7 5 3
5 2 9 3 7 1 4 6 8
7 1 6 4 8 5 9 3 2
3 4 8 6 9 2 5 1 7
9 6 4 2 1 8 3 7 5
2 3 5 7 4 6 1 8 9
8 7 1 5 3 9 2 4 6
```

Solution # 880
```
1 5 3 6 2 9 4 8 7
7 6 2 4 8 1 9 3 5
4 8 9 5 7 3 1 2 6
9 1 7 2 3 6 8 5 4
2 3 5 1 4 8 6 7 9
6 4 8 9 5 7 2 1 3
3 7 4 8 6 2 5 9 1
5 2 1 3 9 4 7 6 8
8 9 6 7 1 5 3 4 2
```

Solution # 881
```
7 4 8 5 6 2 3 9 1
3 6 1 4 8 9 7 2 5
2 9 5 1 3 7 4 6 8
4 7 3 8 2 6 5 1 9
6 1 2 7 9 5 8 3 4
5 8 9 3 4 1 2 7 6
1 3 6 2 5 8 9 4 7
8 2 7 9 1 4 6 5 3
9 5 4 6 7 3 1 8 2
```

Solution # 882
```
8 4 5 6 1 9 7 3 2
9 6 3 7 4 2 5 1 8
1 2 7 5 3 8 4 6 9
5 3 2 8 7 6 9 4 1
6 1 9 4 2 5 8 7 3
4 7 8 1 9 3 2 5 6
2 5 1 3 8 7 6 9 4
7 9 4 2 5 1 3 8 5
3 8 6 9 5 4 1 2 7
```

Solution # 883
```
7 9 6 8 4 3 1 2 5
5 3 1 2 6 7 4 9 8
2 4 8 9 5 1 3 6 7
3 8 9 7 1 6 2 5 4
1 5 4 3 2 9 7 8 6
6 7 2 4 8 5 9 1 3
8 2 5 1 7 4 6 3 9
9 6 7 5 3 2 8 4 1
4 1 3 6 9 8 5 7 2
```

Solution # 884
```
5 4 9 8 6 2 7 3 1
7 8 6 4 1 3 2 9 5
3 1 2 9 7 5 4 8 6
9 2 1 6 3 4 8 5 7
6 3 4 7 5 8 9 1 2
8 5 7 1 2 9 3 6 4
2 7 3 5 8 6 1 4 9
1 9 5 3 4 7 6 2 8
4 6 8 2 9 1 5 7 3
```

Solution # 885
```
6 7 8 3 2 9 1 4 5
9 5 2 1 6 4 8 7 3
1 3 4 8 5 7 9 2 6
8 4 3 5 9 2 6 1 7
5 9 6 4 7 1 3 8 2
7 2 1 6 8 3 5 9 4
2 8 7 9 3 5 4 6 1
4 6 5 2 1 8 7 3 9
3 1 9 7 4 6 2 5 8
```

Solution # 886
```
8 1 6 2 9 5 4 7 3
7 2 5 6 4 3 9 8 1
4 9 3 8 1 7 5 2 6
5 4 2 3 6 9 7 1 8
9 3 1 7 2 8 6 5 4
6 8 7 1 5 4 3 9 2
2 6 9 5 3 1 8 4 7
3 5 8 4 7 2 1 6 9
1 7 4 9 8 6 2 3 5
```

Solution # 887
```
3 4 6 1 8 7 2 5 9
7 1 2 9 3 5 4 6 8
5 8 9 2 4 6 3 7 1
9 2 3 6 7 1 5 8 4
8 6 7 4 5 2 1 9 3
1 5 4 8 9 3 7 2 6
6 7 5 3 1 9 8 4 2
2 3 8 5 6 4 9 1 7
4 9 1 7 2 8 6 3 5
```

Solution # 888
```
2 6 4 3 1 7 9 5 8
3 7 1 9 8 5 4 6 2
5 9 8 4 2 6 3 7 1
4 8 6 2 7 3 5 1 9
1 2 5 6 9 4 7 8 3
9 3 7 1 5 8 6 2 4
8 4 3 5 6 2 1 9 7
6 1 2 7 3 9 8 4 5
7 5 9 8 4 1 2 3 6
```

Solution # 889
```
1 8 5 3 4 2 6 7 9
4 2 6 7 9 8 1 5 3
7 3 9 1 6 5 2 4 8
3 4 7 5 8 6 9 2 1
9 6 8 4 2 1 7 3 5
5 1 2 9 3 7 8 6 4
8 7 1 6 5 3 4 9 2
6 5 4 2 1 9 3 8 7
2 9 3 8 7 4 5 1 6
```

Solution # 890
```
8 7 2 9 4 5 3 6 1
6 5 9 1 2 3 7 8 4
4 3 1 7 8 6 9 5 2
1 2 7 4 6 8 5 3 9
5 8 4 3 9 7 2 1 6
3 9 6 5 1 2 4 7 8
7 4 8 2 5 1 6 9 3
9 1 5 6 3 4 8 2 7
2 6 3 8 7 9 1 4 5
```

Solution # 891
```
4 8 2 6 1 7 5 3 9
9 7 1 3 2 5 6 4 8
6 3 5 9 4 8 7 1 2
3 4 9 5 7 2 1 8 6
1 6 7 8 3 9 4 2 5
2 5 8 4 6 1 9 7 3
5 9 4 1 8 3 2 6 7
8 2 6 7 9 4 3 5 1
7 1 3 2 5 6 8 9 4
```

Solution # 892
```
2 5 4 6 9 7 3 8 1
7 9 6 8 1 3 2 5 4
3 1 8 2 5 4 6 7 9
6 8 5 1 2 9 7 4 3
1 7 3 4 8 6 5 9 2
4 2 9 7 3 5 1 6 8
8 6 1 5 4 2 9 3 7
9 4 7 3 6 1 8 2 5
5 3 2 9 7 8 4 1 6
```

Solution # 893
```
7 5 8 2 6 4 9 1 3
6 4 9 3 7 1 8 5 2
1 2 3 8 9 5 7 6 4
5 3 4 1 2 8 6 7 9
9 8 6 4 5 7 2 3 1
2 7 1 6 3 9 5 4 8
3 9 5 7 1 2 4 8 6
8 1 2 5 4 6 3 9 7
4 6 7 9 8 3 1 2 5
```

Solution # 894
```
5 9 4 6 3 2 1 7 8
1 3 7 9 5 8 4 2 6
8 2 6 4 7 1 3 5 9
4 6 9 2 1 7 5 8 3
2 5 1 8 6 3 9 4 7
3 7 8 5 9 4 6 1 2
9 1 5 7 2 6 8 3 4
7 4 3 1 8 9 2 6 5
6 8 2 3 4 5 7 9 1
```

Solution # 895
```
9 2 3 8 6 7 1 5 4
6 7 1 9 5 4 3 8 2
5 8 4 2 1 3 7 6 9
1 6 8 5 7 9 4 2 3
3 5 7 4 2 6 9 1 8
2 4 9 3 8 1 6 7 5
7 3 5 1 4 8 2 9 6
4 1 2 6 9 5 8 3 7
8 9 6 7 3 2 5 4 1
```

Solution # 896
```
6 3 5 9 2 8 4 1 7
8 4 9 1 7 5 2 3 6
2 1 7 3 4 6 9 5 8
1 7 8 4 6 9 5 2 3
3 9 6 2 5 7 1 8 4
5 2 4 8 1 3 7 6 9
4 5 3 7 8 1 6 9 2
9 6 2 5 3 4 8 7 1
7 8 1 6 9 2 3 4 5
```

Solution # 897
```
4 8 6 2 5 3 7 1 9
1 5 7 6 9 4 2 8 3
2 9 3 8 7 1 4 5 6
9 2 8 3 1 6 5 7 4
5 6 1 7 4 8 9 3 2
3 7 4 9 2 5 8 6 1
7 3 2 1 8 9 6 4 5
8 1 5 4 6 2 3 9 7
6 4 9 5 3 7 1 2 8
```

Solution # 898
```
6 8 1 7 4 9 3 2 5
9 4 3 5 2 8 6 7 1
2 7 5 3 6 1 8 9 4
4 3 8 9 1 6 7 5 2
1 5 2 4 7 3 9 8 6
7 6 9 8 5 2 4 1 3
8 9 6 1 3 5 2 4 7
5 2 7 6 8 4 1 3 9
3 1 4 2 9 7 5 6 8
```

Solution # 899
```
9 7 2 4 5 1 8 3 6
5 3 6 7 8 2 4 9 1
8 4 1 6 3 9 2 5 7
4 9 8 5 1 3 7 6 2
2 6 5 8 7 4 3 1 9
3 1 7 2 9 6 5 4 8
1 2 9 3 4 7 6 8 5
6 5 4 1 2 8 9 7 3
7 8 3 9 6 5 1 2 4
```

Solution # 900
```
8 1 6 4 9 3 5 7 2
7 4 9 2 5 1 8 3 6
3 5 2 7 8 6 9 1 4
4 9 8 5 3 7 6 2 1
5 2 3 6 1 9 4 8 7
6 7 1 8 2 4 3 9 5
9 6 4 1 7 8 2 5 3
1 3 5 9 4 2 7 6 8
2 8 7 3 6 5 1 4 9
```

Solution # 901
```
2 7 4 3 5 6 1 8 9
8 1 6 4 9 7 5 2 3
5 3 9 1 2 8 7 6 4
4 9 5 7 3 2 6 1 8
7 6 1 5 8 9 3 4 2
3 2 8 6 4 1 9 7 5
9 5 7 2 1 4 8 3 6
6 4 3 8 7 5 2 9 1
1 8 2 9 6 3 4 5 7
```

Solution # 902
```
1 3 7 2 6 9 5 4 8
2 9 5 4 8 1 3 6 7
4 8 6 3 5 7 2 1 9
9 6 3 5 1 4 7 8 2
7 1 4 8 3 2 6 9 5
8 5 2 9 7 6 4 3 1
6 4 9 7 2 8 1 5 3
3 2 1 6 9 5 8 7 4
5 7 8 1 4 3 9 2 6
```

Solution # 903
```
5 4 8 1 3 6 7 9 2
3 6 9 8 7 2 4 5 1
2 7 1 4 9 5 3 8 6
4 5 3 6 2 1 9 7 8
9 8 2 7 5 3 1 6 4
6 1 7 9 8 4 2 3 5
1 9 4 5 6 7 8 2 3
7 3 6 2 4 8 5 1 9
8 2 5 3 1 9 6 4 7
```

Solution # 904
```
1 8 3 2 5 7 9 4 6
4 9 7 6 1 8 5 2 3
2 5 6 4 9 3 8 7 1
8 3 4 1 7 5 6 9 2
7 1 2 8 6 9 3 5 4
9 6 5 3 4 2 1 8 7
5 4 8 7 3 6 2 1 9
6 7 9 5 2 1 4 3 8
3 2 1 9 8 4 7 6 5
```

Solution # 905
```
2 3 6 4 9 8 1 7 5
8 7 1 3 2 5 4 9 6
9 5 4 6 7 1 3 8 2
7 9 8 2 1 6 5 4 3
4 1 2 9 5 3 8 6 7
3 6 5 7 8 4 2 1 9
6 4 9 8 3 2 7 5 1
5 2 7 1 4 9 6 3 8
1 8 3 5 6 7 9 2 4
```

Solution # 906
```
3 6 5 4 7 9 1 2 8
2 8 9 3 1 5 6 4 7
7 1 4 2 8 6 5 9 3
1 2 8 9 6 7 3 5 4
6 4 3 5 2 1 8 7 9
5 9 7 8 3 4 2 1 6
9 7 2 6 5 8 4 3 1
4 5 6 1 9 3 7 8 2
8 3 1 7 4 2 9 6 5
```

Solution # 907
```
8 7 4 1 5 3 6 2 9
6 2 5 8 7 9 1 4 3
1 3 9 4 2 6 5 8 7
4 5 7 2 6 8 9 3 1
2 6 3 9 1 5 4 7 8
9 1 8 7 3 4 2 6 5
7 9 6 5 8 4 3 1 2
3 4 1 7 9 2 8 5 6
5 8 2 6 3 1 7 9 4
```

Solution # 908
```
3 7 8 6 2 4 1 9 5
6 1 4 5 3 9 7 8 2
9 2 5 7 8 1 4 3 6
7 6 1 8 4 2 9 5 3
5 4 3 1 9 6 2 7 8
2 8 9 3 7 5 6 1 4
4 9 7 2 5 3 8 6 1
8 5 6 4 1 7 3 2 9
1 3 2 9 6 8 5 4 7
```

Solution # 909
```
4 8 6 3 5 7 1 9 2
1 3 9 8 2 6 4 7 5
5 7 2 1 4 9 6 8 3
3 9 8 4 7 5 2 6 1
2 6 4 9 8 1 5 3 7
7 5 1 6 3 2 9 4 8
9 4 7 2 1 3 8 5 6
6 1 3 5 9 8 7 2 4
8 2 5 7 6 4 3 1 9
```

Solution # 910
```
9 7 4 2 8 1 5 6 3
3 1 5 9 7 6 4 8 2
6 8 2 5 3 4 1 9 7
4 5 3 6 9 8 2 7 1
1 9 8 4 2 7 3 5 6
7 2 6 1 5 3 8 4 9
8 3 1 7 4 9 6 2 5
5 6 7 8 1 2 9 3 4
2 4 9 3 6 5 7 1 8
```

Solution # 911

```
2 7 1 6 9 3 5 4 8
8 9 6 4 7 5 3 2 1
4 3 5 2 1 8 7 6 9
1 6 4 3 5 2 9 8 7
3 2 8 7 6 9 4 1 5
9 5 7 8 4 1 6 3 2
5 4 9 1 2 6 8 7 3
6 1 3 9 8 7 2 5 4
7 8 2 5 3 4 1 9 6
```

Solution # 912

```
3 7 6 9 4 5 8 2 1
4 2 5 8 3 1 6 9 7
8 9 1 2 7 6 5 3 4
5 1 9 3 8 7 2 4 6
6 4 8 1 9 2 7 5 3
2 3 7 5 6 4 1 8 9
7 8 4 6 2 9 3 1 5
9 5 3 7 1 8 4 6 2
1 6 2 4 5 3 9 7 8
```

Solution # 913

```
4 2 8 5 3 9 7 6 1
7 9 3 4 6 1 2 8 5
5 6 1 7 2 8 3 9 4
9 4 6 3 5 7 8 1 2
3 1 2 6 8 4 5 7 9
8 5 7 1 9 2 4 3 6
1 8 4 9 7 5 6 2 3
2 3 9 8 4 6 1 5 7
6 7 5 2 1 3 9 4 8
```

Solution # 914

```
6 3 2 5 9 7 1 4 8
8 9 5 1 4 3 6 7 2
4 7 1 6 2 8 3 9 5
7 6 3 4 8 1 5 2 9
9 5 4 7 6 2 8 1 3
2 1 8 3 5 9 4 6 7
3 2 9 8 1 4 7 5 6
5 4 7 2 3 6 9 8 1
1 8 6 9 7 5 2 3 4
```

Solution # 915

```
3 7 6 1 4 9 5 8 2
8 1 4 3 2 5 9 6 7
2 5 9 7 8 6 4 1 3
4 8 1 2 9 7 6 3 5
9 3 5 8 6 4 7 2 1
6 2 7 5 3 1 8 4 9
1 9 3 4 7 8 2 5 6
7 4 2 6 5 3 1 9 8
5 6 8 9 1 2 3 7 4
```

Solution # 916

```
6 9 1 8 5 3 7 2 4
7 5 4 6 1 2 9 8 3
2 8 3 4 7 9 1 6 5
4 6 8 1 2 5 3 9 7
5 3 7 9 8 6 2 4 1
1 2 9 3 4 7 8 5 6
9 1 5 2 3 4 6 7 8
8 4 2 7 6 1 5 3 9
3 7 6 5 9 8 4 1 2
```

Solution # 917

```
4 2 3 8 9 5 7 1 6
7 1 5 2 6 3 8 4 9
8 9 6 7 1 4 5 3 2
6 4 8 1 5 7 2 9 3
2 3 1 9 4 8 6 5 7
5 7 9 6 3 2 1 8 4
9 8 2 4 7 1 3 6 5
1 5 4 3 2 6 9 7 8
3 6 7 5 8 9 4 2 1
```

Solution # 918

```
7 4 2 5 6 3 1 8 9
1 3 5 9 8 2 7 6 4
6 9 8 4 1 7 3 5 2
3 1 4 8 9 5 6 2 7
9 5 6 7 2 4 8 1 3
2 8 7 1 3 6 4 9 5
4 6 1 3 5 9 2 7 8
8 7 9 2 4 1 5 3 6
5 2 3 6 7 8 9 4 1
```

Solution # 919

```
4 7 1 3 2 6 5 8 9
6 8 3 4 9 5 1 2 7
2 9 5 7 1 8 3 6 4
3 5 6 2 4 7 8 9 1
8 4 9 5 3 1 2 7 6
7 1 2 6 8 9 4 3 5
1 6 7 8 5 2 9 4 3
5 3 8 9 7 4 6 1 2
9 2 4 1 6 3 7 5 8
```

Solution # 920

```
1 3 4 7 5 2 6 9 8
2 5 8 9 6 3 7 1 4
9 7 6 8 1 4 2 5 3
4 9 3 5 2 6 1 8 7
8 1 2 4 9 7 5 3 6
5 6 7 1 3 8 9 4 2
7 2 9 3 4 5 8 6 1
3 8 1 6 7 9 4 2 5
6 4 5 2 8 1 3 7 9
```

Solution # 921

```
2 1 4 7 6 5 8 3 9
6 3 5 9 4 8 7 2 1
8 9 7 1 2 3 5 6 4
9 4 3 5 1 6 2 7 8
5 2 8 4 9 7 3 1 6
1 7 6 8 3 2 4 9 5
3 6 9 2 8 4 1 5 7
4 5 1 3 7 9 6 8 2
7 8 2 6 5 1 9 4 3
```

Solution # 922

```
3 2 4 7 6 5 8 1 9
8 9 5 3 1 2 4 6 7
7 6 1 4 9 8 2 5 3
5 4 7 1 2 6 3 9 8
6 1 2 8 3 9 7 4 5
9 3 8 5 7 4 6 2 1
2 7 3 6 5 1 9 8 4
1 8 9 2 4 3 5 7 6
4 5 6 9 8 3 1 7 2
```

Solution # 923

```
4 1 7 5 9 6 3 2 8
6 5 8 3 4 2 7 1 9
2 3 9 7 8 1 4 6 5
1 4 3 6 2 8 9 5 7
7 2 5 4 3 9 6 8 1
9 8 6 1 5 7 2 3 4
5 9 4 2 1 3 8 7 6
8 7 2 9 6 5 1 4 3
3 6 1 8 7 4 5 9 2
```

Solution # 924

```
7 3 1 6 8 4 2 9 5
4 6 2 1 9 5 8 7 3
5 8 9 7 3 2 4 6 1
2 4 6 3 5 7 1 8 9
8 7 5 4 1 9 3 2 6
9 1 3 2 6 8 5 4 7
1 2 4 5 7 6 9 3 8
6 5 8 9 2 3 7 1 4
3 9 7 8 4 1 6 5 2
```

Solution # 925

```
7 2 4 6 1 9 3 8 5
5 1 8 3 7 2 6 9 4
9 6 3 8 4 5 1 2 7
1 7 5 9 6 3 8 4 2
4 9 6 2 8 1 5 7 3
8 3 2 4 5 7 9 6 1
3 4 9 1 2 8 7 5 6
2 5 1 7 9 6 4 3 8
6 8 7 5 3 4 2 1 9
```

Solution # 926

```
3 6 2 7 8 9 5 4 1
9 4 7 5 3 1 6 2 8
5 8 1 6 4 2 7 9 3
7 5 4 9 1 6 3 8 2
8 3 6 4 2 7 9 1 5
1 2 9 8 5 3 4 6 7
6 1 3 2 9 5 8 7 4
4 7 5 1 6 8 2 3 9
2 9 8 3 7 4 1 5 6
```

Solution # 927

```
1 6 9 4 7 5 3 2 8
3 8 7 2 6 9 1 5 4
4 2 5 1 3 8 9 7 6
8 5 3 6 9 2 4 1 7
7 4 6 3 5 1 2 8 9
9 1 2 7 8 4 5 6 3
6 7 1 9 2 3 8 4 5
5 9 4 8 1 6 7 3 2
2 3 8 5 4 7 6 9 1
```

Solution # 928

```
1 5 8 2 3 4 7 6 9
4 9 7 5 6 1 8 3 2
2 3 6 8 7 9 4 5 1
9 6 1 7 4 5 3 2 8
3 7 5 1 2 8 6 9 4
8 4 2 6 9 3 5 1 7
7 1 4 9 5 6 2 8 3
5 8 3 4 1 2 9 7 6
6 2 9 3 8 7 1 4 5
```

Solution # 929

```
1 2 7 3 9 4 8 5 6
8 5 6 1 2 7 3 9 4
3 4 9 8 6 5 7 1 2
6 8 4 7 3 1 9 2 5
2 3 5 4 8 9 1 6 7
9 7 1 6 5 2 4 8 3
7 6 2 9 4 8 5 3 1
4 9 3 5 1 6 2 7 8
5 1 8 2 7 3 6 4 9
```

Solution # 930

```
5 4 9 8 7 3 6 1 2
7 2 3 6 1 5 4 9 8
6 8 1 4 9 2 5 7 3
2 3 4 5 6 9 7 8 1
9 6 8 7 4 1 2 3 5
1 7 5 3 2 8 9 4 6
8 1 2 9 5 4 3 6 7
4 5 6 1 3 7 8 2 9
3 9 7 2 8 6 1 5 4
```

Solution # 931

```
4 9 6 3 2 7 1 8 5
2 8 7 5 1 6 4 3 9
5 3 1 4 8 9 7 6 2
1 4 8 2 5 3 6 9 7
7 5 9 6 4 8 2 1 3
6 2 3 7 9 1 5 4 8
8 7 5 1 3 4 9 2 6
9 1 2 8 6 5 3 7 4
3 6 4 9 7 2 8 5 1
```

Solution # 932

```
6 3 9 5 1 8 2 7 4
1 2 7 9 4 6 3 5 8
5 4 8 3 2 7 1 6 9
3 9 5 7 6 1 4 8 2
4 8 1 2 5 3 6 9 7
2 7 6 8 9 4 5 3 1
9 5 3 4 8 2 7 1 6
8 1 2 6 7 5 9 4 3
7 6 4 1 3 9 8 2 5
```

Solution # 933

```
9 8 4 2 7 1 6 5 3
5 2 6 4 9 3 1 7 8
3 1 7 8 6 5 4 2 9
1 6 5 7 8 2 9 3 4
7 4 3 6 1 9 2 8 5
2 9 8 3 5 4 7 6 1
8 3 1 9 2 7 5 4 6
6 7 9 5 4 8 3 1 2
4 5 2 1 3 6 8 9 7
```

Solution # 934

```
4 7 6 1 8 9 2 5 3
3 8 1 5 2 6 9 7 4
5 2 9 4 3 7 6 1 8
2 9 7 8 6 1 4 3 5
6 3 4 7 9 5 1 8 2
1 5 8 2 4 3 7 9 6
7 6 2 9 5 8 3 4 1
9 4 5 3 1 2 8 6 7
8 1 3 6 7 4 5 2 9
```

Solution # 935

```
5 4 7 2 1 3 9 8 6
9 1 8 5 6 7 4 2 3
3 2 6 8 9 4 7 5 1
7 6 4 9 2 8 3 1 5
2 9 5 3 4 1 8 6 7
8 3 1 6 7 5 2 4 9
4 8 9 7 5 6 1 3 2
6 7 3 1 8 2 5 9 4
1 5 2 4 3 9 6 7 8
```

Solution # 936

```
4 6 1 9 2 5 8 3 7
5 9 8 4 3 7 6 2 1
3 2 7 8 6 1 5 4 9
1 4 2 3 5 9 7 6 8
7 8 5 2 4 6 1 9 3
9 3 6 1 7 8 4 5 2
6 5 9 7 8 2 3 1 4
2 7 3 5 1 4 9 8 6
8 1 4 6 9 3 2 7 5
```

Solution # 937

```
5 3 7 9 8 4 2 1 6
8 1 4 2 6 3 9 7 5
6 9 2 7 1 5 4 3 8
4 5 6 3 9 7 1 8 2
7 8 9 4 2 1 6 5 3
3 2 1 6 5 8 7 4 9
9 4 5 1 3 2 8 6 7
2 7 3 8 4 6 5 9 1
1 6 8 5 7 9 3 2 4
```

Solution # 938

```
4 6 8 5 9 3 2 1 7
5 9 1 2 6 7 4 3 8
7 2 3 8 1 4 6 5 9
3 8 6 9 2 1 7 4 5
1 7 9 6 4 5 8 2 3
2 4 5 3 7 8 9 6 1
8 1 7 4 5 2 3 9 6
9 3 4 1 8 6 5 7 2
6 5 2 7 3 9 1 8 4
```

Solution # 939

```
3 9 7 2 5 1 6 8 4
1 5 6 3 4 8 2 9 7
2 4 8 9 7 6 1 5 3
4 2 5 8 6 9 7 3 1
7 3 1 4 2 5 8 6 9
6 8 9 7 1 3 4 2 5
9 1 2 5 8 4 3 7 6
5 7 4 6 3 2 9 1 8
8 6 3 1 9 7 5 4 2
```

Solution # 940

```
8 7 2 9 5 3 6 4 1
4 9 6 7 1 2 5 3 8
3 5 1 4 8 6 9 2 7
9 8 5 2 3 4 7 1 6
1 4 3 8 6 7 2 9 5
6 2 7 1 9 5 4 8 3
2 6 4 3 7 8 1 5 9
5 1 8 6 4 9 3 7 2
7 3 9 5 2 1 8 6 4
```

Solution # 941

```
6 2 7 4 1 5 8 3 9
4 5 9 6 8 3 7 1 2
3 8 1 2 7 9 4 6 5
9 4 2 3 5 1 6 8 7
8 1 5 7 6 2 3 9 4
7 3 6 9 4 8 2 5 1
5 6 8 1 2 4 9 7 3
2 7 3 5 9 6 1 4 8
1 9 4 8 3 7 5 2 6
```

Solution # 942

```
2 1 7 4 5 8 6 9 3
8 3 4 2 9 6 5 7 1
6 5 9 1 7 3 2 4 8
5 2 1 8 6 4 7 3 9
4 7 6 3 1 9 8 5 2
9 8 3 7 2 5 1 6 4
3 4 5 6 8 2 9 1 7
7 6 2 9 3 1 4 8 5
1 9 8 5 4 7 3 2 6
```

Solution # 943

```
1 4 2 8 5 7 3 9 6
5 6 3 4 9 1 2 8 7
9 8 7 3 6 2 5 4 1
7 3 9 6 1 4 8 2 5
2 5 4 9 7 8 6 1 3
6 1 8 5 2 3 4 7 9
4 9 1 2 3 6 7 5 8
8 7 6 1 4 5 9 3 2
3 2 5 7 8 9 1 6 4
```

Solution # 944

```
3 2 5 4 1 7 8 9 6
4 8 9 6 3 5 2 7 1
1 7 6 8 9 2 5 3 4
6 9 4 5 2 1 3 8 7
8 3 1 9 7 4 6 2 5
7 5 2 3 8 6 1 4 9
5 4 3 7 6 8 9 1 2
2 6 8 1 4 9 7 5 3
9 1 7 2 5 3 4 6 8
```

Solution # 945

```
4 7 5 9 6 8 2 3 1
9 1 6 7 2 3 8 4 5
8 2 3 4 5 1 7 9 6
7 9 8 5 3 2 1 6 4
3 6 4 1 8 7 9 5 2
1 5 2 6 4 9 3 8 7
5 3 1 8 7 4 6 2 9
6 8 9 2 1 5 4 7 3
2 4 7 3 9 6 5 1 8
```

Solution # 946
```
2 7 5 9 3 1 6 8 4
9 8 6 2 5 4 3 1 7
4 1 3 8 6 7 9 2 5
6 9 7 3 4 2 1 5 8
5 4 8 6 1 9 7 3 2
3 2 1 5 7 8 4 6 9
1 6 9 7 2 5 8 4 3
8 3 2 4 9 6 5 7 1
7 5 4 1 8 3 2 9 6
```

Solution # 947
```
3 1 8 5 4 9 2 6 7
4 5 7 8 2 6 9 1 3
9 6 2 3 1 7 8 5 4
1 3 6 7 9 2 4 8 5
2 8 5 1 3 4 7 9 6
7 9 4 6 5 8 3 2 1
8 4 1 9 7 5 6 3 2
5 2 9 4 6 3 1 7 8
6 7 3 2 8 1 5 4 9
```

Solution # 948
```
2 6 7 1 5 4 9 8 3
5 9 1 2 8 3 4 6 7
4 8 3 9 7 6 2 5 1
9 7 4 5 6 1 3 2 8
8 5 2 4 3 7 1 9 6
1 3 6 8 2 9 7 4 5
7 1 9 6 4 5 8 3 2
3 2 5 7 9 8 6 1 4
6 4 8 3 1 2 5 7 9
```

Solution # 949
```
7 3 2 9 4 1 5 6 8
1 4 8 5 6 3 9 2 7
5 9 6 2 8 7 3 1 4
8 1 3 7 5 6 2 4 9
2 5 7 1 9 4 6 8 3
9 6 4 3 2 8 1 7 5
3 7 5 8 1 2 4 9 6
6 2 9 4 7 5 8 3 1
4 8 1 6 3 9 7 5 2
```

Solution # 950
```
2 3 1 4 6 9 7 8 5
4 9 7 8 1 5 3 2 6
8 5 6 2 7 3 4 9 1
5 4 9 3 8 7 6 1 2
7 6 3 5 2 1 9 4 8
1 8 2 9 4 6 5 3 7
6 7 8 1 3 4 2 5 9
9 2 4 7 5 8 1 6 3
3 1 5 6 9 2 8 7 4
```

Solution # 951
```
7 1 3 8 6 2 4 9 5
9 2 8 4 1 5 6 7 3
5 6 4 9 7 3 1 8 2
4 9 2 7 5 6 8 3 1
8 3 5 2 9 1 7 6 4
1 7 6 3 4 8 5 2 9
3 4 7 5 8 9 2 1 6
2 5 1 6 3 7 9 4 8
6 8 9 1 2 4 3 5 7
```

Solution # 952
```
1 4 8 7 5 6 9 3 2
3 2 6 4 1 9 5 8 7
9 5 7 8 2 3 1 6 4
4 8 5 3 6 1 2 7 9
7 6 9 5 4 2 3 1 8
2 1 3 9 7 8 4 5 6
8 9 2 1 3 7 6 4 5
6 3 4 2 8 5 7 9 1
5 7 1 6 9 4 8 2 3
```

Solution # 953
```
2 1 8 9 7 5 6 3 4
3 7 4 2 6 1 5 9 8
9 5 6 3 8 4 1 7 2
8 4 3 6 1 7 9 2 5
1 9 5 8 2 3 4 6 7
6 2 7 4 5 9 3 8 1
5 3 9 7 4 2 8 1 6
7 6 1 5 3 8 2 4 9
4 8 2 1 9 6 7 5 3
```

Solution # 954
```
4 2 9 8 5 6 7 1 3
1 6 5 9 7 3 2 8 4
7 8 3 1 4 2 6 9 5
8 5 2 6 3 1 4 7 9
9 1 7 5 2 4 3 6 8
3 4 6 7 8 9 5 2 1
2 7 1 3 9 5 8 4 6
5 9 4 2 6 8 1 3 7
6 3 8 4 1 7 9 5 2
```

Solution # 955
```
6 9 3 4 1 5 7 2 8
5 7 1 8 2 6 4 3 9
4 2 8 9 7 3 5 1 6
2 6 5 3 8 7 1 9 4
3 1 4 6 5 9 8 7 2
9 8 7 2 4 1 6 5 3
1 3 9 5 6 8 2 4 7
7 4 6 1 9 2 3 8 5
8 5 2 7 3 4 9 6 1
```

Solution # 956
```
9 1 6 2 7 5 3 4 8
5 3 4 1 8 9 2 7 6
8 7 2 4 3 6 5 9 1
2 5 9 8 4 3 6 1 7
3 6 7 9 5 1 8 2 4
4 8 1 6 2 7 9 3 5
6 4 8 7 9 2 1 5 3
1 2 5 3 6 4 7 8 9
7 9 3 5 1 8 4 6 2
```

Solution # 957
```
1 4 8 5 7 3 6 9 2
3 2 5 9 6 4 8 1 7
9 7 6 2 1 8 5 4 3
2 6 9 4 8 5 3 7 1
7 8 4 1 3 9 2 5 6
5 3 1 6 2 7 4 8 9
4 9 2 3 5 1 7 6 8
6 1 7 8 4 2 9 3 5
8 5 3 7 9 6 1 2 4
```

Solution # 958
```
6 2 3 4 1 9 5 8 7
9 8 1 5 7 2 6 3 4
7 4 5 6 8 3 2 1 9
5 6 7 8 4 1 9 2 3
4 3 8 9 2 6 7 5 1
1 9 2 7 3 5 4 6 8
2 7 6 1 9 8 3 4 5
8 5 4 3 6 7 1 9 2
3 1 9 2 5 4 8 7 6
```

Solution # 959
```
5 6 9 2 3 8 4 1 7
4 3 1 7 9 6 8 2 5
7 2 8 4 1 5 9 3 6
6 8 5 9 2 4 3 7 1
3 9 7 5 6 1 2 8 4
1 4 2 3 8 7 6 5 9
8 1 4 6 5 3 7 9 2
9 7 3 1 4 2 5 6 8
2 5 6 8 7 9 1 4 3
```

Solution # 960
```
8 9 1 5 6 7 4 3 2
2 4 6 9 8 3 1 5 7
3 5 7 2 1 4 8 9 6
5 6 8 1 3 2 7 4 9
4 3 9 6 7 8 2 1 5
7 1 2 4 9 5 6 8 3
9 2 5 8 4 6 3 7 1
1 8 3 7 2 9 5 6 4
6 7 4 3 5 1 9 2 8
```

Solution # 961
```
8 4 2 1 7 3 6 5 9
5 7 6 4 9 2 1 3 8
1 3 9 6 5 8 7 4 2
7 9 5 8 4 1 3 2 6
6 8 1 2 3 5 9 7 4
3 2 4 9 6 7 5 8 1
4 6 3 5 8 9 2 1 7
9 1 7 3 2 4 8 6 5
2 5 8 7 1 6 4 9 3
```

Solution # 962
```
3 6 9 1 4 2 5 8 7
7 4 8 5 9 6 3 1 2
1 2 5 7 8 3 6 4 9
6 8 3 9 5 7 4 2 1
4 9 2 8 6 1 7 3 5
5 7 1 2 3 4 9 6 8
2 1 4 3 7 5 8 9 6
9 3 7 6 1 8 2 5 4
8 5 6 4 2 9 1 7 3
```

Solution # 963
```
3 7 9 6 1 8 5 4 2
1 6 5 2 3 4 7 8 9
8 2 4 7 5 9 1 6 3
2 1 7 5 4 3 8 9 6
4 9 6 8 2 1 3 5 7
5 3 8 9 7 6 2 1 4
9 4 3 1 8 7 6 2 5
7 8 2 4 6 5 9 3 1
6 5 1 3 9 2 4 7 8
```

Solution # 964
```
1 6 2 3 8 7 4 9 5
8 7 9 4 5 6 2 3 1
5 4 3 1 2 9 7 6 8
3 9 5 8 4 2 6 1 7
7 1 4 6 3 5 8 2 9
6 2 8 9 7 1 5 4 3
2 3 7 5 1 4 9 8 6
4 8 6 7 9 3 1 5 2
9 5 1 2 6 8 3 7 4
```

Solution # 965
```
2 7 4 5 9 6 1 8 3
3 9 8 1 7 2 5 6 4
5 6 1 3 4 8 9 2 7
9 4 5 6 3 1 2 7 8
6 8 7 2 5 9 4 3 1
1 3 2 7 8 4 6 5 9
8 1 6 4 2 7 3 9 5
7 2 3 9 1 5 8 4 6
4 5 9 8 6 3 7 1 2
```

Solution # 966
```
4 3 7 1 2 5 8 9 6
8 1 5 9 6 7 2 4 3
6 2 9 3 4 8 1 5 7
7 9 4 6 8 1 5 3 2
3 6 1 5 7 2 4 8 9
5 8 2 4 9 3 7 6 1
9 4 8 7 1 6 3 2 5
2 7 3 8 5 9 6 1 4
1 5 6 2 3 4 9 7 8
```

Solution # 967
```
2 5 4 7 8 6 9 3 1
7 9 6 5 3 1 8 4 2
3 1 8 2 9 4 6 7 5
5 4 2 6 1 7 3 8 9
1 8 9 3 4 5 2 6 7
6 3 7 9 2 8 1 5 4
8 2 3 4 5 9 7 1 6
9 7 5 1 6 3 4 2 8
4 6 1 8 7 2 5 9 3
```

Solution # 968
```
3 2 9 6 7 4 5 1 8
5 6 7 2 1 8 4 3 9
4 8 1 9 3 5 6 2 7
2 1 5 7 9 3 8 6 4
7 9 6 4 8 2 3 5 1
8 4 3 5 6 1 7 9 2
6 3 4 8 2 9 1 7 5
1 5 2 3 4 7 9 8 6
9 7 8 1 5 6 2 4 3
```

Solution # 969
```
8 7 1 4 6 5 2 3 9
5 9 6 3 8 2 4 1 7
2 4 3 1 7 9 5 6 8
7 1 9 5 2 6 3 8 4
4 5 8 7 1 3 9 2 6
3 6 2 9 4 8 7 5 1
9 2 4 6 5 1 8 7 3
1 3 5 8 9 7 6 4 2
6 8 7 2 3 4 1 9 5
```

Solution # 970
```
4 5 7 3 9 8 2 1 6
8 1 9 2 5 6 4 7 3
2 6 3 4 7 1 8 5 9
1 3 8 7 2 9 6 4 5
6 9 2 1 4 5 3 8 7
7 4 5 8 6 3 9 2 1
5 8 1 9 3 4 7 6 2
3 2 4 6 1 7 5 9 8
9 7 6 5 8 2 1 3 4
```

Solution # 971
```
1 7 8 5 6 3 9 4 2
6 5 3 9 4 2 7 8 1
4 2 9 7 8 1 3 5 6
3 4 6 1 7 5 8 2 9
5 9 1 3 2 8 4 6 7
2 8 7 4 9 6 5 1 3
7 6 2 8 5 9 1 3 4
8 1 4 2 3 7 6 9 5
9 3 5 6 1 4 2 7 8
```

Solution # 972
```
8 1 9 6 4 5 2 3 7
4 7 3 2 9 8 5 6 1
5 2 6 3 7 1 9 4 8
1 6 7 5 3 4 8 9 2
9 8 4 7 2 6 3 1 5
2 3 5 1 8 9 6 7 4
3 4 2 8 6 7 1 5 9
6 9 1 4 5 2 7 8 3
7 5 8 9 1 3 4 2 6
```

Solution # 973
```
9 6 1 3 7 2 4 8 5
3 7 4 9 8 5 6 2 1
5 2 8 1 4 6 3 7 9
2 9 5 7 6 1 8 3 4
1 4 3 8 2 9 7 5 6
6 8 7 4 5 3 1 9 2
8 3 2 5 1 4 9 6 7
7 1 6 2 9 8 5 4 3
4 5 9 6 3 7 2 1 8
```

Solution # 974
```
4 2 8 9 6 3 5 1 7
9 7 5 4 1 8 3 2 6
6 3 1 5 7 2 9 8 4
2 6 3 8 9 4 7 5 1
1 5 7 3 2 6 4 9 8
8 4 9 7 5 1 6 3 2
3 9 6 2 8 7 1 4 5
5 1 2 6 4 9 8 7 3
7 8 4 1 3 5 2 6 9
```

Solution # 975
```
6 2 4 1 3 5 9 8 7
9 7 1 2 6 8 4 5 3
5 3 8 7 4 9 2 1 6
1 8 2 3 5 6 7 4 9
4 6 9 8 2 7 5 3 1
7 5 3 9 1 4 6 2 8
3 9 5 6 8 2 1 7 4
8 4 6 5 7 1 3 9 2
2 1 7 4 9 3 8 6 5
```

Solution # 976
```
8 9 3 4 1 5 6 2 7
4 6 2 9 3 7 5 1 8
5 7 1 8 2 6 9 4 3
1 5 6 3 8 4 7 9 2
9 4 8 6 7 2 3 5 1
3 2 7 5 9 1 8 6 4
2 8 9 1 6 3 4 7 5
7 3 5 2 4 9 1 8 6
6 1 4 7 5 8 2 3 9
```

Solution # 977
```
7 6 1 8 5 2 9 3 4
5 3 9 4 7 6 1 8 2
2 4 8 3 1 9 6 7 5
8 9 3 1 6 5 4 2 7
4 7 2 9 8 3 5 6 1
6 1 5 2 4 7 3 9 8
3 5 4 7 9 8 2 1 6
9 8 6 5 2 1 7 4 3
1 2 7 6 3 4 8 5 9
```

Solution # 978
```
9 7 3 5 1 2 4 6 8
8 1 4 9 6 7 2 5 3
6 5 2 3 4 8 9 7 1
3 8 7 6 2 4 1 9 5
5 4 6 1 8 9 3 2 7
1 2 9 7 3 5 6 8 4
4 3 5 2 7 6 8 1 9
7 6 8 4 9 1 5 3 2
2 9 1 8 5 3 7 4 6
```

Solution # 979
```
5 6 9 4 2 7 3 1 8
8 4 3 6 1 5 9 7 2
2 7 1 9 8 3 6 4 5
9 2 8 7 5 1 4 3 6
6 3 4 2 9 8 1 5 7
7 1 5 3 6 4 2 8 9
1 8 2 5 3 6 7 9 4
4 5 6 1 7 9 8 2 3
3 9 7 8 4 2 5 6 1
```

Solution # 980
```
4 5 2 1 9 8 7 3 6
3 9 7 5 6 2 1 8 4
6 1 8 7 4 3 9 2 5
7 8 6 9 1 5 2 4 3
9 2 3 6 8 4 5 7 1
1 4 5 2 3 7 8 6 9
2 6 1 3 7 9 4 5 8
8 7 9 4 5 6 3 1 2
5 3 4 8 2 1 6 9 7
```

Solution # 981

```
2 6 9 5 8 7 3 4 1
3 7 4 6 1 2 5 9 8
1 5 8 3 4 9 6 2 7
6 3 1 8 9 4 7 5 2
9 2 5 1 7 6 4 8 3
4 8 7 2 5 3 1 6 9
7 1 2 4 6 8 9 3 5
8 9 6 7 3 5 2 1 4
5 4 3 9 2 1 8 7 6
```

Solution # 982

```
8 4 9 2 3 7 1 5 6
7 1 5 9 6 8 2 4 3
2 6 3 5 1 4 7 9 8
3 9 8 7 2 5 4 6 1
1 5 7 4 8 6 3 2 9
6 2 4 3 9 1 5 8 7
9 7 2 8 5 3 6 1 4
4 8 6 1 7 2 9 3 5
5 3 1 6 4 9 8 7 2
```

Solution # 983

```
1 3 8 5 4 2 9 7 6
9 5 4 8 6 7 2 1 3
7 2 6 9 1 3 8 5 4
4 9 5 3 7 8 1 6 2
2 6 3 1 5 4 7 8 9
8 1 7 6 2 9 4 3 5
5 4 2 7 8 6 3 9 1
6 8 9 4 3 1 5 2 7
3 7 1 2 9 5 6 4 8
```

Solution # 984

```
6 8 3 4 5 2 1 9 7
2 5 7 1 6 9 8 3 4
1 4 9 8 7 3 2 5 6
5 3 2 7 9 8 4 6 1
8 9 4 3 1 6 7 2 5
7 1 6 5 2 4 9 8 3
4 2 1 9 3 5 6 7 8
3 6 8 2 4 7 5 1 9
9 7 5 6 8 1 3 4 2
```

Solution # 985

```
7 8 4 5 3 1 9 2 6
5 6 9 4 2 7 8 3 1
1 2 3 8 6 9 5 7 4
8 5 7 9 1 2 4 6 3
9 3 1 6 4 5 7 8 2
2 4 6 3 7 8 1 5 9
6 9 8 2 5 4 3 1 7
4 7 2 1 8 3 6 9 5
3 1 5 7 9 6 2 4 8
```

Solution # 986

```
4 9 6 2 3 1 5 8 7
1 8 2 5 7 6 9 3 4
5 7 3 9 4 8 2 6 1
6 4 9 8 1 7 3 2 5
7 3 5 4 2 9 8 1 6
2 1 8 6 5 3 4 7 9
3 2 1 7 9 4 6 5 8
8 5 4 1 6 2 7 9 3
9 6 7 3 8 5 1 4 2
```

Solution # 987

```
2 1 3 7 8 4 9 6 5
9 4 8 6 1 5 3 7 2
5 6 7 3 2 9 4 8 1
6 3 9 4 5 8 1 2 7
7 5 1 9 3 2 6 4 8
4 8 2 1 6 7 5 3 9
1 9 6 2 7 3 8 5 4
3 2 5 8 4 1 7 9 6
8 7 4 5 9 6 2 1 3
```

Solution # 988

```
3 2 4 5 6 7 8 9 1
5 1 8 2 9 3 6 4 7
6 9 7 8 4 1 5 2 3
9 7 6 3 2 4 1 8 5
8 4 2 1 5 6 3 7 9
1 3 5 7 8 9 2 6 4
7 6 1 4 3 2 9 5 8
4 8 9 6 1 5 7 3 2
2 5 3 9 7 8 4 1 6
```

Solution # 989

```
2 9 5 7 8 3 6 4 1
1 7 6 2 9 4 3 5 8
3 8 4 1 5 6 2 7 9
5 3 1 8 4 2 7 9 6
4 2 7 6 1 9 5 8 3
8 6 9 5 3 7 4 1 2
9 5 3 4 2 8 1 6 7
7 1 8 3 6 5 9 2 4
6 4 2 9 7 1 8 3 5
```

Solution # 990

```
3 5 8 9 7 4 2 6 1
7 4 2 3 6 1 5 9 8
9 1 6 2 8 5 7 3 4
5 6 7 8 4 2 3 1 9
1 2 4 6 9 3 8 5 7
8 3 9 1 5 7 4 2 6
6 7 1 5 3 8 9 4 2
4 9 3 7 2 6 1 8 5
2 8 5 4 1 9 6 7 3
```

Solution # 991

```
3 1 6 2 8 4 5 9 7
8 9 5 6 7 1 2 3 4
4 2 7 5 9 3 1 6 8
2 8 9 7 1 5 6 4 3
1 5 4 3 6 9 8 7 2
7 6 3 8 4 2 9 5 1
6 3 8 9 2 7 4 1 5
5 4 2 1 3 6 7 8 9
9 7 1 4 5 8 3 2 6
```

Solution # 992

```
4 3 1 6 8 9 2 7 5
9 8 7 5 2 4 6 1 3
2 5 6 3 1 7 4 9 8
6 7 9 2 4 3 5 8 1
1 2 5 9 6 8 3 4 7
8 4 3 7 5 1 9 6 2
5 1 2 8 9 6 7 3 4
7 9 4 1 3 5 8 2 6
3 6 8 4 7 2 1 5 9
```

Solution # 993

```
4 2 5 8 9 7 6 1 3
7 8 3 6 5 1 9 4 2
6 1 9 2 4 3 8 5 7
3 9 2 4 1 6 7 8 5
5 6 4 9 7 8 2 3 1
1 7 8 3 2 5 4 6 9
9 5 7 1 6 4 3 2 8
8 4 1 7 3 2 5 9 6
2 3 6 5 8 9 1 7 4
```

Solution # 994

```
7 2 9 6 4 8 1 3 5
1 5 6 2 3 9 4 8 7
4 3 8 7 5 1 2 6 9
9 6 7 4 8 2 5 1 3
2 1 4 5 6 3 7 9 8
5 8 3 9 1 7 6 2 4
3 9 5 1 7 6 8 4 2
6 7 2 8 9 4 3 5 1
8 4 1 3 2 5 9 7 6
```

Solution # 995

```
8 6 2 7 5 3 1 9 4
4 5 3 6 1 9 8 2 7
9 1 7 8 2 4 6 3 5
5 8 1 4 3 6 9 7 2
6 2 9 1 7 5 4 8 3
7 3 4 9 8 2 5 1 6
3 9 5 2 4 1 7 6 8
1 4 8 3 6 7 2 5 9
2 7 6 5 9 8 3 4 1
```

Solution # 996

```
1 8 7 5 9 6 4 2 3
5 3 4 7 2 8 9 6 1
9 2 6 4 1 3 7 8 5
4 7 8 2 6 5 3 1 9
2 5 3 1 7 9 6 4 8
6 1 9 8 3 4 2 5 7
7 4 2 3 5 1 8 9 6
3 6 5 9 8 2 1 7 4
8 9 1 6 4 7 5 3 2
```

Solution # 997

```
3 8 6 2 7 9 4 1 5
2 7 4 1 5 6 3 9 8
5 9 1 4 8 3 2 6 7
7 4 5 3 9 2 6 8 1
6 3 9 5 1 8 7 2 4
1 2 8 6 4 7 5 3 9
4 5 2 9 3 1 8 7 6
9 6 7 8 2 5 1 4 3
8 1 3 7 6 4 9 5 2
```

Solution # 998

```
9 5 7 8 3 2 6 4 1
2 8 4 6 7 1 5 3 9
3 6 1 5 4 9 2 8 7
7 9 8 3 6 5 1 2 4
1 2 6 4 9 8 3 7 5
5 4 3 2 1 7 8 9 6
6 3 5 9 8 4 7 1 2
8 7 9 1 2 6 4 5 3
4 1 2 7 5 3 9 6 8
```

Solution # 999

```
6 2 9 7 5 1 3 8 4
3 4 5 8 9 6 2 7 1
8 1 7 2 4 3 6 5 9
2 3 6 5 8 4 9 1 7
1 7 4 9 6 2 5 3 8
5 9 8 1 3 7 4 2 6
9 8 3 6 1 5 7 4 2
4 6 2 3 7 8 1 9 5
7 5 1 4 2 9 8 6 3
```

Solution # 1000

```
5 3 9 1 7 8 2 6 4
6 1 7 3 2 4 8 9 5
4 8 2 5 6 9 7 1 3
2 4 5 7 9 6 1 3 8
9 6 3 8 5 1 4 2 7
8 7 1 2 4 3 6 5 9
1 9 4 6 3 7 5 8 2
7 5 6 9 8 2 3 4 1
3 2 8 4 1 5 9 7 6
```

Solution # 1001

```
2 4 3 1 8 7 6 5 9
1 7 9 5 4 6 2 3 8
5 6 8 2 3 9 1 4 7
3 5 6 4 9 2 7 8 1
8 2 1 7 5 3 4 9 6
7 9 4 6 1 8 3 2 5
4 8 2 9 6 1 5 7 3
6 3 5 8 7 4 9 1 2
9 1 7 3 2 5 8 6 4
```

Solution # 1002

```
8 4 9 6 7 1 5 3 2
5 2 3 4 9 8 1 7 6
6 1 7 5 3 2 4 8 9
4 9 5 3 2 7 6 1 8
3 6 8 9 1 4 2 5 7
2 7 1 8 6 5 9 4 3
9 5 4 2 8 3 7 6 1
7 3 6 1 4 9 8 2 5
1 8 2 7 5 6 3 9 4
```

Solution # 1003

```
6 5 8 4 2 1 7 3 9
7 3 2 8 9 6 5 4 1
4 9 1 3 5 7 8 2 6
3 2 6 5 4 9 1 7 8
9 1 5 7 8 2 4 6 3
8 7 4 1 6 3 2 9 5
5 8 7 6 3 4 9 1 2
2 4 3 9 1 5 6 8 7
1 6 9 2 7 8 3 5 4
```

Solution # 1004

```
6 1 3 4 8 7 9 5 2
2 9 4 5 6 1 7 3 8
8 7 5 9 3 2 1 4 6
5 3 2 7 9 4 8 6 1
7 6 1 2 5 8 4 9 3
4 8 9 6 1 3 5 2 7
1 2 8 3 4 9 6 7 5
3 4 6 8 7 5 2 1 9
9 5 7 1 2 6 3 8 4
```

Solution # 1005

```
6 9 3 8 1 5 2 7 4
7 5 4 6 9 2 1 8 3
2 8 1 7 4 3 6 9 5
8 3 5 2 6 7 4 1 9
4 2 6 9 5 1 8 3 7
9 1 7 4 3 8 5 6 2
1 6 9 3 2 4 7 5 8
5 4 8 1 7 9 3 2 6
3 7 2 5 8 6 9 4 1
```

Solution # 1006

```
2 9 7 6 1 3 5 4 8
8 4 6 2 9 5 1 7 3
1 5 3 7 4 8 2 6 9
6 1 5 8 2 9 4 3 7
3 2 9 4 6 7 8 5 1
4 7 8 3 5 1 9 2 6
5 8 4 9 7 6 3 1 2
9 6 1 5 3 2 7 8 4
7 3 2 1 8 4 6 9 5
```

Solution # 1007

```
2 8 4 9 5 6 3 1 7
9 7 5 2 3 1 8 6 4
1 3 6 4 7 8 9 2 5
4 5 2 7 6 3 1 8 9
6 1 7 8 4 9 2 5 3
3 9 8 1 2 5 4 7 6
5 2 9 6 8 4 7 3 1
7 4 3 5 1 2 6 9 8
8 6 1 3 9 7 5 4 2
```

Solution # 1008

```
5 1 8 3 2 9 6 4 7
3 6 7 1 4 5 2 9 8
4 9 2 8 6 7 1 5 3
8 7 4 2 1 6 5 3 9
2 3 1 9 5 8 4 7 6
9 5 6 4 7 3 8 2 1
7 2 9 6 8 4 3 1 5
1 8 5 7 3 2 9 6 4
6 4 3 5 9 1 7 8 2
```

Solution # 1009

```
2 8 4 3 5 9 6 7 1
1 3 7 2 6 4 9 8 5
6 9 5 8 7 1 4 2 3
8 6 2 9 3 7 5 1 4
7 5 1 4 2 6 8 3 9
9 4 3 1 8 5 2 6 7
3 7 6 5 4 8 1 9 2
4 1 8 7 9 2 3 5 6
5 2 9 6 1 3 7 4 8
```

Solution # 1010

```
6 8 1 4 2 5 3 9 7
3 5 7 9 8 1 4 6 2
2 9 4 3 7 6 1 5 8
7 4 5 1 9 3 8 2 6
8 1 6 5 4 2 9 7 3
9 3 2 8 6 7 5 1 4
5 6 9 7 3 4 2 8 1
1 2 3 6 5 8 7 4 9
4 7 8 2 1 9 6 3 5
```

Solution # 1011

```
9 6 1 8 7 5 4 3 2
8 4 7 3 2 6 9 5 1
2 3 5 4 9 1 6 8 7
3 1 9 2 4 8 5 7 6
7 5 6 1 3 9 8 2 4
4 2 8 5 6 7 3 1 9
5 7 4 9 8 2 1 6 3
6 8 3 7 1 4 2 9 5
1 9 2 6 5 3 7 4 8
```

Solution # 1012

```
4 5 9 3 8 7 2 1 6
1 7 6 2 5 4 9 3 8
2 8 3 6 1 9 7 4 5
7 9 8 5 6 3 1 2 4
3 2 1 7 4 8 6 5 9
6 4 5 1 9 2 8 7 3
8 1 4 9 7 5 3 6 2
9 6 2 4 3 1 5 8 7
5 3 7 8 2 6 4 9 1
```

Solution # 1013

```
6 7 4 2 8 9 5 1 3
3 2 5 7 4 1 9 6 8
9 1 8 3 6 5 7 2 4
2 8 9 4 1 6 3 5 7
7 6 3 5 9 2 4 8 1
4 5 1 8 3 7 6 9 2
8 9 6 1 7 4 2 3 5
1 3 2 6 5 8 4 7 9
5 4 7 9 2 3 1 8 6
```

Solution # 1014

```
1 8 7 9 2 4 6 5 3
5 9 6 3 7 8 4 1 2
4 2 3 5 1 6 9 7 8
3 5 4 2 9 7 1 8 6
8 6 1 4 5 3 7 2 9
2 7 9 8 6 1 3 4 5
7 3 2 1 8 9 5 6 4
6 4 8 7 3 5 2 9 1
9 1 5 6 4 2 8 3 7
```

Solution # 1015

```
3 5 2 9 8 4 7 1 6
7 9 6 2 1 3 5 4 8
4 1 8 5 6 7 2 9 3
6 7 5 3 2 1 9 8 4
1 4 3 7 9 8 6 5 2
2 8 9 4 5 6 3 7 1
8 3 4 6 7 5 1 2 9
9 6 7 1 4 2 8 3 5
5 2 1 8 3 9 4 6 7
```

Solution # 1016
```
7 6 8 2 5 9 3 4 1
1 3 2 4 7 8 9 5 6
5 4 9 6 3 1 7 2 8
9 7 4 5 1 3 8 6 2
2 8 5 7 9 6 1 3 4
3 1 6 8 4 2 5 7 9
4 9 3 1 6 5 2 8 7
6 2 1 3 8 7 4 9 5
8 5 7 9 2 4 6 1 3
```

Solution # 1017
```
7 8 4 1 5 6 9 3 2
3 1 5 7 9 2 6 4 8
2 6 9 3 8 4 7 1 5
5 7 8 2 4 1 3 9 6
6 4 3 8 7 9 2 5 1
9 2 1 6 3 5 8 7 4
8 3 6 5 1 7 4 2 9
4 5 2 9 6 3 1 8 7
1 9 7 4 2 8 5 6 3
```

Solution # 1018
```
2 8 5 7 9 6 4 3 1
1 3 7 4 2 5 8 9 6
4 9 6 3 8 1 7 5 2
9 1 2 5 4 8 3 6 7
5 6 3 9 1 7 2 4 8
8 7 4 6 3 2 5 1 9
7 5 9 2 6 4 1 8 3
3 2 1 8 5 9 6 7 4
6 4 8 1 7 3 9 2 5
```

Solution # 1019
```
4 2 6 9 5 7 3 1 8
3 5 1 4 2 8 9 7 6
8 9 7 1 3 6 5 2 4
5 3 9 6 8 2 7 4 1
6 7 8 3 4 1 2 9 5
2 1 4 5 7 9 8 6 3
1 8 3 2 9 4 6 5 7
9 6 5 7 1 3 4 8 2
7 4 2 8 6 5 1 3 9
```

Solution # 1020
```
1 6 7 9 4 8 5 3 2
2 4 5 3 1 6 9 8 7
9 3 8 7 2 5 4 1 6
7 8 1 6 5 9 2 4 3
4 2 6 8 3 1 7 5 9
5 9 3 4 7 2 1 6 8
8 5 2 1 9 6 3 7 4
6 7 4 5 9 3 8 2 1
3 1 9 2 8 4 6 7 5
```

Solution # 1021
```
6 3 8 1 2 5 4 9 7
9 1 7 8 4 3 2 6 5
2 5 4 6 9 7 1 8 3
8 4 9 2 7 1 3 5 6
5 7 2 3 8 6 9 1 4
3 6 1 9 5 4 7 2 8
7 2 5 4 1 8 6 3 9
4 9 6 5 3 2 8 7 1
1 8 3 7 6 9 5 4 2
```

Solution # 1022
```
5 8 3 9 1 4 7 6 2
4 9 1 2 7 6 5 8 3
7 2 6 3 5 8 9 1 4
2 1 8 7 4 5 6 3 9
3 7 5 8 6 9 4 2 1
6 4 9 1 2 3 8 7 5
1 6 4 5 8 2 3 9 7
8 3 7 4 9 1 2 5 6
9 5 2 6 3 7 1 4 8
```

Solution # 1023
```
6 1 2 9 4 8 7 3 5
7 3 9 6 5 1 2 4 8
8 5 4 7 2 3 9 6 1
2 6 1 3 8 5 4 7 9
9 4 5 2 6 7 8 1 3
3 8 7 1 9 4 5 2 6
5 7 8 4 3 6 1 9 2
4 2 3 8 1 9 6 5 7
1 9 6 5 7 2 3 8 4
```

Solution # 1024
```
2 4 5 3 1 9 8 6 7
9 8 7 2 6 4 1 5 3
3 1 6 7 5 8 4 2 9
1 9 3 8 4 6 5 7 2
6 2 4 1 7 5 9 3 8
7 5 8 9 2 3 6 1 4
8 7 2 6 9 1 3 4 5
5 6 9 4 3 7 2 8 1
4 3 1 5 8 2 7 9 6
```

Solution # 1025
```
9 5 6 2 8 1 4 3 7
7 3 8 5 4 6 2 1 9
4 2 1 9 7 3 5 6 8
1 6 9 4 3 2 8 7 5
2 7 4 6 5 8 3 9 1
3 8 5 1 9 7 6 2 4
5 4 2 7 6 9 1 8 3
6 9 3 8 1 5 7 4 2
8 1 7 3 2 4 9 5 6
```

Solution # 1026
```
3 6 2 1 9 7 8 4 5
5 7 9 2 4 8 3 1 6
8 4 1 5 6 3 9 2 7
1 9 3 7 8 6 2 5 4
2 5 6 9 1 4 7 3 8
4 8 7 3 5 2 6 9 1
7 1 8 4 2 9 5 6 3
6 2 5 8 3 1 4 7 9
9 3 4 6 7 5 1 8 2
```

Solution # 1027
```
8 4 5 3 9 6 2 1 7
6 7 9 2 4 1 3 5 8
3 1 2 5 8 7 4 9 6
7 5 4 6 3 8 1 2 9
9 3 1 4 7 2 8 6 5
2 6 8 9 1 5 7 4 3
5 9 7 1 2 3 6 8 4
4 2 3 8 6 9 5 7 1
1 8 6 7 5 4 9 3 2
```

Solution # 1028
```
7 9 6 3 8 5 2 1 4
1 5 3 4 2 6 8 9 7
4 8 2 1 9 7 3 6 5
9 6 4 8 5 2 1 7 3
5 1 7 9 4 3 6 8 2
2 3 8 6 7 1 5 4 9
8 2 5 7 1 4 9 3 6
3 7 9 5 6 8 4 2 1
6 4 1 2 3 9 7 5 8
```

Solution # 1029
```
1 2 6 4 8 3 9 5 7
8 3 9 2 5 7 4 6 1
4 5 7 9 1 6 2 3 8
6 8 4 1 7 5 3 2 9
5 7 1 3 2 9 6 8 4
2 9 3 6 4 8 1 7 5
3 1 5 8 6 4 7 9 2
9 4 8 7 3 2 5 1 6
7 6 2 5 9 1 8 4 3
```

Solution # 1030
```
6 8 3 4 9 2 7 1 5
4 1 9 7 8 5 3 6 2
7 2 5 6 1 3 4 8 9
9 7 8 1 2 4 6 5 3
5 4 6 9 3 7 1 2 8
2 3 1 5 6 8 9 7 4
8 9 7 3 5 6 2 4 1
1 6 2 8 4 9 5 3 7
3 5 4 2 7 1 8 9 6
```

Solution # 1031
```
3 8 9 6 7 4 1 2 5
1 7 4 5 9 2 8 3 6
6 5 2 1 8 3 9 4 7
2 1 7 9 6 5 4 8 3
9 6 3 4 1 8 5 7 2
8 4 5 3 2 7 6 1 9
4 3 1 2 5 9 7 6 8
7 9 6 8 3 1 2 5 4
5 2 8 7 4 6 3 9 1
```

Solution # 1032
```
9 6 4 1 5 7 8 2 3
7 8 2 4 3 6 5 9 1
3 1 5 2 9 8 6 7 4
1 9 8 5 4 2 7 3 6
5 4 7 3 6 9 2 1 8
6 2 3 8 7 1 9 4 5
8 3 6 7 1 4 3 5 2
2 5 1 9 8 3 4 6 7
4 7 9 6 1 5 3 8 2
```

Solution # 1033
```
5 2 4 6 1 8 9 3 7
8 7 6 4 9 3 5 1 2
9 3 1 2 7 5 8 4 6
3 4 9 1 8 2 7 6 5
2 1 5 3 6 7 4 9 8
6 8 7 9 5 4 3 2 1
7 6 3 8 2 9 1 5 4
1 9 8 5 4 6 2 7 3
4 5 2 7 3 1 6 8 9
```

Solution # 1034
```
1 2 5 9 8 3 7 4 6
8 6 7 4 2 1 3 9 5
9 4 3 7 6 5 2 8 1
5 7 2 8 1 9 4 6 3
6 1 8 3 7 4 9 5 2
3 9 4 6 5 2 8 1 7
4 5 1 2 9 7 6 3 8
2 3 6 5 4 8 1 7 9
7 8 9 1 3 6 5 2 4
```

Solution # 1035
```
6 2 5 8 3 4 7 9 1
9 3 8 1 5 7 4 6 2
1 4 7 9 2 6 5 3 8
4 1 9 3 6 5 2 8 7
8 5 2 4 7 9 3 1 6
3 7 6 2 1 8 9 4 5
2 8 3 7 9 1 6 5 4
7 6 4 5 8 3 1 2 9
5 9 1 6 4 2 8 7 3
```

Solution # 1036
```
9 2 7 6 5 8 3 4 1
8 5 4 1 7 3 2 9 6
3 1 6 4 9 2 8 7 5
4 9 2 8 1 6 5 3 7
1 7 5 3 2 9 4 6 8
6 8 3 5 4 7 1 2 9
2 3 9 7 8 1 6 5 4
7 4 1 2 6 5 9 8 3
5 6 8 9 3 4 7 1 2
```

Solution # 1037
```
4 3 5 9 1 7 2 8 6
6 2 9 8 4 5 3 1 7
8 7 1 2 6 3 4 9 5
3 1 4 6 5 2 9 7 8
5 8 7 3 9 4 6 2 1
2 9 6 1 7 8 5 4 3
7 5 3 4 2 1 8 6 9
9 4 8 7 3 6 1 5 2
1 6 2 5 8 9 7 3 4
```

Solution # 1038
```
6 2 3 7 8 9 4 1 5
8 5 9 6 4 1 2 3 7
7 4 1 5 3 2 8 6 9
3 1 8 4 2 5 9 7 6
4 7 2 9 1 6 3 5 8
5 9 6 8 7 3 1 2 4
1 6 5 3 9 8 7 4 2
9 3 4 2 5 7 6 8 1
2 8 7 1 6 4 5 9 3
```

Solution # 1039
```
4 6 8 9 1 3 5 7 2
5 2 3 7 6 8 4 9 1
1 7 9 2 4 5 8 6 3
7 5 6 1 3 4 9 2 8
9 8 1 5 2 7 3 4 6
3 4 2 6 8 9 1 5 7
8 3 7 4 5 2 6 1 9
2 1 5 3 9 6 7 8 4
6 9 4 8 7 1 2 3 5
```

Solution # 1040
```
3 8 6 4 9 7 5 2 1
9 4 7 5 1 2 6 3 8
1 5 2 3 8 6 4 7 9
8 2 5 1 7 3 9 4 6
4 6 3 2 5 9 1 8 7
7 1 9 6 4 8 2 5 3
2 9 1 8 3 5 7 6 4
5 7 8 9 6 4 3 1 2
6 3 4 7 2 1 8 9 5
```

Solution # 1041
```
2 5 4 6 7 9 1 3 8
3 8 9 5 2 1 7 6 4
6 7 1 3 4 8 9 2 5
9 6 5 2 1 7 4 8 3
8 2 7 4 9 3 5 1 6
4 1 3 8 6 5 2 9 7
1 9 8 7 3 4 6 5 2
5 4 6 9 8 2 3 7 1
7 3 2 1 5 6 8 4 9
```

Solution # 1042
```
2 7 6 1 8 3 4 5 9
5 4 3 7 2 9 1 6 8
8 9 1 4 6 5 2 7 3
4 8 9 2 5 7 6 3 1
1 3 7 8 4 6 5 9 2
6 5 2 9 3 1 8 4 7
7 6 8 3 1 4 9 2 5
3 1 5 6 9 2 7 8 4
9 2 4 5 7 8 3 1 6
```

Solution # 1043
```
4 2 8 3 6 5 7 9 1
7 3 5 8 1 9 4 2 6
1 9 6 2 4 7 5 8 3
5 1 4 6 7 2 9 3 8
6 8 2 4 9 3 1 7 5
9 7 3 1 5 8 6 4 2
2 5 1 9 8 4 3 6 7
3 6 9 7 2 1 8 5 4
8 4 7 5 3 6 2 1 9
```

Solution # 1044
```
1 6 7 2 4 3 9 5 8
9 5 2 6 8 1 3 4 7
4 8 3 9 5 7 6 2 1
5 2 4 8 6 9 1 7 3
3 7 8 1 2 5 4 6 9
6 9 1 7 3 4 2 8 5
7 1 5 4 9 6 8 3 2
8 3 6 5 1 2 7 9 4
2 4 9 3 7 8 5 1 6
```

Solution # 1045
```
7 8 9 4 6 1 2 3 5
4 1 3 5 2 8 6 9 7
6 5 2 9 7 3 1 8 4
9 2 4 7 8 6 3 5 1
5 6 1 3 4 9 7 2 8
3 7 8 1 5 2 9 4 6
2 4 7 6 9 5 8 1 3
8 3 6 2 1 4 5 7 9
1 9 5 8 3 7 4 6 2
```

Solution # 1046
```
2 1 7 8 5 6 3 9 4
3 8 9 2 1 4 5 7 6
4 6 5 3 9 7 8 2 1
1 2 6 4 7 3 9 8 5
5 4 8 9 2 1 7 6 3
9 7 3 6 8 5 4 1 2
7 5 4 1 6 8 2 3 9
8 9 1 5 3 2 6 4 7
6 3 2 7 4 9 1 5 8
```

Solution # 1047
```
5 8 6 7 3 4 1 2 9
9 3 4 2 6 1 7 8 5
7 1 2 5 8 9 4 3 6
4 6 7 8 2 3 9 5 1
3 5 1 4 9 6 2 7 8
8 2 9 1 5 7 3 6 4
2 9 8 3 1 5 6 4 7
6 4 5 9 7 2 8 1 3
1 7 3 6 4 8 5 9 2
```

Solution # 1048
```
5 9 2 4 7 3 1 8 6
3 6 7 1 2 8 4 5 9
4 1 8 9 5 6 2 7 3
6 7 4 5 9 1 3 2 8
1 8 5 7 3 2 6 9 4
9 2 3 8 6 4 5 1 7
2 3 1 6 8 7 9 4 5
8 5 6 2 4 9 7 3 1
7 4 9 3 1 5 8 6 2
```

Solution # 1049
```
9 8 7 6 5 2 3 1 4
4 5 1 8 7 3 2 6 9
2 6 3 4 9 1 7 8 5
3 4 2 5 8 7 6 9 1
6 7 9 1 3 4 5 2 8
5 1 8 2 6 9 4 3 7
7 2 6 9 4 8 1 5 3
1 9 4 3 2 5 8 7 6
8 3 5 7 1 6 9 4 2
```

Solution # 1050
```
9 1 8 3 2 4 7 5 6
4 7 6 8 5 9 3 2 1
3 2 5 7 6 1 4 8 9
8 6 9 2 4 3 5 1 7
5 4 7 6 1 8 2 9 3
1 3 2 5 9 7 8 6 4
6 5 1 4 3 2 9 7 8
7 9 4 1 8 5 6 3 2
2 8 3 9 7 6 1 4 5
```

www.ingramcontent.com/pod-product-compliance
Lightning Source LLC
Chambersburg PA
CBHW080826220526

45467CB00008B/2209